BRAINS TOP DOWN

Is Top-Down Causation Challenging Neuroscience?

editors

Gennaro Auletta
Pontifical Gregorian University, Italy
University of Cassino, Italy

Ivan Colagè
Pontifical Gregorian University, Italy

Marc Jeannerod
Claude Bernard University, France

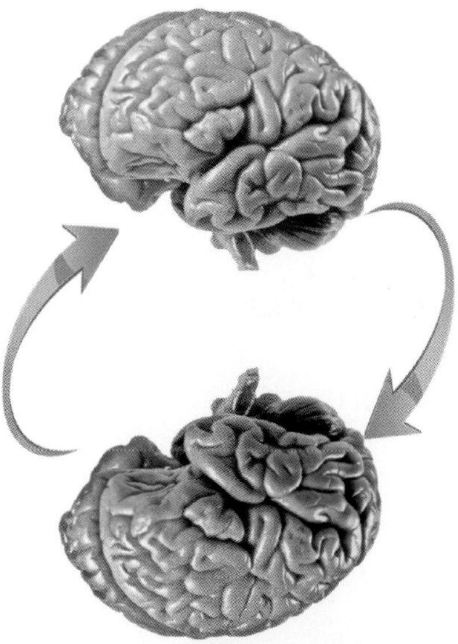

BRAINS TOP DOWN

Is Top-Down Causation Challenging Neuroscience?

World Scientific

NEW JERSEY • LONDON • SINGAPORE • BEIJING • SHANGHAI • HONG KONG • TAIPEI • CHENNAI

Published by

World Scientific Publishing Co. Pte. Ltd.
5 Toh Tuck Link, Singapore 596224
USA office: 27 Warren Street, Suite 401-402, Hackensack, NJ 07601
UK office: 57 Shelton Street, Covent Garden, London WC2H 9HE

Library of Congress Cataloging-in-Publication Data
Brains top down : is top-down causation challenging neuroscience? / [edited by] Gennaro Auletta, Ivan Colagè, Marc Jeannerod.
p. ; cm.
Includes bibliographical references and index.
ISBN-13: 978-9814412452 (hardcover : alk. paper)
ISBN-10: 9814412457 (hardcover : alk. paper)
I. Auletta, Gennaro, 1957– II. Colagè, Ivan. III. Jeannerod, Marc.
[DNLM: 1. Brain--physiology. 2. Consciousness--physiology. WL 300]
QP376
612.8'2--dc23
2013007836

British Library Cataloguing-in-Publication Data
A catalogue record for this book is available from the British Library.

Typeset by Stallion Press
Email: enquiries@stallionpress.com

Printed in Singapore by Mainland Press Pte Ltd.

List of Contributors

Michael A. Arbib
USC Brain Project, University of Southern California, USA

Gennaro Auletta
Specialization in Science and Philosophy, Pontifical Gregorian University, Italy; University of Cassino, Italy

Jeffery Burgdorf
J. P. Scott Center for Neuroscience, Northwestern University, USA

Jean-Pierre Changeux
Institut Pasteur, Paris, France

Ivan Colagè
Specialization in Science and Philosophy, Pontifical Gregorian University, Italy

Charles R. Crowell
Department of Psychology, University of Notre Dame, USA

Antonio R. Damasio
USC Brain and Creativity Institute, University of Southern California, USA

Pierre Jacob
Institut Nicod, CNRS-EHESS-ENS, Paris, France

Marc Jeannerod (†)
Claude Bernard University, Lyon, France

Joseph R. Moskal
Biomedical Engineering Department, Northwestern University, USA

Jaak Panksepp
Center for Neuroscience, Mind & Behavior, Bowling Green State University, USA

Marcus E. Raichle
School of Medicine, Washington University in St Louis, USA

Juan Josè Sanguineti
Faculty of Philosophy, Pontifical University of the Holy Cross, Italy

Wolf Singer
Department of Neurophysiology, Max Planck Institute for Brain Research, Germany

From the left to the right: Marcus Raichle, Wolf Singer, Jean-Pierre Changeux, Pierre Jacob, Hanna Damasio, Antonio Damasio, Gennaro Auletta, Marc Jeannerod, Tomasz Trafny, Charles Crowell, Juan José Sanguineti, Paolo D'Ambrosio, Ivan Colagè, Erwin Guerrero, Carly Andrews, Michael A. Arbib.

Contents

CHAPTER 1

Introduction: Consciousness as a Top-Down Causal Agency

*Gennaro Auletta and Marc Jeannerod**

1.1. BASIC NOTIONS

We would like to approach the significance and role of consciousness (especially its causal downwards significance, if any) starting from its biological background, following Damasio's contribution. It is indeed a difficult matter to understand this issue, and it becomes impossible if biological and evolutionary considerations do not form an appropriate background.

1.1.1. The Biological Self

What is characteristic of life is its ability to shield its internal processes against the disruptive action of the environment and to preserve and develop order,

*Note by G. Auletta: The following introduction was written at a time in which Marc Jeannerod was already very sick (*see Postface* and Appendix I). An initial draft dealing with the main theoretical questions was written by me and approved by Marc. The deal was that he would write a sort of comment on the different contributions to insert at the end of this introduction, but unfortunately he was hampered by his ill health. By reading the whole manuscript, I became aware that it would be much better to insert the comments on the different contributions to this volume in the body of the text and not at the end, and I decided to reformulate the draft in this way. This means that the critical examination of the contributions has not been checked by Marc. However, I have conserved his name as co-author because I have found this joint effort in any case as a remarkable example of dialogue and convergence between two minds, or two brains. It is clear that I alone am responsible for any shortcomings or mistakes.

apparently against the Second Law of thermodynamics. This is illusionary, since any organism discharges in the environment at least as much entropy as it gets free energy, so that the equilibrium (homeostasis) is maintained. However, what is not illusionary is the ability to keep the metabolic exchanges with the environment under control (Auletta 2011b). If this were not the case and if the organism were under the control of the environment, the possibility that the latter spontaneously produced a configuration that would result in promotion of the internal order of the organism is a probability so tiny as to be very near to zero. No life could exist and endure for millions and billions of years in such conditions. This is not the right context for examining which are the mechanisms allowing this controlled exchange. We only wish to point out that the necessary condition for this to happen is that the organism constitutes itself as a *biological self*. As far as we can understand, this notion corresponds to the concept of *proto-self* introduced in Parvizi and Damasio (2001) and is central also to Antonio Damasio's contribution, Chap. 4 (it is possible that is finally an issue of nomenclature). With the notion of *biological self* we intend the complex of all functions and processes of a living organism that for their execution owe much more to its internal constitution than to the exterior.

Such a fundamental distinction between what pertains to the organism (the self) and what does not (the non-self) is basilar to virtually all biological functions and activities, and to many subsystems, like the immune system or the brain. J. Monod defined the organism as an object endowed with a project represented by its structures and realized through its performances (Monod 1970, Chap. 1). The same question has been dealt with by W. Wundt, along with W. James, the father of modern psychology, who said that, even if at the lowest levels of life the processes of consciousness are confined within extremely narrow limits, and the will is determined by the universal organic impulses only in the very simplest way, it is also true that, even among the lowest protozoa, the manifestations of life are explicable only upon the hypothesis that they possess a mind (Wundt 1893, p. 29). Here, the words "mind" and "consciousness" are probably not entirely appropriate. However, they capture the fundamental and basic fact of life that we have called "biological self", and any shortcoming is likely to be of a terminological nature.

1.1.2. The Concept of Emergence

For many people, it is often already difficult to accept this basic level of the discussion. It is indeed a great leap to begin with brute, mechanical matter and to end with something like the self. How can a self have an existence at all in a world in which it is assumed that the only existing things are atoms, molecules and at most chemical reactions? So, the problem is not with human consciousness, as many believe, but must rather be settled already at a biological level. We can very well understand how a ball sets another ball in motion by impinging on it, but how can we grasp the constitution of a self controlling metabolic exchanges with the environment? Should not the environment, on the contrary, act in a mechanical way on a self, and that is all? We have already mentioned that precisely this cannot account for the distinctive way organisms are able to endure.

Therein is a deep misunderstanding among many biologists and neurologists about the concept of matter. Indeed, the understanding of matter of a couple of centuries ago is no longer the way in which we scientifically consider matter today. The structure of matter (atoms, molecules, chemical reactions) is an emergence from quantum mechanics (Auletta *et al.* 2009, Chaps. 11–12; Atkins and De Paula 2006). By *emergence*, we mean here the fact that matter is ruled by quantum-mechanical laws, but since those laws are only probabilistic, one cannot predict which specific characters this matter organization will take. This implies that, although phase transitions (from a liquid to a solid state, for instance) is well defined in general terms (the phase that minimizes the free energy is selected), the specific parameters that describe a phase transition of a concrete chemical (like water) cannot be deduced *a priori* but need to be extracted from empirical data and therefore the transition simulated. Moreover, water is not constituted by H_2O molecules but rather by dynamic aggregates $H_{2n}O_n$, where n is often much larger than unity. Finally, we cannot say in any literal sense that a molecule is formed by atoms: it is rather a new structure built of elemental centers. (The individual atoms are no longer preserved as such, as is evident in benzene, where electrons are shared by the whole molecule, i.e., all the so-called π bonds constitute a single giant quantum system.) Then what is a molecule? It is an integrated *system* of physical processes encompassing both local interactions and long-ranging correlations.

The issue of emergence is a very difficult one, and it is not easy to find a balanced position that takes into account what is scientifically known thus far. This has been especially the concern of Michael Arbib's contribution (Chap. 3) with reference to Sperry's work. Arbib rightly points out that emergence of new levels of reality cannot imply that we have removal of lower levels. We may certainly assume that the level of electrons no longer plays a crucial role when speaking, e.g., of brain excitation patterns, so that there is a certain loss of relevance of low levels the more and more we mount the ladder of complexity. Nevertheless, following Arbib's suggestions, two caveats seem important here:

- The low level that directly underlies an emerged level is crucial for everything that happens here, since it contributes with its dynamism and variety, which represents the necessary conditions for any possible event or process at this high level. For instance, there cannot be any rational planning without involving neural circuitry.
- Any fundamental rearrangement in the organism's and the brain's functionalities must necessarily imply also a rearrangement (and even a rebuilding) of the underlying structures, like organs, brain areas or neural connections, and so on. In other words, we cannot have changes in functionality without appropriate structural circuitries supporting them.

1.1.3. Associationism and Life

Let us now apply the previous considerations to the issue of the organism. What then is an organism? It is an integrated system of physical-chemical processes but also of biological functions that are needed for survival (like the metabolic ones). This is true for the biological self or proto-self (without which an organism could not exist at all) but also for the higher levels like the core self (in Damasio's terminology) and the personal self, as we shall see: both are integrated systems of functions. Therefore, the organisms both continuously change and preserve a certain systemic organization. In this sense, brain excitation patterns are, in principle, not very different from those (chemical-genetic) patterns that are built during epigenetic processes. Both patterns cannot be considered in themselves (it makes even no sense to consider them in themselves) but have a function in the interaction of the organism with other organisms or objects. (For instance, consider how Damasio (Chap. 4) depicts the way in which a proto-self builds a representation of an object.) The same is true *mutatis mutandis* for the personal self, even if at a higher level of integration. Indeed, both the biological and psychological-personal self are nothing more than this interactive system able to build an integrated but transient unity. If, with the nice words of Herbart (1816) [see also De Biran (1803), pp. 97–99], we say that any perception is a dynamic reaction against an external perturbation that needs to be integrated, we may also add that any self as such is preserved in an itinerant dynamical process through reaction and integration.

Here, we touch on a very important point. Can we explain life by starting from molecules and by aggregating them piecemeal in order to finally get an organism? It is not impossible that a superior unity can emerge through aggregation of more elementary units. It is quite likely that in putting such units together several interconnections are spontaneously established so that we have indeed the emergence of the whole. As a matter of fact, and physically speaking, when you put more and more systems (or units) together, the maximal entropy attainable grows exponentially (due to the combinatorial explosion of all the possible configurations). Now, thermodynamics only forbids that the resulting entropy be lower than the sum of the entropies of the systems as they were at the start (that is, separated). This entropy is, however, much lower than the maximal entropy attainable when the systems or units are put together (Auletta 2011a, Sec. 9.9). This allows for spontaneous emergence of order and it indeed often occurs in nature. Therefore, there is nothing wrong, in itself, with the assumption that organisms can arise through such a combinatorial assembly.

What is wrong, however, is that in this way we can adequately explain how self-organization processes spontaneously arise in the organism but not the distinctive way in which organisms deal with their environment and endure. Suppose, indeed, that you were able to build an organism and its resulting modality of action by putting together more and more pieces, and therefore by assembling its behavior purely bottom-up starting from elementary inputs coming from the outside (as associationism depicts the story). Then the organism would be enslaved by the environment and no subsequent change or mutation could eliminate the basic fact that it is a mechanical putting-together of elements or units that are not under its control. It is clear that it is one thing to linearly assemble and another to nonlinearly assemble, as occurs in all spontaneous self-organization processes in nature, as stressed by W. Singer in his contribution (Chap. 7). Nevertheless, spontaneously arising prebiotic self-organized systems like Bénard cells show no autonomous activity in themselves (they are not able to provide for their own self-maintenance) and therefore rely completely on external inputs (the amount of heat pumped into the system).

The situation is similar with physicalist attempts that are linearity-based, as seems to be the case for the approach endorsed by Pierre Jacob in his contribution (Chap. 11). This is why one of us has tried to build a consistent alternative to traditional associationism (Jeannerod 1988). The question is: Can we do that without falling into some form of Intelligent Design (God or some Designer intervening in order to design organisms with a specific constitution) or even vitalism? We see this as a very serious issue. One of us has tried to develop the idea that (i) a spontaneous process of aggregation of biomolecules out of an initial variety (a bottom-up process), resulting essentially in some piece of RNA insulated through some form of membrane, together with (ii) a systemic rearrangement of the way in which this complex system deals with information (a top-down processes) that (iii) has been selected and fixed through natural selection could be a viable answer at a biological level (Auletta 2011a, Chaps. 7–8). It could be interesting to see whether or not on this point there could be a certain agreement with other approaches presented in this volume. We have already remarked on the examination of the biological self or proto-self undertaken by A. Damasio. We also mention here that in the second session of the Overall Discussion (Chap. 12), Jean-Pierre Changeux has vindicated a non-reductionist point of view.

This examination can also have some interesting consequences for the issue of the brain. In his contribution (Chap. 8), Marcus Raichle has indeed strongly stressed (and, thanks to comparative studies on the brain's energy consumption during the day and night, shown remarkable empirical evidence for) the endogenous character of most of the brain's activity, a point on which there is full agreement from both of us. Moreover, what is remarkable with the brain is that its excitation patterns are transient (a point stressed in W. Singer's contribution, Chap. 7, in demonstrating how the brain is not a computer, since it lacks the basic distinction between hardware and software). However, what is also interesting is that the brain shows continuity in its operations and even a cognitive-operative coherence that is already evident in mammals, never mind primates and humans. It is difficult to explain this by starting in a bottom-up way only from the excitation patterns themselves: how could a stability be built? By pure association with other excitation patterns? But this would again be an association pattern, a token so to speak, created itself in a bottom-up way in a certain specific environmental context and having certain stimuli. In the potent words employed by A. Damasio at the end of the discussion in Chap. 4, the highest level of the brain's organization is not the "theatre" of the mind; it is the *maker* of the "theatre" of the mind.

1.1.4. Top-Down Causation

The previous examination brings us to a preliminary consideration of top-down causation. Ultimately, there seems to be a wide agreement among the participants concerning this concept (see for instance the discussion following M. Raichle's talk and the first session of the Overall Discussion, Chap. 12), with the apparent exception of Pierre Jacob. Even Wolf Singer, who had expressed some important concerns, finally agrees about the relevance of the concept. Let us first of all clarify an important point: Although with different words, both Arbib and Changeux agree with us (Auletta *et al.* 2008; see the literature quoted in the Auletta and Colagè's contribution, Chap. 10, and also the Appendix II) on the fact that we cannot understand this concept as meaning that there would be a form of dynamical-causal determination of what happens at a lower level by an upper level. This would not only contradict the closure of the physical laws but would also be a chimera from a philosophical point of view (Auletta 2008). This is why, in his contribution (Chap. 3), M. Arbib speaks of "downward constraints" and, for example, admits that once several perceptual schema instances are active, they may instantiate others in a "hypothesis-driven" way. Jean-Pierre Changeux preferred to speak of regulation or regulative influence of the top level (see the first session of the Overall Discussion, Chap. 12). We believe that, despite the usage of different terms and perhaps a variation in emphasis, all of the contributors to the volume agree on this.

One of the most interesting ways to consider this problem is to see an entrenchment between bottom-up and top-down processes, as pointed out by Changeux in his own talk (see Chap. 10), and expressed by W. Singer in the third session of the Overall Discussion (Chap. 12). The latter has pointed out that it is a resulting global state of the system that is also able to influence in a top-down way the respective local processes in a distributed way. These local processes being influenced by the global state will in turn generate the next global state, and so there is a continuous alternation between top-down causation of the global state that influences local processes, and local processes being changed in this way that generate other global states; this is the passage from top-down to bottom-up causation and *vice versa*, determining an intertwined alternation between local and global. We think that this is extremely helpful; let us understand how tangled these two forms of causation are, as we would expect in a true complex system. It is a sort of new form of feedback, since it does not occur at the same level but among different levels of complexity.

In a similar way (although using a different approach), A. Damasio stressed in the discussion at the end of Chap. 4 that the higher level of the brain hierarchy gives the possibility of relatively simultaneous reactivation of information that is not contained within itself but within cortices such as those of the temporal region or the prefrontal territory. In humans, this is an organization that will momentarily bring information about one's past lived events and one's constructive plans for the future, one's identity and personhood. Therefore, by having a structure such as this, there is the possibility of reactivating within a short time-window activity that is very diverse and could not possibly be recorded within one simple structure, or center, but is likely to be highly distributed.

As mentioned, the approach of P. Jacob seems to be different. Both in the discussion following C. Crowell's contribution (Chap. 9) and in his own contribution (Chap. 11), Jacob emphasized the temporal component of any causal process and its effectiveness in producing results. In the context of top-down causation, Jacob considers a single event eventually able to cause another single event at an upper level (according to a traditional mechanical understanding of causality). The problem is that we do not have such linearity at any level of reality even in the case when pure bottom-up effects are being considered. The reason is simple: any bottom-up cause must be combined with upper-level factors to produce an effect at that level. For instance, electrons cannot alone produce chemical reactions without specific chemical constraints concerning permissible building and breaking of molecular connections. Since Jacob considers the problem in a linear and mechanical way, he arrives quite naturally at the conclusion that the upper event or properties are supervenient on the lower event or properties. In other words, they can be ontologically reduced to the latter. In this sense, there is *a fortiori* no reason whatsoever to speak of top-down causation. This conclusion, although accepted by many philosophers, does not seem sufficiently justified. Our concern here seems also to be the essence of the comment made by M. Raichle in the discussion following Crowell's contribution (Chap. 9): it is useless to speak of top-down causation without specific top-level functionalities having emerged, and this explains why the example of the switch mechanism (used by Crowell himself during his presentation) is not only not very helpful but can even turn out to be misleading.

Another point is whether it is appropriate to speak of top-down *causality* or not. We will not concern ourselves with terminology here. What is important is that top-down processes can be relevant causal *factors*, that is, they can contribute to the final outcome of certain operations. The contributions of Arbib, Damasio, Singer and Auletta in this volume may help to overcome misunderstandings about the role (namely, its character and extent) played by those factors.

1.2. EVOLUTION, AWARENESS AND CONSCIOUSNESS

1.2.1. What is the Brain for?

Once that we have preliminarily clarified some basic notions, the issue of the fundamental function of the brain can be properly addressed. In other words, what is the brain for? This is fundamental for our enquiry; otherwise we could not capture the specificity of consciousness either. It is well known that bacteria like *E. coli* are able to survive and develop several reactions as to a number of environmental stimuli. They are even able to form elementary models of the environment (Auletta 2011b). Eukaryotic unicellular organisms are also able to develop actions on their own (the cytoskeleton and pseudopods allow them to actively and endogenously search for food resources). Therefore, a brain is not necessary for acting on the environment, nor for having a model of it. In order to understand its function, we need to consider creatures that are in some sense at the limit of brain-endowed organisms: the sea squirt, a primitive creature in the class Ascidiacea, is a sac-like sessile marine being that develops a nervous system (of 300 neurons) for only one day during its larval stage to migrate to a new place (Llinás 2001, pp. 15–17). We can therefore assume that the evolutionary significance of the brain is to *connect* motion with representation and therefore to monitor the *effects* of the action of the organism on the environment in order to establish feedback circuits allowing for more error correction and thus more control (Auletta 2011a, Sec. 12.2–12.3).

Then, we can assume that multicellular organisms endowed with a brain are already somehow capable of developing even an elementary *awareness of their environment* (Lloyd Morgan 1896, pp. 126–131, 147; Lloyd Morgan 1900, pp. 100–106). We think that this stage of evolution can correspond to what in Parvizi and Damasio (2001) has been called the core self (see also Damasio's contribution, Chap. 4). In this case, the animal shapes its actions in cognitive *consistency* with past experiences, as experiments on the delayed reaction procedure have shown (Tinklepaugh 1928; Tinklepaugh 1932). The biological origin of this awareness is to be found in the fact that many vertebrates develop sophisticated mechanisms of proprioception that allow for the continuous checking of internal and external states and therefore for an increasing control on behavior. These mechanisms can initially be reflex-like, but their composition can give rise to new functions having cognitive consistency. Thanks to this awareness of the environment, the individual actions of animals always constitute a history not only for themselves but also for their offspring as well as for other species (and therefore even for whole ecosystems). The importance of this is due to the mentioned circumstance that animals can only exist as interactive systems maintaining a transient (itinerant) dynamics in coping with the environment (finally establishing interdependencies, which is the result of a process of co-adaptation). Obviously, there are different levels at which this can be done, and what matters here is the complexity of the organisms and of the web of relations that they establish.

We would like to stress a point here that has rather general value. As mentioned, this level of awareness (or the core self) shows how relevant is the co-evolution or co-adaptation between organism and environment (other species included). W. Singer has stressed this point in his contribution (Chap. 7). He points out that evolution can be considered as a cognitive process through which the evolving organisms acquire knowledge about those features of the world whose consideration facilitates survival and reproduction. Knowledge acquired through evolution is stored in the genes, expressed in the functional architecture of the brain and then complemented by developmental processes and learning. In a certain sense, this allows us to fully solve a traditional philosophical problem, that of the possible connection between knowing subject and object. As a matter of fact, we construct our objects through our perception (we recall Arbib's stress on hypotheses-building), but nonetheless those objects are not deprived of objectivity or reality since such a construction is functional to survival; and this also explains why higher organisms can have a richer understanding of objects than lower organism have (Auletta 2011a, Sec. 4.4). It suffices to consider here that when we speak of external reality we are not always aware that such a world (we mean the Earth) has been shaped by the biological species themselves by a great deal. However, Singer's stress on a Kantian view could give the feeling that the way in which we consider the world is purely illusionary or at least subjective, while the constraints exerted on cognition are not only dependent on the "subject" but also on the way it interacts (and its or his ancestors have interacted in the past) with an external environment. It is perhaps appropriate to recall that even attempts at naturalizing Kantism, like those of K. Lorenz, have been criticized for their tendency to underestimate both development and interaction with the environment, so that one sometimes has the feeling that these inborn schemata (imprinting) are a sort of re-edition of the *homunculus*, since the genome is thought to host a perfect image of the mature organism, including its behavior (Auletta 2011a, Sec. 15.2).

1.2.2. Mammals' Awareness

It can therefore be enlightening to make another evolutionary jump and to consider a higher level of awareness, namely the specific kind of mammal awareness. As is well known, mammals are characterized by some important changes relative to reptiles and birds: they take care of infants (nursery), are able to give rise to interactive sensorimotor schemata, and play (MacLean 1990; Panksepp 1998; Auletta 2011a, Secs. 12.7, 14.2 and 16.4). All these changes are due to the establishment of a true emotional system in the brain that is able to put part of the metabolic system under control of, or at least in connection with, the brain. Damasio (in the first session of the Overall Discussion, Chap. 12) distinguishes between emotions and feelings. This could be helpful in clarifying that lower forms of life have still different forms of reactivity even though they lack the mentioned integrated limbic system and (probably for the same reason) a true psychology. Ultimately, it is a matter of terminology provided that we agree on the main purpose of such distinctions.

As matter of fact, this new neural organization allows for a completely new kind of awareness in mammals: understanding the consequences of their own actions *on other organisms.* Mammals are indeed organisms that are much more tuned to expectation and anticipation (included simulation of actions) than any other biological classes. For this reason, they have developed characteristic error-correction mechanisms based on both feed-forward and feedback circuits (Freeman 2000a; Freeman 2000b). There is a huge amount of ethological research on all forms of conflict and competition between conspecific males for territory, females, and leadership of the group. What is common to all these very different forms of behavior is that one competitor takes into account the possible reactions of its opponent in its own actions. This demands the ability to understand the effect of an action on the competitor and therefore how the opponent will likely react. Such an incredible ability to understand a wide range of different effects allows mammals to develop a rudimentary understanding of causal relations, which is a known characteristic of rodent behavior. In other words, it is clear that an emotional system can have deep consequences for both cognition and rational plans or decisions (Damasio 1994; Damasio 1999; Damasio 2003; Damasio 2010).

1.2.3. Primates as Highly Social Animals

What has been said thus far allows us to familiarize with some basic forms of awareness before dealing with human consciousness. In order to understand the latter a little more, we need to look at our kin, the animals that have developed the behaviors and conditions out of which humanity has emerged: primates. Primates (as well as cetaceans) give rise to fusion-fission societies in which small groups, short-lived coalitions and not long-running alliances are constituted that confer high plasticity to their social organization. As such, these parties (comprising 4 to 10 members) are continuously formed and dissolved. Studies on chimpanzees in the wild show that intergroup policy is determined by several hierarchies (Boesch and Boesch-Achermann 2000): (a) the social dominance system, (b) the displacement leader, and (c) the hunt leader. For Humphrey (1976 and 1983) the natural equivalent of a laboratory test of intelligence is society. The game of social plotting cannot be played merely on the basis of accumulated knowledge acquired by chance. This raises completely new perspectives in learning and behavior. Indeed, subsistence techniques that are common among other mammals do not require truly creative intelligence, whose chief role is to hold a dynamic society together, since any social interaction here becomes a *transaction* between social partners. The main function of a primate society is to act as a polytechnic school for teaching to deal with highly sophisticated subsistence technology in two ways: (i) by allowing a period of relatively long immaturity of the young animals during which they are free to experiment and to explore, and (ii) by bringing them into contact with older members of the community. In such a complex social environment, primates are able to communicate signs about specific dangers (e.g., leopards and eagles) to their conspecifics. This communication system is already very complex since we have here a sign about (referred to) an event or object for a specific receiver that the primate is addressing (Cheney and Seyfarth 1990). For this reason, primates have also developed a remarkable ability to read and interpret the behavior (but not necessarily the intentions) of conspecifics (and even of humans) with a significant capacity for emulation, also showing a certain affective interaction with other primates (Auletta 2011a, Secs. 18.2, 18.6, 20.2 and 21.2). Summarizing, primates have developed a higher form of awareness: the awareness of the consequences of the actions of other primates (including humans), which allows for ternary social relations, and perhaps a rudimentary awareness of their own body.

Awareness of the environment, of the effects of our actions on others, and of the effects of others' actions do not demand a particular ability apart from an executive or at most a program-based motor system able to give rise to sequences of coordinated actions (see Fig. 2.2, a remake of a scheme by M. Jeannerod). This is completely understandable in terms of the so-called intentions in action and therefore of ordinary excitation patterns of the brain. A paradigmatic aspect of this is represented by mirror neurons

(Rizzolatti *et al.* 1990). These are neurons that discharge when an action is performed and also when it is only observed. Mirror neurons are therefore active when primates perform certain tasks, but they also fire when they watch another primate performing the same specific task. Arbib has particularly stressed, in his contribution (Chap. 3) and elsewhere (see the literature quoted by him), the crucial relevance of mirror neurons for the emergence of language in humans. Nevertheless, although understanding the action of another in such a reflexive way is impressive, it does not demand a special super-control instance for directing those motor processes. What is likely to be at the threshold between biological-neural awareness and, let us say, mental consciousness is the so-called mind reading, which is the awareness of others' minds (more correctly it is the guessing of others' intentions). It is a matter of debate whether non-human primates are able to read the behavior of others or also the mind. However, we shall not deal with this problem here.

1.3. HUMAN CONSCIOUSNESS

The preliminary issue as to human consciousness is a methodological one. Traditionally, in psychology and philosophy the methods employed for dealing with consciousness were phenomenological introspection and subjective report. Recently (Dehaene and Changeux 2011, p. 201), it has been shown that subjective and objective methods arrive essentially at the same results, which sets this whole matter on firmer grounds. Conscious processes can indeed be studied experimentally by comparison with closely matched unconscious processes, an approach that Baars (1988; 2002) has called "contrastive analysis." As to this general methodology, we like to quote Baars words (1988, Sec. 1.1):

> Our view here is that *both* conscious and unconscious processes involve inferences from publicly observable data. Thus conscious and unconscious events reside in the same domain of discourse — the domain of inferred psychological events. From this perspective William James was wrong to insist that all psychological events must be reduced to conscious experiences, and behaviorists were equally wrong to insist that we cannot talk about consciousness at all. Once we accept a framework in which

we simply try to understand the factors underlying the observations in exactly the way geologists try to understand rocks — that is to say, by making plausible and testable inferences about the underlying causes — the way becomes much clearer.

1.3.1. The Problem of Human Consciousness

Relative to the previously examined problems, a different issue is whether humans have consciousness or self-consciousness. As Marcus Raichle recalls in his contribution (Chap. 8), there is neural evidence for the existence of an interpreter in the human brain (in the left hemisphere), i.e., of a system for integrating facts in a unitary manner. This phenomenon is proved in patients who present a dissociation of the left and right hemispheres (Gazzaniga *et al.* 1998, pp. 672–675). Following W. James, we can take consciousness to be a sort of attentional-supervisor (James 1890, Chap. 5; Dehaene and Changeux 2011, p. 209). Damasio (especially in the discussion at the end of Chap. 4) has also spoken of a super-binder, although probably referring to another brain region. Let us therefore provisionally assume that there is a consciousness. The only character that reasonably does not seem to be reducible to the biological-neural processes that we have sketched so far is that of a reflexive consciousness, that is, self-consciousness. To be very specific, let us distinguish between two different kinds of self-consciousness. The first is objective self-awareness, which is a refinement of the biological self and encompasses the sense of one's own boundaries as well as, in several forms, considering the self as an object (physically, socially, psychologically). The second is subjective self-awareness, which requires identification with a certain class of stimuli and has some cognitive components such as personal memories (together with a sense of one's own continuity), representation of oneself (and a sense of one's own distinctiveness), theories about oneself and one's own agency power. This corresponds to the third level of selfhood pointed out by Damasio, that is, the autobiographical self. Three basic requirements for speaking of a personal self have also been distinguished (Blanke and Metzinger 2009): sense of the whole body, space-time localization, and first-person perspective. The first two are the basis of objective self-awareness, the third of the subjective one. These two varieties together constitute self-consciousness.

1.3.2. Unconscious and Conscious Processes

Let us first recall the main differences between unconscious and conscious processes. Unconscious processes are highly efficient in their own tasks, whereas conscious processes seem inefficient. Whilst unconscious processors have a limited range over time and each one is relatively isolated and autonomous (i.e., modular), conscious ones have different contents over time and can relate both different contents and conscious events to their contexts. By context we understand, along with Baars (1988, Chap. 4), a large cognitive envelope arising from a subjacent unconscious information-processor, which without being conscious itself, can shape or contribute to shaping conscious experience. For this reason, conscious processes show a fundamental limitation not only in the number of independent features that can in principle be processed (just four to seven) but also in the number that can be simultaneously attended to (just one), that is, discriminated within a single conscious state without interfering with the integration and the coherence of that state (Dehaene and Changeux 2011, pp. 209–210). Unconscious processes, instead, may work in parallel. We may assume that unconscious contexts are built from specialized processors when several of them cooperate in determining a routine and in shaping our conscious experience without being conscious at the same time. Here, a remarkable level of integration of different inputs can be found (see Fig. 10.5 in Auletta and Colagè's contribution). These contexts include unconscious expectations and goals. However, unconscious contextual assumptions can become consciously accessible (and therefore conscious contexts) when they are violated or when one learns a skill (a form of decontextualization): problems represent sources of special attention towards the implicit premises of activity.

Summing up:

- Unconscious processors are various, can operate in parallel and together have a great capacity, whilst
- Conscious ones have internal consistency, seriality and limited capacity.

The limited capacity, the serial succession, and the sensibility to interference of conscious processes are the price paid for their integration capability. Instead, unconscious processes are blind and proceed by ignoring changes in task and context. Specialization is good for routine situations where one applies a well worked-out algorithm, but this is tempered by a loss in flexibility (Baars 1988). Conscious processes are context-sensitive, since they are always consistent (as concepts are), whilst the unconscious ones may also be inconsistent (due to the absence, or at least the low level, of integration). Indeed, consciousness can bend or shrink but does not tolerate breaks in coherence or continuity. Interestingly, people who suffer from hemineglect (the failure to be aware of objects located on the left or on the right of a visual field) deny that anything is wrong with them. It is reasonable to assume that neural processes underlying consciousness also show integration and differentiation. For these reasons, when dealing with consciousness, the hypothesis of a global workspace has been introduced (Baars 1988; Changeux and Dehaene 2008); see also Auletta and Colagè's contribution to this volume, Chap. 10 (a point also supported by Singer).

Therefore, consciousness is sometimes taken to be a slower and less reliable process than unconscious processes occurring in the brain. Indeed, if we consider the survival value of consciousness, we may arrive at the conclusion that it is not adaptive, since it may be a disadvantage when dealing with quick choices or competition in a non-social environment. However, when considering the social aspects of the primate life, consciousness turns out to be of highest relevance, since it has co-evolved with a highly sophisticated social environment that is finally evolved into a cultural one. Humans are indeed fully social animals. Our symbolic systems, like language, are constituted in a social exchange. They are indeed conventional and presuppose a pragmatics in order to share common meaning (Putnam 1988). Moreover, it is well known that when children learn to speak, they are able to follow the gaze of the mother in the search for a shared intentionality towards a third party or object (Tomasello 1999; Tomasello 2003).

1.3.3. Symbolic Systems and Language

We may take for granted that, when we perform habitual actions, we do this in an unconscious way and that consciousness is only activated when dealing with new and especially problematic tasks (Dehaene and Changeux 2011, p. 208). As we have seen, unconscious processes proceed in parallel. We may assume that a hierarchy of tasks and unconscious contexts is necessary for a conscious being showing a supervising instance. This is rooted in the symbolic nature of the world in which humans live. Arbib, in his contribution (Chap. 3), has stressed the tight connection between language and consciousness by saying that human consciousness seems qualitatively different in nature, since it incorporates the functions of the brain that express language.

When we speak of language, it seems quite normal to consider the Chomskian school, probably the most relevant one in cognitive studies to date. Indeed, Pierre Jacob has devoted his contribution to this issue. It is well known that Chomsky's program is the naturalization of the mind and an attempt at treating the language system avoiding traditional dualistic and intentional assumptions. However, Chomsky's program was essentially a methodological naturalism. Indeed, as Jacob recalls, in his later work Chomsky stressed that his research program tries to go beyond the traditional monism/dualism antinomy and in particular tries to avoid a strict reduction to a form of physicalism affirming that physical theories are essentially influent on such a research program. Taking the opposite position, Jacob stresses that Chomsky's epistemic goal of theoretical unification in cognitive science could be better served by monist physicalism than by Cartesian substance dualism, thus putting the problem in ontological terms. This seems to be not completely in agreement with another central tenet of Jacob, the idea that the epistemic unification between computational theories of the human language faculty and neuroscientific theories of the human brain may well involve and presuppose an *ontological* distinction between computational *properties* and lower-level (presumably physical) *properties* of the human brain. Indeed, if we would like to reduce everything to a physical level, should not a computational process itself be understood in causal-physical terms? This is certainly what happens in a machine like a computer (as stressed by Fodor, again quoted by Jacob), so the individuation of a computational level as distinct from causal-mechanical explanations (see the fifth session of the Overall Discussion, Chap. 12) seems to be not completely justified. This would be justified only if we maintained that the mind and the language system, in their computational processes, showed aspects or properties that cannot be reduced to the physical and mechanical-causal level of explanation. As a matter of fact, the only sentence explicitly examined in Jacob's contribution (Chap. 11) is a piece of evidence showing that linguistic reference is an intentional action. However, if we introduce the notion of intentionality, we are already using a language that is likely to be not naturalistic in the sense of

Chomsky or Jacob. Moreover, Chomsky seems to assume a basic distinction between tokens and types. We are not able to judge whether or not this distinction is consonant with the naturalistic program of Chomsky himself, but certainly it does not fit stronger versions of physicalism. It is likely that the way to deal with this problem in truly naturalistic terms is to start with functional equivalence classes as something having a biological support, as stressed in Auletta *et al.* (2008) and in Crowell's contribution (Chap. 9), as well as in several interventions of the latter, especially in the fifth session of the Overall Discussion (Chap. 12). This is likely to be the reason why Changeux (in the same discussion, for instance) and others have contested the possible relevance of Chomsky's positions with regard to the issue at stake in this workshop.

The latter approach can be furthered by considering language as embedded in the general symbolic capacity of human beings, because this makes the problem manageable from the evolutionary, neural and cultural perspectives, without putting these approaches in conflict. A common characteristic of all symbolic systems is that they show information codification. Letters, numbers, gestures and other symbolic elementary units are combined (in natural languages, in mathematics, in liturgy, and so on) according to syntactic rules in a way that is both linear and hierarchical, with a partial independence of each level, allowing in this way the recursivity of any symbolic system (Auletta 2011a, Chap. 19). Thus, symbolic systems can give rise to items that have no representational counterpart (e.g., concepts like infinity, science, truth, imaginary numbers), which are so relevant to any human culture (and especially to cognition) and constitute the unique human ability to invent objects and even to reinvent the environment. These concepts without representational counterpart are so relevant because they allow management, inference and processing of aspects of our world that are hidden to our direct experience (like the concept of force). Symbols are indeed organized in huge hierarchical semantic networks according to both thematic (such as causal, spatially or physically belonging, and so on) and taxonomic (levels of conceptual generality, conceptual belonging) structures (Auletta 2011a, Secs. 20.4–20.6).

This information codification cannot be accomplished without having a *finite* number of such units. When W. Singer says in his contribution (Chap. 7) that information codification is ensured in the brain by the presence of 10^{11} neurons, we would say that this is the surest mark of the fact that there

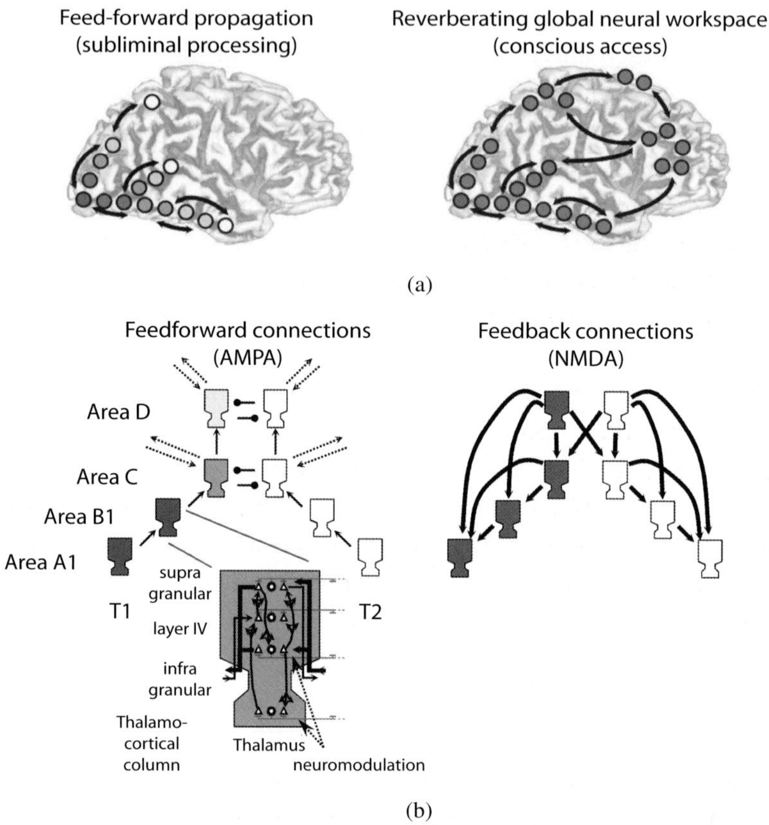

Fig. 1.1 (a) A bottom-up unconscious process is shown on the left whilst the top-down conscious effect of ignition is shown on the right. (b) The feedforward propagation through different levels of the global workspace is shown on the left whilst the feedback reverberating connections are shown on the right. (Adapted from Dehaene and Changeux 2011).

is no codification of the excitation patterns, because such an astronomic number would immediately lead to a combinatorial explosion and therefore to the impossibility of performing any search (among memories, percepts, and so on). Obviously, the single neurons are codifying units (and also the information in the peripheral sensory system is codified), but when information is combined in the brain through self-organizing excitation patterns (an issue very much stressed by Singer himself; see again Fig. 1.1), it can no longer preserve any trace of codification. Patterns arise and are combined with other patterns without any syntax ruling these combinations.

Symbolic systems can arise even in the presence of this fundamental limitation of the brain because humans have the unique capacity to make use of bodily or external (at least, extraneural) physical items (like vocalizations, gestures, stones, computers, and so on) as units of an informational game, as codifying units. Therefore, it is not the brain's excitation patterns being codified, but such bodily or external elements are used for this purpose. It is the grandest trick of human intelligence. Another aspect is the social dimension: symbols can only exist as long as they are socially shared. This explains also why dead languages are easily forgotten if nobody uses them.

Symbols are crucial for dealing with conscious processes. Indeed, due to its globality, the mind needs to have a common language (a *lingua franca*), while this is not necessarily the case for ordinary neural unconscious processes, as clearly stressed by Baars (1988, Sec. 1.5). So, there is no universal language of the brain as such (we have indeed said that symbols can exist and be managed only thanks to extraneural resources). Then, such a *lingua franca* is necessarily symbolic and mental.

1.3.4. Unreflexive Consciousness

In the current literature, one often takes into account only reflexive consciousness or self-consciousness as crucial (Libet 2004). To a certain extent it is necessary, since we have seen that it is the only variety of awareness that we assume to be essentially different from the kinds present in other animals. However, when focusing on reflexive consciousness, it is likely that some of its fundamental mechanisms are obscured. On the other hand, it is also likely that the awareness of the environment (and of others) is crucially different for a being that is endowed with reflexive consciousness and symbolic systems. Therefore, it could be interesting to have a look at some basic mechanisms of non-reflexive consciousness in humans (Auletta 2011a, Sec. 21.3). In one of his interventions following his own contribution (Chap. 5), M. Jeannerod seems to hint at this distinction when he says that the self-consciousness that you get after a failure of the unconscious-automatic system is not at the same level as when in the conscious system the state prior to the action and the state after the action are compared. In other words, the self-consciousness at the first level is not of the same nature of that reached when the other system is at play. In fact, this is the reason why Jeannerod makes the distinction between actor and author. In the first case, self-consciousness is transient, discontinuous, and vanishes as soon as one passes to the next step, whereas in the second, self-consciousness has a temporal window that exceeds the action duration and is continuous over successive episodes. Let us therefore consider here the first kind of awareness and in the subsequent section, the latter.

The conceptual hierarchy that is typical of symbolic systems is relevant to this enquiry, since our guess is that it is precisely this kind of organization that makes all forms of human awareness different relative to those characterizing other animals and gives us a good starting point. The fact that this hierarchy is modularized allows different unconscious processes to go on at different levels without giving rise to conflicts (and this is likely to be true also for other animals). However, when in our experience some problems arise, when some incongruence comes out, we may assume that it has a destabilizing effect. If such an effect can again be dealt with thanks to small local rearrangements and reorganization (where thematic relations predominate), this can still be done in fully unconscious ways by essentially preserving the previous hierarchical order of contexts and tasks. This is why M. Jeannerod has stressed in his contribution (Chap. 5) that not every error correction needs to be conscious. For instance, it is well known that trajectory corrections during a prehension act are mostly unconscious (Milner and Goodale 1995; Auletta 2011a, Sec. 5.1). However, sometimes the destabilizing effect determines a disruption of contexts that can no longer be managed with ordinary non-conscious means. In this case, it is necessary that a global action will be undertaken with the result giving rise to new hierarchies of tasks and contexts (heavily affecting taxonomic relations). The necessary condition is that previous (unproblematic) unconscious contexts become consciously (and attentionally) accessible in order that new connections between (even distant) contexts are established [see also Dehaene *et al.* (2006)]. It is here that the mentioned computer-simulation based hypothesis of a global workspace could become helpful, as recalled in Auletta and Colagè's contribution [Chap. 10; see also Changeux and Dehaene (2008); Fig. 1.1]. It is not unusual in the natural sciences that qualitative theory often precedes quantitative or formal theory (Baars 1988, Sec. 1.3). Nevertheless, we raise here the question whether the model should not be already integrated with a more systemic view of the brain in which the three major systems and their

relations are considered, namely the sensory one (located mainly in the posterior part of the brain) that processes input information, the emotional one (located at the base of the brain) that deals with emotionally-charged events and experiences, and the motor-decisional one (located in the prefrontal part of the brain) that represents the governing entity for motor outputs (Auletta 2011a, Chap. 13). The neural connections governing each of these circuits are well established. In themselves, these systems are empty or only provide the context and not the content (Changeux 2002, p. 79). Then, the decisional system in itself can only give rise to attention as the empty *alert state* thanks to which we enhance perception and the further elaboration of certain items and decrease or suppress elaboration of other items; the emotional system can only produce a generic *arousal state*; the sensory system can produce a default state of information-processing. When cooperating, these systems are able to provide a *supplementary* selection of a stimulus or neural event and to produce the three main functions of memory, learning and behavior (Auletta 2011a, Chaps. 14–17; see also Auletta and Colagè's contribution to this volume, Chap. 10). Our main point here, therefore, is that it is worth further studying how consciousness is related to, or supported by, an interconnection of these diverse systems and functions as intermediary instances between specific excitation patterns and a global state or workspace.

We can summarize the preceding discussion by saying that an external, very dissonant stimulus can break the ordinary unconscious information-processing activity, being the source of some conflicts that need to be solved. It seems that already Pani has hypothesized this mechanism as basilar for consciousness, and Baars calls it deautomatization or decontextualization (Baars 1988, Secs. 1.4 and 4.1): strong violations of contexts can become consciously accessible. Although the main elements of this explanation are due to Baars, it remains true that when he says (Baars 1988, Sec. 5.1) that any learnable task goes from context-creation to conscious information and then to redundancy, he seems to reverse the natural order, since we shall assume that unreflexive consciousness follows damage to, or disruption of, contexts, its effect being instead the establishment of new relations among different contexts, whilst reflexive consciousness (likely to be equivalent to Baars' conscious information) will precede the updating or even the creation of new contexts out of this cross-contextual experience (Auletta 2011a, Sec. 21.3). Therefore, in view of the data about motor cognition to be analyzed below, we would not say that the function of consciousness is also to recruit a goal context in order to help maintain relevant information on the workspace (Baars 1988, Sec. 5.2), since most of the goal contexts are fully unconscious. Nevertheless, Baars (1988, Sec. 5.1) acknowledged that consciousness may have the function to create new contexts. As a consequence, a conscious state arises that, having taken notice of the problem, starts a jump-like conscious search of new possible solutions. Due to the global interconnections established during the conscious state, consciousness is likely to bring together unconscious contexts that were previously unrelated. If this represents a good, if not optimal, solution to the problem, it will give rise to a new series of unconscious actions (which are in general not related with the series of actions that was performed before the destabilizing stimulus) able to operatively deal with the previous problem. If this solution is considered particularly new and enlightening, it is what we ordinarily call an insight. This explains the evolutionary significance of consciousness (because it allows us to solve otherwise intractable problems), and its rooting in the social behavior of other primates (since it is here that particular difficult problems arise that demand very flexible, novel solutions).

1.3.5. Reflexive Consciousness

Let us now consider a very simple example. Suppose that somebody drives a car. If the path is known (e.g., between home and office), he or she will perform this without taking care of the executed operations and therefore in an unconscious modality of action. However, if a danger is suddenly present, the driver will become immediately conscious and a quick search will give rise to a new series of actions that are again performed unconsciously but are unrelated to the previous ones, like suddenly turning the steering wheel or braking.

Now suppose that so far the solution that has been found is a good one, even an insight. This may turn out to be a surprise also for the person to whom it occurred. Indeed, this is the result of a jump from one context to another or of putting together previously unrelated contexts. Then, the issue is how, starting with this insight, we are able to restructure the hierarchical space of contexts. This is also known as ignition (see Fig. 10.5), that is, the self-reverberating of this conscious state (Dehaene and Changeux 2011,

p. 211). In other words, this new solution needs to be assimilated and incorporated in the web of previous relations, and this will likely be done in two steps. The first step is a conscious one; it is even what ordinarily is taken as consciousness. This is *reflexive* consciousness. Here, we take the solution (and not the external stimulus) as an input and undertake a similar process as that described for unreflexive consciousness. Since this solution, especially when it is an insight, is something new that still has a destabilizing power as the original external stimulus had, the issue here is to search for connections between contexts allowing the right contextualization of this novelty. This shows that a solution of this kind is not optimal for the organism and is searched only insofar as it is an obligatory way out. Conversely, as mentioned, this shows how powerful consciousness can be for solving problems that are otherwise intractable both for a machine and for an unconscious animal: in the first case, since this modality of action allows for quick solutions even when not all the initial conditions and data for solving a certain problem are available; relative to an unconscious animal, since it allows for solutions that, although apparently illogical from a biological point of view, are able to save our lives in many concrete situations. It is very probable indeed that no other animal (in the fictional situation in which it would be a driver) would react like a human when facing, for example, a truck about to collide with the car: the animal would jump out of the car, or at least try to and not try, instead, to get the car under the control in order to avoid the collision.

Now, the business of the reflexive consciousness is to give rise to new connections between contexts able to deprive the insight of its potentially disruptive power. But this is only one side of the problem. If the insight or solution was good, we need to be able to use it in other concrete situations, which implies that we need to integrate and to generalize it in order that becomes itself part of the hierarchical, web-like organization of our mind. If this reflexive operation is successful, this means that we have began to contextualize this novelty. Now, the process can go on in an unconscious way (this is the second step) in terms of tightening certain connections in order that this solution becomes full available for further situations. This shows that reflexive consciousness has the same modality of operation as other forms of awareness. The bridge is provided by the unreflexive human awareness of the environment that is both similar to the awareness of other species and unique in its ability to jump contexts.

1.4. INTENTIONS IN ACTION AND PRIOR INTENTIONS

1.4.1. Action and Consciousness

The biggest difficulty here is how to relate an active role of consciousness with the fact that it does not seem to be a causal agency at all when motor acts are involved. Indeed, being aware of a goal is not being aware of how it is reached. This is the main subject of M. Jeannerod's contribution (Chap. 5). Visuomotor representations lack conceptual content and are not conscious (Jacob and Jeannerod 2003; Jeannerod 2006). Interesting studies show that intentional and conscious processes tend to temporally shorten the time delay between the onset of the cause (for instance, pressing a certain key) and the onset of the effect (for instance, an auditory signal) (Haggard *et al.* 2002); see Fig. 2.4. In a series of experiments originally performed by Nielsen (1978) [see also Jeannerod (2009), pp. 185–202], subjects believed that they were following the motion of their own hand, while in fact (thanks to a reflection in a mirror) they were following the movements of the hand of an experimenter, which deviated more or less from the subjects' intended motions. In order to compensate for these deviations, the subjects adjusted their own movements in a direction opposite to that of the hand that they could see. However, they were unaware of their compensatory movement. These experiments have been further developed by Daprati *et al.* (1997) in order to ascertain self-recognition and ownership (see also Fig. 5.1). A subject was asked whether the hand he/she saw was his/her own or not. Three different images could be shown in each trial: (i) his or her own hand, (ii) the experimenter's hand performing a different movement, and (iii) the experimenter's hand performing the same movement. Test subjects performed well in cases (i) and (ii) but much worse in case (iii). Therefore, in this case, subjects acknowledged actions, rather than hands or other parts of the body, as their own. Since the sense of discrepancy in the attribution of movement is fundamental here, all these data could be sufficiently accounted for in terms of a default configuration in which motion is self-attributed if the effects are exactly those that have been expected given those decisions. However, when there is a certain discrepancy, the subject must search for another explanation.

Other experiments showed that subjects may be conscious of a goal but not of motor performances and therefore they would not be conscious of the way to achieve it or of the performance that is needed (Fourneret and

Jeannerod 1998). Slachewsky *et al.* (2001), pursuing this line of study, showed that subjects may become aware of an abnormal deviation from the willed trajectory (due to distortion effects) only when the discrepancy passes a certain threshold, for instance an angle greater than 14° when tracing lines with a pen on paper. This confirms the fact that consciousness is always bound to some form of failure originating in the sensory data or in the accomplishment of a movement. Conscious experience about one's movement deals with the later matching of two sorts of data: central and peripheral signals. In conclusion, some have assumed that consciousness comes too late to be a causal agency (Wegner 2002) [see Haggard (2008) and Schultz *et al.* (2004) for criticism].

1.4.2. Plans and Prior Intentions

The previous examples also show that the subject may become conscious when something goes wrong, accordingly to our previous analysis of unreflexive consciousness. Using the words with which Jeannerod makes this point in his contribution (Chap. 5), when an action that was unfolding automatically and outside the subject's awareness fails because of some unexpected event, this automatic action is stopped and remains incompletely executed. As a consequence of the mismatch between the desired motor output and its observed result, the action is deautomatized, delayed, and will continue under conscious control.

Moreover, the distinction between primary or prior intentions (expressing long-term planning) and intentions in action (the motor intentions producing movements in due course) could allow for a different evaluation of consciousness even in its *anticipatory* quality (Searle 1983, pp. 79–111). Indeed, if consciousness is bound to prior intentions and long-term plans, it could represent an integration system able to frame different specific actions in a wider context of planning. If so, it may be able to determine the general conditions in which a specific act (and therefore an intention in action) comes into play. Top-down causation of endogenous cognitive processes requires that some representation of the goal precedes the action that is aiming at achieving the goal, as suggested by studies showing a hierarchy of the body motions, such that the sub-segments are nested in larger motor segments, like Russian dolls (Bernstein 1967; Jeannerod 1988). As Jeannerod says in his contribution (Chap. 5), all of us may occasionally experience feelings of concentration, fatigue or even exhaustion when faced with a difficult problem, e.g., a multiple choice decision, even when no action is actually planned. These feelings show many similarities with physical fatigue when fighting against a resistance. Therefore, prior intentions are able to influence motor processes in a top-down fashion. Indeed, Jeannerod says that when "the information transferred, say, from the cognitive network to the motor network undergoes transformations, such that the output of the motor network will not be a simple copy of the input received from the cognitive network, it will represent new information where the action will have acquired new characteristics as a result of this top-down control. In turn, after completion of the action, these new characteristics will become observable and will affect the content of the cognitive network" in a bottom-up way. The process described by Jeannerod may show some important connections with the previously mentioned self-reverberating process of ignition introduced by Changeux and coworkers.

Therefore, by making use of a sort of two-way dialogue between different levels, the will could determine the (functional or motor) equivalence class to which several concrete actions (as well as intentions in action) and means belong. This equivalence class is determined by the general goal of the primary intentions and by its subgoals. Indeed, when dealing with goals and intentions, the way in which they are executed does not matter, but only their final outcome, as a representation of the goal. Thus, the specific way in which an action is performed is not controlled by consciousness, but its functional value is (see Auletta's contribution, Chap. 2). If this is correct, consciousness could still be a causal agency of actions but only in an indirect and constraining sense, and not in a direct and dynamic sense as J. Searle still assumed (Searle 1983, pp. 94–95). Evidence for this is the circumstance that, when consciousness is at play, primary motor areas are not involved but only high-level decisional areas located in prefrontal and lateral cortices (angular gyrus) (Dehaene and Changeux 2011, p. 207).

So, in what way can consciousness exert this causal influence? If it is a framing, rather than a direct, dynamical action, it is likely that its function is rather more negative than positive. Indeed, interdependencies in general (even at a physical level) do not possess the effectiveness to start actions but can very well canalize courses of action by blocking inappropriate actions already started without conscious control. This is precisely what Libet (2004) assumed [see also James (1890), Vol. 1, p. 140]. We do not need to concern ourselves with all the specifics of his model (there is a detailed analysis of Libet's experiment and of some of the further experiments performed by Haggard and others in Crowell *et al.*'s contribution to this volume, Chap. 9), but the idea of consciousness as a sort of instance of veto is very promising. Indeed, a series of appropriate blockages can in principle redirect a course of action along a planned path.

It seems to us that, in his praise of the self and of the mind as a cause (see Chap. 6), Sanguineti, as he explicitly admits, touches many issues of great relevance that would certainly need a further deepening in a larger context. Such further work is very welcome if we desire that some of the distinctions or considerations that he introduces can be made operative and therefore helpful for making further advances in this field.

1.4.3. Failure and Systemic-Functional Reality of the Self

If the formulated hypothesis is correct, how can we deal with the previously described failures? Let us go back to some very general systemic and functional considerations. A system like an organism is characterized by the specific cluster of relations, correlations and/or interdependencies that holds among its subsystems; in brief, let us say, by its configuration. The same subsystems arranged in different configurations generally provide to the whole system different properties, different (potential, and not merely possible) behaviors, different potential answers or reactions to stimuli, perturbations or actions, different capabilities (even endogenous ones). If we understand both the biological and the personal self as integrated systems in dealing with the environment, this implies that any organism both accommodates itself to the external environment and assimilates the latter to itself (consider what bacteria or plants do). Both aspects constitute what we call adaptation (Auletta 2011a, Sec. 8.4). This is a difference relative to any prebiotic physical system. Also, these systems reach an equilibrium with the environment, only that in this case we can speak neither of accommodation nor of assimilation. The split of the "equilibrium-reaching" process in these two sides is precisely due to the dividing line between what is external (non-self) and what is internal (self). Moreover, this gives rise to the fact that any organism tries to preserve its organization not in spite of change but through change (with an itinerant dynamics). Given these considerations, we expect that the self, at any level of complexity, is subjected to errors or functional anomalies precisely because any self continuously changes and adapts due to the modifications of the contexts in which it works and deals. We could even say that errors and dysfunctions are the surest mark of the reality of consciousness and not only of a sort of phenomenological, ghostly existence.

Therefore, the previously described failures do not imply a denial of the self or of the consciousness, since the business of consciousness is to be a higher (global) control (and therefore also integration) instance intervening when there are characteristic discrepancies between what one expects and what is the case. Then, it is obvious that the self does not apperceive errors under a certain threshold (apart from pathological cases). In this sense, consciousness arises and plays a role only *before* (when planning) and *after* certain very important discrepancies have been produced, but never in the course of the action itself. An old-fashioned view of planning (still determining the research in AI until the 1990s) assumes that plans deal with all the details of the planned action. This is precisely what rational plans do not and cannot do (Suchman 1987). A very common experience is that decisions are taken simply to diminish the dimensions or variables of a given problem and therefore in order to take further decisions. It is obvious that, when starting with the mentioned wrong assumption, once that we discover that intentions in actions are not controlled by consciousness, we also deny any causal value to consciousness. Nevertheless, our projections and our retrospective reconstructions are the principal means by which we catch hold of a situated action and reason about it, while a situated action in itself, by contrast, cannot be fully grasped by us as actors (we live here *within* the intentions in action). However, the present course of choice and action to a certain extent is *per se* an abstraction that always involves both reconstructions and projections. Indeed, there is a continuous advancing horizon of (prior) intentions. Furthermore, there is also a continuously receding horizon of understanding to be accounted for. For this reason, deciding how the background premises of action should be concretely enumerated remains an *ad hoc* procedure (Garfinkel 1967). Instead, a background assumption is generated by the need to account for an action when the premise of the action is called into question.

Obviously, the continuity of the author (in Jeannerod's terminology), that is, of an agent that lasts not only for the time interval of a given motor performance but for a long time span allowing both predictions and retrodictions, could be questioned. As a matter of fact, there are characteristic failures also in this case (Jeannerod 2009, pp. 225–269). Nevertheless, again consciousness, and its activity of planning, does not need to be, and even cannot be, infallible. What really matters is that, in certain situations, it can causally influence the course of action.

1.5. NEURAL ARCHITECTURE

There is neural evidence for the existence or role of consciousness. Obviously, we can find neural correlates here, which is not proof in a strong sense. However, the findings are amazing. Indeed, fMRI techniques show that ordinary unconscious processing areas are activated more intensively, as well as new ones being activated, when conscious processes go on (Dehaene and Changeux 2011, pp. 202–207). We can therefore assume that consciousness is correlated with a sort of global state of the brain in which distant areas are connected and there is also an increased coherence between distal regions (Tononi and Edelman 2000). Such a state fits very well with the idea that the primary function of consciousness is to provide connections between distant contexts that, at least in many cases, are likely to be processed in brain areas located far away from each other. Moreover, it is likely that the brain state corresponding to consciousness is a metastable one, that is, a state of equilibrium that can easily be disrupted through a small perturbation (Dehaene *et al.* 2003), as also stressed in Singer's contribution (Chap. 7). This explains why consciousness can last for a short time and is followed by an unconscious modality of action. Moreover, since any action, at least in mammals, is strongly connected with anticipation, which implies that energy has already been assigned to the movement before execution, a block or veto only has to redirect part of this energy along another path.

This global state is what allows the integration function of consciousness. Integration can be defined in terms of mutual information between cerebral subsystems and in particular as the difference between the sum of the entropies of the single brain components independently considered and the entropy of the whole system. We can refine this concept and define the neural complexity characterizing consciousness in terms of a cluster index (Tononi and Edelman 1998; Edelman and Tononi 2000). The cluster index for each set of subsystems is the ratio between the integration of this set and the mutual information between this set and the rest of the brain. Anatomically, we can indeed distinguish between three systems: (i) a meshwork of segregated yet integrated circuits constituting the thalamocortical loops (probably the basis of the different unconscious contexts), (ii) a set of parallel unidirectional chains that link the cortex with specialized structures like the cerebellum, and (iii) a large set of connectives resembling a large fan, influencing billions of synapses, which allows for the distributed nature of consciousness and therefore also for its integration capability. A crucial concept for understanding these kinds of connections is reentry (Edelman 1987). Reentry is very common between different brain areas and neuron groups. It is especially relevant when information already processed in higher areas feeds back into primary receiving areas (for instance, from visual integration areas back to V1 and V2), allowing in this way the reverberating top-down effects of consciousness on lower levels of the neural hierarchy, specific contexts and local information processors. It does not simply consist of feedback (that occurs along fixed loops using previous instructionally derived information for control and correction) but occurs in selection systems across multiple parallel paths where information is not specified in advance.

This shows again that consciousness cannot be understood without taking into account the characteristic hierarchy of levels of the brain. This is an issue that many have pointed out in their contributions and interventions. If we speak of a hierarchy, it is clear that we also expect to find some areas in the brain that sit at the top. Essentially, we have two main proposals in this sense: one is mainly due to Jean-Pierre Changeux and puts the stress on the prefrontal areas (Dehaene *et al.* 1998; Changeux and Dehaene 1998); the other, to Damasio and puts the stress on the posterior parts of the brain, with a shift of stress, over time, from the brain stem with its nuclei (Parvizi and Damasio 2001; Parvizi and Damasio 2003) towards the posterior medial cortices (PMC) and in particular the posterior cingulate area and the region called precuneus (Parvizi *et al.* 2006). It is interesting to observe that Baars' work (1988), which is at the origin of a more developed workspace hypothesis, is also crucial for the latter developments, since it individuated the so-called Extended Reticular-Thalamic Activating System (ERTAS), including the traditional reticular formation (essentially connected with the brain stem), the thalamus (that Auletta and Colagè's contribution (Chap. 10) considers crucial for the global workspace), and cortical regions that, at least in part, cover the PMC.[1] It is quite remarkable that the prefrontal area deals more with decisional and motor aspects, while the PMC seems to deal more with integration of sensory information. Then, an integration of these two views could be possible, which would also be in accordance with the three-system view previously presented if also the emotional component mainly concerning the central part of the brain is considered [see Chap. 10 and also Dehaene and Changeux (2011), p. 214]. Indeed, this is the focus of Raichle's contribution (Chap. 8). He deals with the problem from the perspective of the default mode of the brain function, that is, the baseline, endogenous activity that goes on in the brain when a human subject rests quietly but is simultaneously awake with his/her eyes closed [Raichle *et al.* 2001; see also Hagmann *et al.* (2008)]. This possible integration of these perspectives may even be considered as one of the most qualified and qualifying results presented in this volume. Moreover, this approach raises the very interesting question whether or not these two areas are also able to establish a connection in an active way, i.e., to dialogue directly through the kinds of connections pointed out by Raichle. Since the PMC and the prefrontal cortex are the top levels of ascending inputs coming from without and of descending outputs coming from within, respectively, this could explain a strict interrelation between prefrontal areas and unreflexive consciousness as well as between PMC and reflexive consciousness, respectively. These processes should be seen in sequence, so that a conscious decision (starting in the prefrontal part) or step could give rise to a reflexive re-elaboration of certain experiences that is likely to be more associated with the posterior parts of the brain (PMC). It is not by chance that the circuit more associated with these areas is considered as expressing self-referential processing and consciousness (Hagmann *et al.* 2008; see also Auletta and Colagè's contribution (Chap. 10) to this volume).

---

[1] In Auletta (2011a, Sec. 21.3), Baars' argument is still considered as a formulation of the classical reticular formation.

1.6. SOCIETY OF MINDS

In the Overall Discussion (Chap. 12), the issue of the society of brains emerged. In particular, W. Singer stressed that there could also be a top-down process going from the global state represented by the interaction among different brains to the global states of the individual brains and from here, in a bottom-up fashion, again towards the social global network. This is likely to be much more than a metaphor. As a matter of fact, as Singer always stresses, many of our behaviors, beliefs and attitudes can only be understood if we take this social globality seriously into account. Furthermore, due to the tools of language, of shared knowledge, of formal tools like mathematics, as well as collective experiences and experiments, such a collectivity represents the result of a true convergence process towards ideas, values, strategies, and so on, that have a general value (potentially able to attract other people, as we hope that this volume will do) if not potentially universal, at least in the intentions of the proponents.

This is made possible by the mentioned fact that we are symbolic animals, which means being able to share intentionality, information, to speak about what is not immediately present in our experience, and so on. However, this is something that other (non-symbolic) species cannot do. They can constitute groups, even social groups with the high complexity of primates, but the motivations are ultimately of a biological nature. In contrast, humans can meet and decide to freely submit themselves to a *tour de force* only for the sake of the enquiry. This is why we would prefer to speak of a society of minds rather than a society of brains, stressing here what is really universal and unique to our species.

Since we were, in fact, able to constitute such a community (as we become aware especially during the third session of the Overall Discussion), we think that it would be nice with this volume to encourage a free network of exchanges at this level, following a suggestion of Crowell and, if we are not wrong, also of Changeux.

REFERENCES

Atkins PW, De Paula J. (2006) *Physical Chemistry*. Oxford University Press, Oxford.

Auletta G. (2008) How Many Causes Are There? *21mo secolo. Scienza e tecnologia* **5**: 41–48.

Auletta G. (2011a) *Cognitive Biology: Dealing with Information from Bacteria to Minds.* Oxford University Press, Oxford.

Auletta G. (2011b) Teleonomy: The feedback circuit involving information and thermodynamic processes. *J Mod Phys* **2**(3): 136–145.

Auletta G, Ellis G, Jaeger L. (2008) Top-down causation by information control: From a philosophical problem to a scientific research programme. *J R Soc Interface* **5**: 1159–1172.

Auletta G, Fortunato M, Parisi G. (2009) *Quantum Mechanics.* Cambridge University Press, Cambridge.

Baars BJ. (1988) *A Cognitive Theory of Consciousness.* Cambridge University Press, Cambridge.

Baars BJ. (2002) The conscious access hypothesis: Origins and recent evidence. *Trends Cogn Sci* **6**: 47–52.

Bernstein NA. (1967) *The Coordination and Regulation of Movements.* Pergamon, Oxford.

Blanke O, Metzinger T. (2009) Full-body illusions and minimal phenomenal selfhood. *Trends Cogn Sci* **13**: 7–13.

Boesch C, Boesch-Achermann H. (2000) *The Chimpanzees of the Tai Forest: Behavioural Ecology and Evolution*, Oxford University Press, Oxford.

Changeux J-P. (2002) *The Physiology of Truth: Neuroscience and Human Knowledge.* Harvard University Press, Cambridge, MA.

Changeux J-P, Dehaene S. (1998) Hierarchical neural modeling of cognitive functions: From synaptic transmission to the tower of London. *CR Acad Sci III-Vie* **321**: 241–247.

Changeux J-P, Dehaene S. (2008) The neuronal workspace model: Conscious processing and learning. In: R. Menzel (ed.), *Learning Theory and Behavior.* Elsevier, Oxford, pp. 729–758.

Cheney DL, Seyfarth RM. (1990) *How Monkeys See the World: Inside the Mind of Another Species.* University of Chicago Press, Chicago.

Damasio AR. (1994) *Descartes' Error: Emotion, Reason, and the Human Brain.* (Last ed. 2000) Quill, New York.

Damasio AR. (1999) *The Feeling of What Happens: Body and Emotion in the Making of Consciousness.* (Last ed. 2000) Harcourt, New York.

Damasio AR. (2003) *Looking for Spinoza: Joy, Sorrow and the Feeling Brain.* Harcourt, New York.

Damasio AR. (2010) *Self Comes to Mind: Constructing the Conscious Brain.* Pantheon, New York.

Daprati E, Franck N, Georgieff N, *et al.* (1997) Looking for the agent: An investigation into consciousness of action and self-consciousness in schizophrenic patients. *Cognition* **65**: 71–86.

De Biran M. (1803) *Influence de l'habitude sur la faculté de penser.* Henrichs, Paris. Reprinted in: M De Biran. (1841) *Œuvres philosophiques* (ed. V Cousin), Vol. 1. Ladrange, Paris.

Dehaene S, Changeux J-P. (2011) Experimental and theoretical approaches to conscious processing. *Neuron* **70**: 200–227.

Dehaene S, Kerszberg M, Changeux J-P. (1998) A neuronal model of a global workspace in effortful cognitive tasks. *Proc Natl Acad Sci USA* **95**: 14529–14553.

Dehaene S, Sergent C, Changeux J-P. (2003). A neuronal network model linking subjective reports and objective physiological data during conscious perception. *Proc Natl Acad Sci USA* **100**: 8520–8525.

Dehaene S, Changeux J-P, Naccache L, *et al.* (2006) Conscious, preconscious, and subliminal processing: A testable taxonomy. *Trends Cogn Sci* **10**: 204–211.

Edelman GM. (1987) *Neuronal Darwinism: The Theory of Neuronal Group Selection.* Basic Books, New York. (2nd ed. 1989, Oxford University Press, Oxford.)

Edelman GM, Tononi G. (2000) *A Universe of Consciousness.* Basic Books, New York.

Fourneret P, Jeannerod M. (1998). Limited conscious monitoring of motor performance in normal subjects. *Neuropsychologia* **36**: 1133–1140.

Freeman WJ. (2000a) Brains create macroscopic order from microscopic disorder by neurodynamics in perception. In: P Arhem, C Blomberg and H Liljenström (eds.), *Disorder versus Order in Brain Function: Essays in Theoretical Neurobiology.* World Scientific, Singapore, pp. 205–219.

Freeman WJ. (2000b) *Neurodynamics: An Exploration in Mesoscopic Brain Dynamics.* Springer, Berlin.

Garfinkel H. (1967) *Studies in Ethnomethodology.* Prentice Hall, Englewood Cliffs, NJ.

Gazzaniga MS, Ivry RB, Mangun GR. (1998) *Cognitive Neuroscience: The Biology of the Mind.* W. W. Norton, New York. (2nd ed. 2002.)

Haggard P. (2008) Human volition: Towards a neuroscience of will. *Nat Rev Neurosci* **9**: 934–946.

Haggard P, Clark S, Kalogeras J. (2002) Voluntary action and conscious awareness. *Nat Neurosci* **5**: 282–285.

Hagmann P, Cammoun L, Gigandet X, *et al.* (2008) Mapping the structural core of human cerebral cortex. *PLOS Biol* **6**: 1479–1493.

Herbart JF. (1816) *Lehrbuch zur Psychologie.* Reprinted in: G Hartenstein (ed.), *JF Herbart (1850–1851)*, Vol. 5. L. Voss, Leipzig, pp. 1–187.

Humphrey NK. (1976) The social funtion of intellect. In: PPG Bateson and RA Hinde (eds.), *Growing Points in Ethology*. Cambridge University Press, Cambridge, pp. 303–317.

Humphrey NK. (1983) *Consciousness Regained*. Oxford University Press, Oxford.

Jacob P, Jeannerod M. (2003) *Ways of Seeing: The Scope and Limits of Visual Cognition*. Oxford University Press, Oxford. (2nd ed. 2004; 3rd ed. 2006.)

James W. (1890) *The Principles of Psychology*. H. Holt. (Reprinted: Dover, New York, 1950.)

Jeannerod M. (1988) *The Neuronal and Behavioural Organization of Goal-Directed Movements*. Oxford University Press, Oxford. (2nd ed. 1990.)

Jeannerod M. (2006) *Motor Cognition: What Actions Tell the Self*. Oxford University Press, Oxford.

Jeannerod M. (2009) *Le cerveau volontaire*. Odile Jacob, Paris.

Libet B. (2004) *Mind Time: The Temporal Factor in Consciousness*. Harvard University Press, Cambridge, MA.

Lloyd Morgan C. (1896) *Habit and Instinct*. E. Arnold, London.

Lloyd Morgan C. (1900) *Animal Behavior*. E. Arnold, London.

MacLean PD. (1990) *The Triune Brain in Evolution: Role in Paleocerebral Functions*. Plenum, New York.

Milner AD, Goodale MA. (1995) *The Visual Brain in Action*. Oxford University Press, Oxford. (2nd ed. 1996; 3rd ed. 2000.)

Monod J. (1970) *Le hasard et la nécessité*. Seuil, Paris.

Nielsen TI. (1978) *Acts: Analyses and Syntheses of Human Acting, Concerning the Subject and from the Standpoint of the Subject*. Dansk Psykologisc Forlag, Copenhagen.

Panksepp J. (1998) *Affective Neuroscience: The Foundations of Human and Animal Emotions*. Oxford University Press, Oxford.

Parvizi J, Damasio AR. (2001) Consciousness and the brainstem. *Cognition* **79**: 135–160.

Parvizi J, Damasio AR. (2003) Neuroanatomical correlates of brainstem coma. *Brain* **126**: 1524–1536.

Parvizi J, Van Hoesen G, Buckwalter J, Damasio A. (2006) Neural connections of the posteromedial cortex in the macaque: Implications for the understanding of the neural basis of consciousness. *Proc Natl Acad Sci USA* **103**: 1563–1568.

Putnam H. (1988) *Representation and Reality*. MIT Press, Cambridge, MA. (2nd ed. 1991; 3rd ed. 2002.)

Raichle ME, MacLeod AM, Snyder AZ, *et al.* (2001) A default mode of brain function. *Proc Natl Acad Sci USA* **98**: 676–682.

Rizzolatti G, Gentilucci M, Camarda RM, *et al.* (1990). Neurons related to reaching-grasping arm movements in the rostral part of area 6 (area 6aβ). *Exp Brain Res* **82**: 337–350.

Schultz J, Sebanz N, Frith CD. (2004) Conscious will in the absence of ghosts, hypnotists, and other people. *Behav Brain Sci* **27**: 674–675.

Searle JR. (1983) *Intentionality: An Essay in the Philosophy of Mind.* Cambridge University Press, Cambridge. (2nd ed. 1997.)

Slachewsky A, Pillon B, Fourneret P, *et al.* (2001) Preserved adjustment but impaired awareness in a sensory-motor conflict following prefrontal lesions. *J Cogn Neurosci* **13**: 332–340.

Suchman LA. (1987) *Plans and Situated Actions: The Problem of Human/Machine Communication.* Cambridge University Press, Cambridge. (2nd ed. 1994.)

Tinklepaugh OL. (1928) An experimental study of representative factors in monkeys. *J Comparat Psychol* **8**: 197–236.

Tinklepaugh OL. (1932) The multiple delayed reaction with chimpanzees and monkeys. *J Comparat Psychol* **13**: 207–s43.

Tomasello M. (1999) *The Cultural Origins of Human Cognition.* Harvard University Press, Cambridge, MA. (2nd ed. 2003.)

Tomasello M. (2003) *Constructing a Language: A Usage-Based Theory of Language Acquisition.* Harvard University Press, Cambridge, MA.

Tononi G, Edelman GM. (1998) Consciousness and complexity. *Science* **282**: 1846–1851.

Tononi G, Edelman GM. (2000) Schizophrenia and the mechanisms of conscious integration. *Brain Res Rev* **31**: 391–400.

Wegner D. (2002) *The Illusion of Conscious Will.* MIT Press, Cambridge, MA.

Wundt WM. (1873–74) *Grundzüge der physiologischen Psychologie.* (2nd ed. Engelman, Leipzig, 1880.)

Wundt WM. (1893) *Principles of Physiological Psychology.* Allen, London. [English translation of Wundt (1873–74), Vol. 1. New York, 1904.]

CHAPTER 2

A Constraining Role of the Mind?

Gennaro Auletta

In starting my studies, actually through the work of Marc Jeannerod (1988; 2006), the concept of motor cognition soon became very important to me. As far as I can understand, in the context of neurocognitive studies, this notion is rooted in the work of Bernstein, on the one hand, and in that of Hebb on the other. Bernstein (1967) was aware of the fact that goal-directed movements cannot reflect a one-to-one relationship between the specific patterns of motor-neuron firing (or indeed the forces generated by the muscles that you use) and the actual movement produced. Hebb (1949) speaks about the different motor actions as schemes that share important general characteristics, even if performed in different ways. The fact is that single motor acts may be embedded in more complex ones in both a serial and a hierarchical-modular sense. In this way, the different parts can be assembled moment by moment when a certain need or stimulus is present, giving rise to a class of similar whole structural patterns that we call movements. Since within a certain tolerance window, many minute differences in the execution of the subsegments do not matter at all, we need to consider a whole movement as an equivalence class of *chains* of motor segments able to obtain the same result. In other words, several individual motor segments are (within certain limits) appropriate for combining in a pattern to generate a certain result, thereby giving rise to different possible chains that can be considered instances of the same movement, thus constituting a motor equivalence class. Different motor performances that can lead to the same result are then said to be motor equivalent, which is therefore a functional equivalence. This

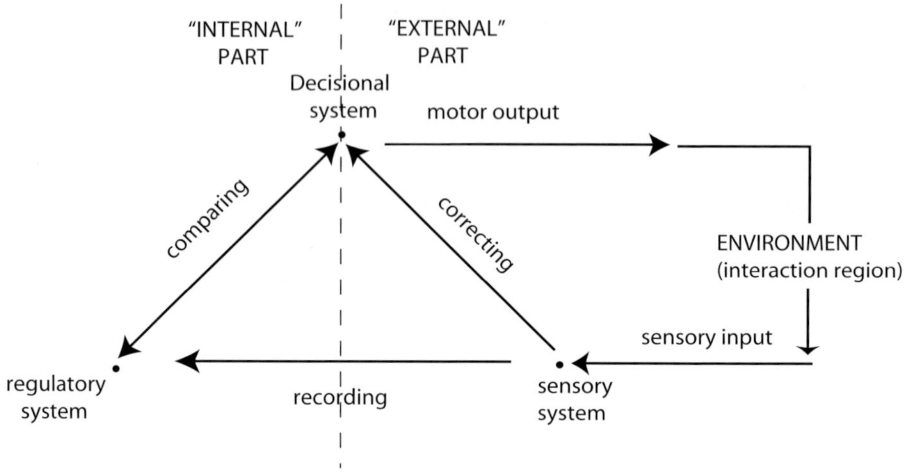

Fig. 2.1 Bernstein's model simplified. The sensory system produces allocentric maps, the reafferent information from the sensory system (consisting in visuomotor representations) goes to the regulatory system in order to be stored and to the decisional system (where goals are established) for quick corrections of motor outputs. (Adapted from Auletta 2011, p. 147.)

notion will prove very important and we will later stress the specific concept of functional equivalence class.

If we take the representation of the general way in which the motor system can work, in Bernstein's original formulation, we see that we have an internal and an external part (as shown in Fig. 2.1) and three main neural centers: the commanding or decisional system, the regulatory system and the sensory system. A sensory input coming from the environment is acquired, a sort of recording goes to the regulatory system, and another signal reaches the commanding system for comparison and correction. Obviously, the motor output coming from the commanding system contributes to a certain interaction with the environment in order to produce a new sensory input; hence, this is also a feedback circuit. Now, Bernstein acknowledged that the relationship between movement and the innervational impulse that evokes it was not univocal. A given impulse can produce completely different effects under different conditions, because the response will depend on the initial position of the body and on the external force fields in which the movement is executed. Now, in my opinion, the interesting point with regard to highlighting the relevance of the functional equivalence involved here is to consider how the organism reacts when there is an unexpected event that could block or impair the movement's execution, that is, the way in which there is *error correction.* I stress that in general error correction is really very important to understanding such matters. According to Bernstein, when an action encounters surprises (i.e., unexpected effects or events), it is impossible or irrelevant to reestablish the initial plan of action. In other words, the action must be *reorganized* according to unforeseen events. In this way, a motor equivalence is also somehow established between the *original* movement and the *corrected* one. This means that in the hierarchical-modular organization of the movement we can substitute few or several subsegments in order to partially correct the trajectory and converge with the previous one at a certain stage. Then, the reason for such structural (and global) properties of goal-directed movement is the following: Reaching movements, for instance, cannot be carried out as a series of sequential steps but present a hierarchical structure. Neural commands forwarded to muscle groups may be generated in parallel even though the overt movement appears sequentially.

Allow me now to come back to the issue of equivalence and recall what is the notion of equivalence in general. Equivalence class, a rather formal concept that can be commonly found in many fields, has the following properties:

- *Symmetry*: If item *a* is equivalent to item *b*, then *b* is also equivalent to *a*.
- *Reflexivity*: *a* is equivalent to *a*.
- *Transitivity*: If *a* is equivalent to *b*, and *b* to *c*, then *a* is also equivalent to *c*.

I have stressed that we are dealing here with a very specific meaning, actually a biological meaning, of equivalence, that is *functional* equivalence (Auletta *et al.* 2008). What does functional equivalence mean? The point is that since operations (for instance, motor actions or segments) equivalent with respect to a certain function are not automatically equivalent for other functions, it is important to unequivocally identify the function concerned. Thus, when we deal with functional equivalence, even at the biomolecular level (for instance, splicing can be performed in different ways), the important point is to understand what the function concerned is, because we may have structures or motor segments that are equivalent in certain contexts for certain functions, but in different contexts they may be not equivalent. In other words, the criterion by which items are judged to be or not to be members of such classes is a specific function. Therefore, although functional equivalence still fulfills the above criteria of formal equivalence, it does this not in general or under every circumstance, as is true for mathematical and logical entities, but only in a conditioned and partial form. In this way, functional equivalence classes are characterized by a part-whole relation. Only specific aspects of two operations will really matter for judging whether or not they are functionally equivalent. In particular, the most relevant aspect for movements is their outcome or the final result that represents the final goal of a motor sequence. Therefore, with respect to equivalence classes, the outcome of the operation performed in a certain context may be considered as a sign (the part) of the function (the whole) being deployed successfully or not, and hence that the operation really is or is not a member of the equivalence class. I believe that this is the way in which the brain evaluates different operations as well as specific motor subsegments and is therefore able to correct movements in due course.

According to Jeannerod (1999a; 1999b), the whole decisional system, at least in humans, and in a very rudimentary form in other primates, can be represented as a planning-programming-executing system, as shown in Fig. 2.2. The upper level represents planning; the middle level, programming; and the lower one, executing. Cognitive constraints are exerted on planning (in the secondary sensorimotor prefrontal integration system), while emotional inputs contribute to the intermediate level: programming.

Programming happens in the premotor cortex and inputs are received from three main sources:

(i) the motor nuclei in the ventro-anterior and ventro-lateral thalamus, which receive inputs from the basal ganglia and the cerebellum; the cerebellum provides the basic maps that are used for forming segmental motor schemes.
(ii) the primary somatosensory cortex and the parietal association cortex; this is the reafferent information about the ongoing motor response as well as from the emotional sphere.
(iii) the prefrontal association cortex, the upper-level, planning system.

It is very important to understand that we have *regulation* between the executing and the programming levels, emotional components intervene at the intermediate level, whilst control comes from above and sensory

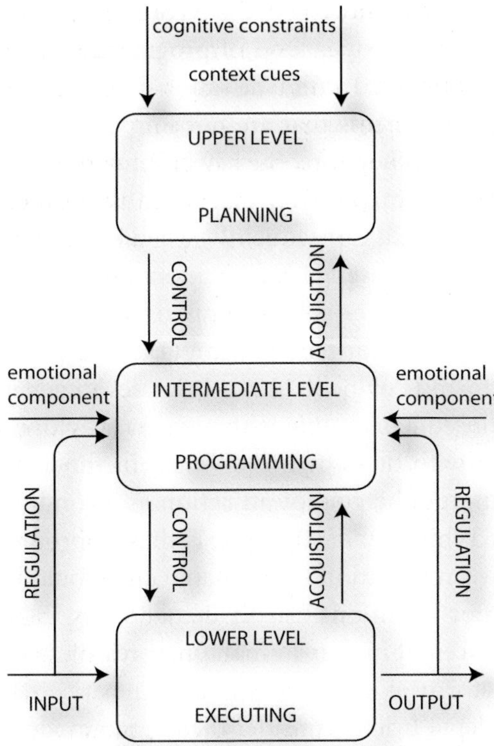

Fig. 2.2 Jeannerod's distributed and hierarchical model for generations of actions. Note that the higher instance controls the lower instance whilst acquiring information from the latter. (Adapted from Auletta 2011, p. 446.)

inputs from below (peripheral sensory system). In addition, we have cognitive constraints on the upper level and environmental cues are also used.

As is well known, a very important distinction between prior intention and intentions in action has been introduced (Anscombe 1957; Searle 1983). We may further distinguish between goals, purposes, intentions in action, and prior intentions. Goals only demand the self-maintenance of the system and can be found at any level of life. This is purely the executing level. Intentions in action are those goals that tend to give rise to a coordinated sequence of actions or single movements with a practical and immediate result (we may assume that reptiles are fully able to work at this level). The expression of purposeful and interactive behavior requires the mammal level of complexity. This is the proper level of programming. At a further level, prior intentions are connected with true rational *plans* and demand, in general, abstract aspects and items that are not immediately (neither spatially nor temporally) present. Therefore, the key point with regard to planning is that objects that are not immediately present in the perceptive field of the agent are considered as well. Consequently, when one plans, one formulates guesses and tries to extract hidden information out of the natural or social environment. For instance, to establish causal connection between events that are spatially and temporally far removed is a task that non-human primates usually cannot perform (Povinelli 2000). Primates in general only observe behavior and its consequences; this is the reason why one speaks of redoing the same action with one's own individual methods and distinguishes this from understanding the goal of an action as a condition for true imitation (Galef 1992; Whiten *et al.* 1996). It is then appropriate to introduce the distinction between imitation (in humans) and emulation (also in other primates). Thus, prior intentions usually demand this abstract component, and this level is not present in non-human primates, or can be found only in a very rudimentary form.

Now, I would like to briefly consider the refined model of Wolpert *et al.* (1995), as shown in Fig. 2.3, where the executive level is that of the means and of the intentions-in-action. The high-level of the decisional system is the planning level and that of prior intentions. Without a proper comparator there would be no connection between these two levels, and the forward and inverse models would be impossible. The forward model is crucial in that it ensures the capability of prediction and, in particular, of anticipating feedback effects (and therefore not only dealing with actual feedback).

It is now time to turn to the role of consciousness. We can take for granted the following result of recent studies [see Blakemore and Frith (2003) as well as Jeannerod (2006) for a summary]: Being aware of a goal is not being aware of how it is reached. Visuomotor representations lack conceptual content and are not conscious. Interesting studies show that intentional (conscious) processes tend to temporally shorten the time delay between the onset of the cause (for instance, pressing a certain key) and the onset of the effect (for instance, an auditory signal), as shown in Fig. 2.4. In the case of a conscious experience (middle layer of Fig. 2.4), the time of the movement

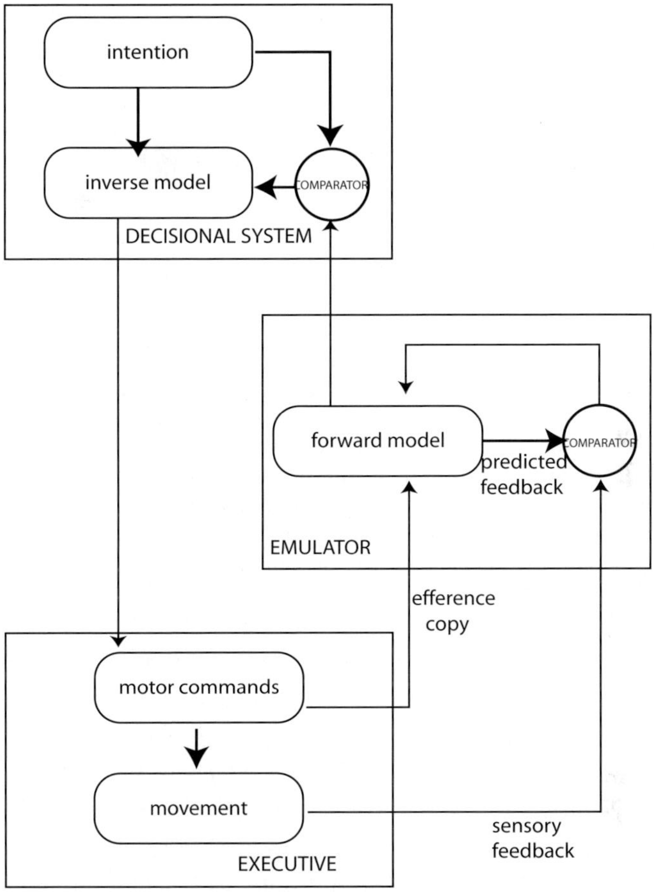

Fig. 2.3 Wolpert *et al.*'s model. The executive level is that of the intentions in action. The decisional system is the planning level, i.e. that of prior intentions. Without a comparator there would be no connection between these two levels. It provides for simulation and testing of external reality. (Adapted from Auletta 2011, p. 540.)

execution and of the onset of the acoustic signal are estimated earlier and later, respectively (relative to their real timing, shown in the upper layer), so that the time interval is narrowed (or widened, in the case of unconscious experience) by about 60 msec. This is the consequence of over-attributing external events (the tone) to oneself when they are considered as consequences of self-produced actions. We have quite the opposite situation when the subject is asked to estimate the time interval in the case of involuntary

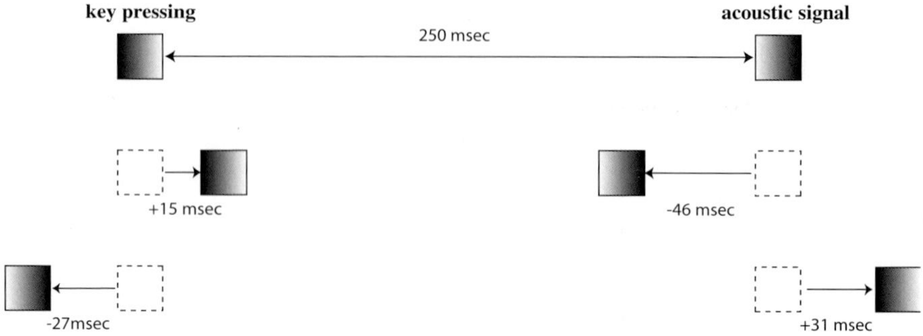

Fig. 2.4 Conscious processes tend to temporally shorten the time delay between the onset of the cause and the onset of the effect (here represented by pressing a key and an acoustic signal, respectively). In the case of a conscious experience (middle layer), the time of the movement execution and of the onset of the acoustic signal are estimated earlier and later (relative to their real timing, shown in the top layer), respectively, so that the time interval is shrunk by about 60 msec. This is the consequence of overattributing external events (the tone) to oneself when they are considered as consequences of self-produced actions. We have quite the opposite situation (bottom panel) when the subject is asked to estimate the time interval in the case of an involuntary act (the time interval is widened by about 60 msec). (Adapted from Auletta 2011, p. 542.)

acts. Experiments performed by Nielsen (1978) and by Fourneret and Jeannerod (1998) show that subjects may be conscious of a goal but not of the way to achieve it or of the required performance. Slachewsky *et al.* (2001), pursuing this line of study, showed that subjects may become aware of an abnormal deviation from the willed trajectory (due to distortion effects) only when the discrepancy passes a certain threshold (angle $>14°$ when tracing lines with a pen on paper, as shown in Fig. 2.5). This shows that consciousness is always bound to some failure originating in the sensory data or in the accomplishment of a motion. Then, conscious experience about one's movement is due to the mismatch between two sorts of reafferent data: central and peripheral signals, i.e., between what we expect and what actually occurs. Therefore, the issue of whether or not consciousness is a causal agency may be raised (Jeannerod 2009). As a matter of fact, motor programs and motion are executed in complete absence of consciousness.

The above distinction between primary or prior intentions and intentions in action allows for a different evaluation of consciousness even in its *anticipatory* quality. Indeed, if the consciousness is bound to prior intentions and long-term plans, it presents itself as an integration system able

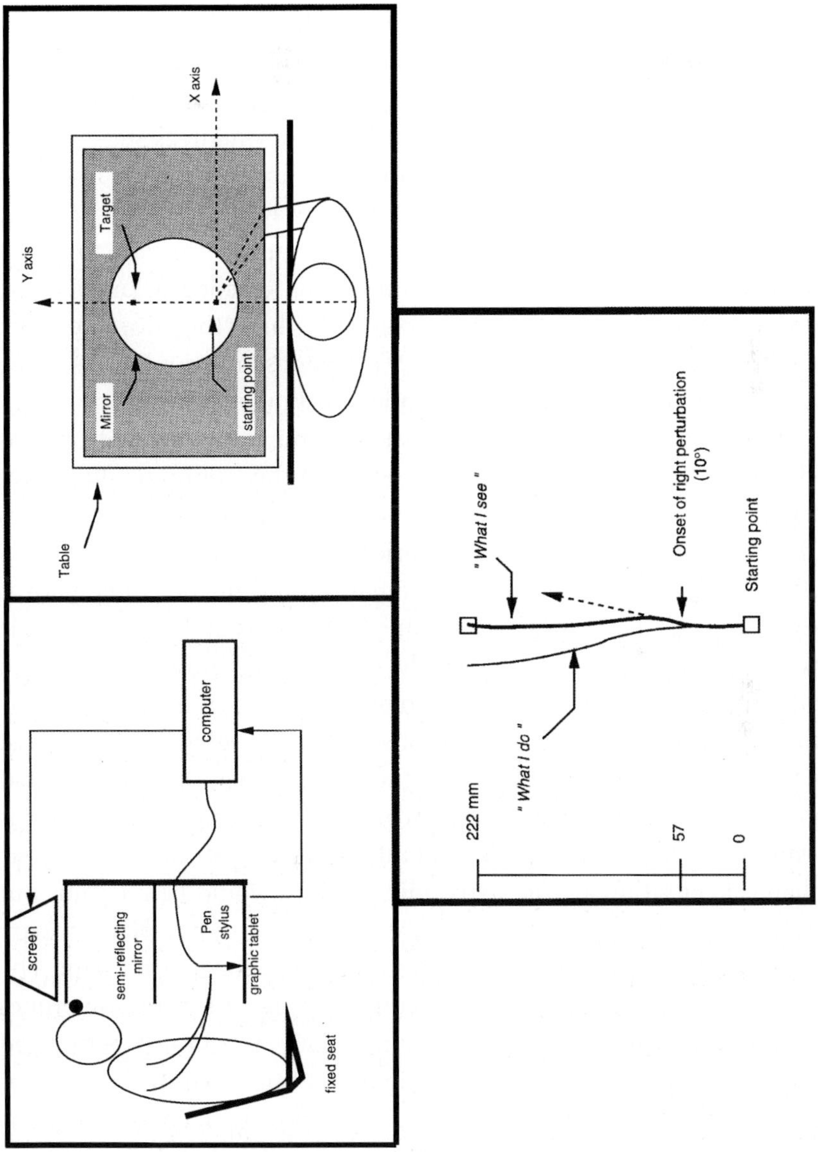

Fig. 2.5 The discrepant-line experiment.

to frame different specific actions into a wider context of planning, which is again a functional equivalence class but at a higher level than the equivalence established among different movement executions. If so, it may be able to determine the general conditions in which a specific act (and therefore an intention in action) comes into play. Top-down causation (Auletta *et al.*, 2008) of endogenous cognitive processes requires that some representation of the goal precedes the action aimed at achieving it, as suggested by the hierarchy of the body motions, such that the subsegments are nested in larger motor sequences (like Russian dolls). In this way, consciousness can (or could) determine (through prior intention) the equivalence class to which several concrete actions characterized by intentions in action belong. Let me give a simple example. If I would like to go to Paris, this is the prior intention. Obviously, the way in which I can reach Paris depends on many factors, some external (like having a means of locomotion or buying an airplane ticket, or having the money or the time, and so on) and some internal or connected with the self (like being healthy, preferring this option to alternative ones, especially in changing conditions and so on). Some segments of this whole decisional path, leading from the prior intention up to the realization of the final goal, are executed in an automatic way, but the whole path is consciously and rationally chosen. The equivalence class of different possible plan executions (e.g., going by plane or by car, staying for a week or for ten days, lodging in Boulevard Saint-Germain or Saint-Michel) is determined by the general goal of the primary intentions and by its subgoals. Recall that when dealing with goals and intentions, the way in which they are executed does not matter, but only their final outcome representing the goal (this is called *equifinality*). This is very important, because if we had only a sort of mechanical causation at play here, only the path would be relevant. But dealing with equivalence classes and the execution of motor plans, we are concerned with the final outcome, and this is what is ultimately relevant. The way in which one arrives at the final outcome is far less relevant, sometimes not relevant at all. Thus, the *specific way* in which an action is performed is not controlled by consciousness (consciousness cannot control the motor segments as such), but its *functional value* is. If I am right, consciousness is indeed a causal agency of actions but only in an indirect and constraining sense. "Constraining" here means the capability of applying conditions that are able to somehow canalize the lower-level motor segments. Let us consider how this may happen.

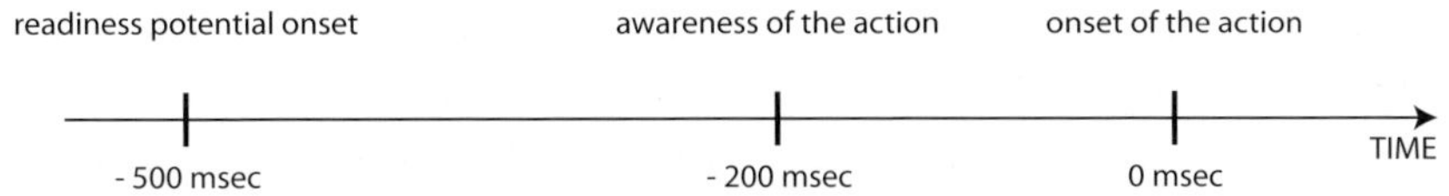

Fig. 2.6 Libet's temporal model of actions.

The way in which the conscious mind could exert a constraining influence on behavior is by means of *action suspension*. We know of data that can open up the possibility of such a constraining influence. When an activity persists longer than 500 msec, according to Libet (1985; 2004), it may become conscious (Fig. 2.6). Then, even if any action begins involuntarily and after an involuntary process has already started, the voluntary act still appears 150–200 msec before the beginning of the muscular movements themselves (actually, only 100 msec, since 50 msec are necessary for the primary motor cortex to activate the spinal motor nerve cells). Libet assumes that during this time interval (100 msec), one may block an unconscious process and so avoid a certain act. Thus, consciousness is not able to produce a motor segment but can still block or suspend it (a negative and not a positive action). This would make it viable, from a neuronal viewpoint, to speak of the constraining influence of the will on the brain, and therefore on actions and behavior (Auletta 2011, Chs. 18 and 24). In this way, the mind would be able to canalize the brain activity, and consequently the behavior, towards the chosen purpose. It is like a ball rolling down a hill; the ball will possibly follow very different paths, but if there are two parallel walls the ball could be canalized along the path confined by those walls. Therefore, thanks to subsequent blocking and suspensions of action, the conscious mind is able to canalize a series of motor acts, without violating the physical laws: to have such a constraining influence does not imply playing the role of a dynamical causal factor. Indeed, it suffices to redirect in due course part of the energy and the connected neural activity that is already spontaneously started about 400 msec before the onset of the movement. Therefore, such a constraining influence of the mind (which could be interpreted as the highest brain's control instance) is likely to be found in the interrelations between automatic intentions-in-action and the fully conscious prior intentions framing the former. Moreover, I have already recalled that consciousness comes into play exactly when the goal at which an action is aimed fails to be achieved.

From my point of view, consciousness comes into play each time we have problems. Such problems in certain circumstances may suggest blocking actions. By blocking the action we are able to canalize and to frame different motor segments. This situation is somewhat analogous to what happens in a cell. In a cell a huge number of chemical interactions are in principle possible, but the number of the reactions occurring is a small subset of those abstractly possible (in the simplest organisms like bacteria, about 700 biomolecules are present but only 1,100 chemical reactions occur, which is a very tiny fraction of the set of possible ones); this is again due to a canalizing and framing influence.

DISCUSSION

Michael Arbib: I have two questions. (i) I am at the restaurant, I look at the menu and I choose the dish. It seems to me that just saying that there is an equivalence class of actions is not really a solution of what is going on in that situation. In particular, the fact that there is a canalization or constraint seems to me almost irrelevant. Of course, it is true that there is canalization — I am in a situation when I choose — but the subconscious processes entering into this are so numerous that I do not see how top-down causation can be useful here as an explanatory tool. (ii) Now, the second, harder question is that, at this table we have people that have chosen to live a secular life and people that decided to enter the Church. This seems to me to be an extremely important act of the will, but again I do not see the value in analyzing this at the simple level of saying that choosing a secular life is an equivalence class of action and entering the Church is another equivalence class. Instead, it seems to me that we have to go into an incredible array of life experience to understand how at a crucial moment the decision is taken.

Gennaro Auletta: Thank you. Obviously, for me it is very important to pick the right equivalence class; picking an irrelevant one is indeed not very useful. When I go to the restaurant, the question is, am I really hungry? Suppose I am. Probably, I will choose any restaurant that I can find. In this case, this would be an equivalence class, as here the specific restaurant is irrelevant. But, if you do not have a lot of money, then you will choose among a subset of the available restaurants. This will again be an equivalence class. Let me take another example. Suppose I utter the letter "A" 1,000 times. Each time I will pronounce it in a slightly different way. And different people will pronounce this in different ways. But "A" is an equivalence class and this fact does not seem to me to be totally irrelevant. As to your second question, the point is this: People can, say, enter the Church in a lot of different ways and your personal path is what matters here. The specific way in which any future priest enters the Church is not equivalent from a personal point of view, obviously. But the final result, the outcome, is what matters here; and from this point of view you establish an equivalence class. If you consider the specific modalities in which you do things, obviously the differences are relevant, because you are on the level of motor execution. If you consider the outcome, and therefore the function of the motor act, the way in which it is performed does not matter at all. Only the result is relevant. Note that also equivalence classes are nested at different levels of generality.

Jean-Pierre Changeux: I have a couple of objections too. First of all, you have considered the Libet experiment. Many experiments on the line of that of Libet have been performed and they are highly controversial; there have been errors and misinterpretations. Secondly, in the first part of your presentation, the term "mind" is not present and you did not need it. You introduce this concept just in the last part. But as soon as you introduce it, you replace it with the brain control instance. Now, I agree with the latter terminology, as it is not clear to me what the concept of mind means.

Gennaro Auletta: I am not able to judge whether the Libet experiment is appropriate or not; I hope that further discussions will cast more light on that. Anyway, being controversial is one thing and being wrong is another. Let me focus specifically on your second question, that of mind. I am not a dualist in any sense of the word. I want just to attract your attention on this specific point (and this is a question rather than an answer). The brain does not codify information at all, because you have coding at the peripheral sensory organs, but not in the brain. The brain deals with information but does not codify it for the reason that to have information codification you need elementary elements to be combined, but you do not have these elements in the brain. You likely have infinite possible excitation patterns (which are tokens and not types) that are always slightly different to one another, even if they refer to the same thing. Now, the interesting point is that the brain does not codify information, while the mind does, and this is a difference.

Jean-Pierre Changeux: Do you think that your statement that the brain does not codify information is a scientific statement?

Gennaro Auletta: Yes, I do. Please, if I am wrong, show me that I am. I think that any attempt at finding information codification in the brain will fail.

Antonio Damasio: In my opinion, the question is not if you are wrong or not, but what you mean by information codification. What do you exactly mean when you say that "the brain does not codify information"?

Gennaro Auletta: Well, in order to codify information you need, first of all, a set of elementary units (like the letters of the alphabet or the DNA bases) that can be combined according to syntactical rules. Moreover, in my modest opinion, you also necessarily need at least two codes and some translation rules between them. The brain satisfies only a part of these requirements. You have excitatory patterns, even self-organizing patterns, and some distributed network of interconnections, but this is not codified information.

Michael Arbib: Before I talk of the brain, you have this strange distinction, I think, between the brain not coding information and the mind coding information. So I ask you as a parallel: would you agree that an electron cannot orbit the sun but only a planet can orbit the sun?

Gennaro Auletta: My point is that quantum systems do codify information but you do not have information codification at the atomic, molecular or chemical level, not at all.

Michael Arbib: No, this was not my question. The planet can orbit the sun, but the planet is made of electrons and protons. So, do you agree that a comet can orbit the sun?

Gennaro Auletta: Yes, sure, I see no problem in that.

Michael Arbib: Well, would you now agree that the planet could not go around the sun if not for the motions of the individual particles that form the planet?

Gennaro Auletta: Yes. The issue is only to understand if the planet is simply a bag in which you put the particles or it is an organized entity. Well, this is obvious also for the mind. I repeat, the mind, for me is not an entity; it is not a substance. What I am trying to say is that the mind must be understood as a cluster of processes strictly related to the brain. The mind is able to codify information only because it is able to combine external or bodily physical elements like phonemes, gestures or stones. Without such a combination of physical elements this would be impossible.

Michael Arbib: Let us assume that neurons are these elementary components. Then we put them together in layers that, to my mind, do codify information. The retina is a special form of organization of matter that can respond to light intensity and the rest of the brain is able to take that peripheral sensory code of the light patterns and determine the three-dimensional disposition of objects so that we can, for example, move the hand to grasp this object. In some cases we are able to see in great detail how the code changes from the retinal code to the motor neurons firing, which finally triggers the muscles to contract. It seems to me unhelpful to say that none of that is codified in the brain and that only the mind can code. Now, where I have disagreed with my learned colleague from Paris is that I think that the word "mind" is a useful shortcut for some brain processes that are of interest to us. But I do not see authoritative reasons to draw a sharp distinction between the brain that cannot code information and the mind that can. I think this is a misuse of that useful shortcut.

Jean-Pierre Changeux: I think you have said, at the very end, that the combination of physical elements allows the mind to codify information. So, I could say that the combination of physical elements allows the conscious workspace. Therefore, you are saying that in the brain, in its highest form of organization which is consciousness (which, in turn, can be implemented in neural terms), there is a way to combine physical elements, and therefore to codify information. Then, why you are using the word "mind" is a total mystery.

Gennaro Auletta: I totally agree with you, Jean-Pierre. You are the specialist. Children learn language exactly when they learn to combine elements. My only point is that in order to combine physical element so that they can represent information codification you need a sort of symbolic activity. Well, animals do not have that. Primates, which are very intelligent, do not display symbolic activity. So, there is something that distinguishes humans, even though this does not allow anybody to introduce some kind of dualism. Now, Michael, I am sure that the peripheral sensory system codes information. But when this information is combined in the brain, it is not combined in a way in which information is codified. There are studies showing that information can be combined in two different ways: according to information codification (in a digital way, as words are combined), or according to a compositional method (as when you assemble a puzzle). In the latter case, there is no codification. In my opinion, it is exactly in this second way that the brain combines excitation patterns; therefore, they are not codified but simply combined.

Pierre Jacob: I think you have tried to establish the causal role of the prior intentions. The way in which you did this is through the important notion of equivalence class. I want to suggest that we should make a distinction between causation and explanation. I think that there are very important scientific explanations that are not causal at all. Maybe, you have actually confused explanation with causation and taken that all kinds of explanations must be causal.

Gennaro Auletta: It is clear to me that there are explanations that are not causal. My field is quantum mechanics and here you have a lot of correlations without any causes at all. But I am suggesting discriminating the concept of cause: there are things that have causal power (let me call this dynamical causes), and there are things (in general, structures) that do not have a dynamical causal power but are nonetheless causally relevant. Think about the structure of a forest, i.e., the disposition of the trees. Is this real or not? According to a reductionist point of view only trees are real, of course. Obviously, the structure of a forest is passive; it does not act in any way. But suppose that you have a dynamical agent like fire or wind. The structure may well canalize it, or even block it; this depends on the structure. So, what is here not able to produce any effect alone somehow concurs in the determination of a specific effect in conjunction with dynamical causes. In this way, it is causally relevant. This is the way in which I mean that the mind is causally relevant. This is, in the end, the reason why I have spoken of suspension of actions. The brain is a causal agent in the dynamical sense; for the mind I have rather spoken of a constraining influence. In quantum mechanics, when we have entanglement, the results that you obtain when you measure are quite different from the case where there is no entanglement. In this sense, entanglement, even if it is not a dynamical cause, is relevant. To come back to your question, I wish to distinguish between the "genetic" and the functional. With respect to the "genetic," everything comes bottom-up. But, once a network, or some other kind of formal constraint, is established, will this network react and/or influence, in return, the lower level from which it has emerged? My answer will be that the structure is able to have such an influence.

REFERENCES

Anscombe GEM. (1957) *Intentions*. Blackwell, Oxford.

Auletta G, Ellis G, Jaeger L. (2008) Top-down causation by information control: From a philosophical problem to a scientific research programme. *J R Soc Interface* 5: 1159–1172.

Auletta G. (2011) *Cognitive Biology: Dealing with Information from Bacteria to Minds.* Oxford University Press, Oxford.

Blakemore S-J, Frith CD. (2003) Self-awareness and action. *Curr Opin Neurobiol* 13: 219–224.

Bernstein NA. (1967) *The Coordination and Regulation of Movements.* Pergamon, Oxford.

Fourneret P, Jeannerod M. (1998) Limited conscious monitoring of motor performance in normal subjects. *Neuropsychologia* 36: 1133–1140.

Galef BG. (1992) The question of animal culture. *Hum Nat* 3: 157–178.

Hebb DO. (1949) *The Organization of Behavior: A Neuropsychological Theory.* Wiley, New York. (2nd ed. Erlbaum, London, 2002.)

Jeannerod M. (1988) *The Neuronal and Behavioural Organization of Goal-Directed Movements.* Oxford University Press, Oxford.

Jeannerod M. (1999a) The cognitive way to action. In: RJ Russell, N Murphy, T Meyering and A Arbib (eds.), *The Neurosciences and the Person: Scientific Perspectives on Divine Action.* Vatican Observatory Publications and CTNS, Vatican City and Berkeley, CA, pp. 57–66.

Jeannerod M. (1999b) Are there limits to the naturalization of mental states? The cognitive way to action. In: RJ Russell, N Murphy, T Meyering and A Arbib (eds.), *The Neurosciences and the Person: Scientific Perspectives on Divine Action.* Vatican Observatory Publications and CTNS, Vatican City and Berkeley, CA, pp. 121–128.

Jeannerod M. (2006) *Motor Cognition: What Actions Tell the Self.* Oxford University Press, Oxford.

Jeannerod M. (2009) *Le cerveau volontaire.* Odile Jacob, Paris.

Libet B. (1985) Unconscious cerebral initiative and the role of conscious will in voluntary action. *Behav Brain Sci* **8**: 529–539.

Libet B. (2004) *Mind Time: The Temporal Factor in Consciousness.* Harvard University Press, Cambridge, MA.

Nielsen TI. (1978) *Acts: Analyses and Syntheses of Human Acting, Concerning the Subject and from the Standpoint of the Subject.* Dansk Psykologisc Forlag, Copenhagen.

Povinelli DJ. (2000) *Folk Physics for Apes: The Chimpanzee's Theory of How the World Works.* Oxford University Press, Oxford.

Searle JR. (1983) *Intentionality: An Essay in the Philosophy of Mind.* Cambridge University Press, Cambridge. (2nd ed. 1997.)

Slachewsky A, Pillon B, Fourneret P, *et al.* (2001) Preserved adjustment but impaired awareness in a sensory-motor conflict following prefrontal lesions. *J Cogn Neurosci* **13**: 332–340.

Whiten A, Custance DM, Gomez J-C, *et al.* (1996) Imitative learning of artificial fruit processing in children (*Homo sapiens*) and chimpanzees (*Pan troglodytes*). *J Comparat Psychol* **110**: 3–14.

Wolpert DM, Ghahramani Z, Jordan MI. (1995) An internal model for sensorimotor integration. *Science* **269**: 1880–1882.

CHAPTER 3

Neurons, Schemas, Persons and Society — Revisited

*Michael A. Arbib**

3.1. THE CHALLENGE

Roger Sperry spoke of *downward causation* from the mental to the physical level. He said that the ontology of mind is nothing without the brain but adds that the emergence of the mental supersedes the laws of the physical parts of the brain:

Evolution keeps complicating the universe by adding new phenomena that have new properties and new forces and that are regulated by new scientific principles and new scientific laws... [T]he old simple laws and primeval forces of the hydrogen age never get lost or cancelled in the process of compounding the compounds. They do, however, get superseded, overwhelmed and outclassed by the higher-level forces as these successively appear at the atomic, the molecular and the cellular and higher levels... When it comes to brains,... the simpler electric, atomic, molecular, and cellular forces and laws, though still present and operating, have been superseded by the configurational forces of higher-level mechanisms. At the top, in the human brain, these include the powers of perception, cognition, reason, judgment, and the like, the operational, causal

*The title refers to that of an earlier conference paper addressing somewhat related themes (Arbib 1991).

effects and forces of which are equally or more potent in brain dynamics than are the outclassed inner chemical forces (Sperry 1964).

On this account, the "neuron-talk" ceases to be relevant once it has yielded the emergent properties of conscious activity. This is the view of *top-down causation* I will be refining in this chapter. My counterclaim is that no matter how far a mind-level psychology or a theology of the human spirit progresses, there will be phenomena that cannot be explained simply in terms of causation at this top level. The low-level details will continue to be relevant. However, in presenting this argument, I will be fully in sympathy (though implicitly) with Sperry's (1980) rejection of "various mentalist-related concepts and dualist beliefs in the supernatural, the paranormal and in unembodied forms of conscious existence that receive no logical support from neuroscience." Just as for Sperry (1980), the mind-brain model advanced here — linking neural networks, schemas in the head and social schemas — is "fundamentally monistic and not only fails to support dualism, but serves to further discount fading prospects for finding dualist forms or domains of conscious experience not embodied in a functioning brain."

3.2. TWO-WAY REDUCTION

It is a general property of the universe that matter organizes itself into more and more embracing entities whose dynamics can be described approximately without reference to lower levels; but subtleties of these dynamics and understanding of their breakdown can only be understood by analysis of constituent systems. Conversely, we may not come to understand the full dynamics of lower-level systems unless we can place them in the context of larger systems.

As a simple example of this, consider a spoked wooden wheel, rolling down a hill. If the wheel is intact, its dynamics can be readily computed in terms of its diameter, moment of inertia and other macro-variables. In this way, we may imagine that top-down causation has taken control of the movement of all the molecules in the wheel. But suppose the wheel hits a rock. Will it shatter? In what way? Here, only a micro-analysis of the wheel's structure will explain what happens.

In the same way, an individual's consciousness or will or soul (still poorly understood concepts for the neuroscientist seeking a bridge to philosophy or theology) can be seen as a high-level, but not necessarily accurate, précis of the detailed workings of his brain. At times of conscious, rational decision making, we may feel that our consciousness exerts downward causation on our behavior. Yet we often act in ways that are not premeditated and whose causes are to be found in patterns of neural activity dependent on neural traces of memory to which we have no conscious access. Even if our course of action is what we intended, the details of its execution (including errors, and the degree of success) will depend on these subconscious subtleties. And once we turn to mental disease and neurological disorders, we recognize that the downward causation has been distorted in ways that depend on neuronal and hormonal changes whose cause lies outside the domain of the will. Therefore, my theme is that downward causation is not the right notion for a research program for neuroscience. Instead, I would rather talk about the "two-way reduction" (Arbib and Hesse 1986) of an inter-level conversation, where a higher level may express a dynamics that can, for long stretches of time, render consideration of the lower levels superfluous. However, (a) this higher-level dynamics is an expression of the statistical regularities emergent from lower-level processing. For example, statistical mechanics explains bulk properties of matter by appropriate averaging over the interactions of myriad constituent particles but this process of explanation is guided by phenomena of thermodynamics, observed at the macro level. (b) The macro-dynamics can be disrupted by circumstances not captured by the high-level description (a region of structural weakness of the wheel; the atypical geometry of the rock that encounters that weakness to break the wheel). (c) Moreover, understanding of when and how the high-level dynamics holds may help us ask the "right questions" to discern more about the low-level dynamics and determine appropriate ways of aggregating that dynamics to both yield the default macro-dynamics and illuminate the cases when it breaks down. For example, the photoelectric effect defied explanation in terms of pre-quantum microphysics, and so demanded redefinition of physical theory at the micro level.

The aim of *two-way reduction* is thus not only to explain the macro level in terms of the micro level but also to redefine our theories of the micro level as we seek to make sense of phenomena that we observe at the macro level. In the same way, the experiments that led to the development of quantum physics all involved macro level phenomena observable by humans and were contrary to the Newtonian view of macro level interactions. In the process, we came to understand why in many cases the non-relativistic motion of large objects could be understood in Newtonian terms. We would not use quantum mechanics to compute a control strategy for a spaceship or satellite. Yet, at the same time, we would not (or, at least, I would not) say that the Newtonian variables exert downward causation on the quantum wave function.

Mary Hesse, as a philosopher of science, studied the way in which data cannot in themselves determine a theory. Rather, given a basic set of observations, a scientific community must assess those data to decide which are important and then decide what theory adequately meets these data in order to be accepted as an appropriate theory. Let me place this in a perspective we developed on this basis (Arbib and Hesse 1986, Ch. 1):

There is certainly instrumental progress in science in that the pragmatic possibilities of predicting and controlling empirical events have vastly increased. When a new theory replaces the ontology of an old theory, it may also change the set of phenomena we choose to consider important. But for those phenomena addressed by both theories, the observation language of the new theory must be able to represent the observations made within the old theory, and the new theory must explain the phenomena so represented. In particular, a successful new theory should help us understand where the old theory succeeded even while replacing the ontology of the old explanation. A given theory lets us develop an observation language (based in part on the scientific instruments made possible by that theory), which enables us confidently to specify the nature of experiments and the results to be obtained under specific conditions. This procedure not only enables us to express the success of the theory but also to represent the discrepancies that provide the "error signals" for what Peirce characterized as the self-corrective feedback procedure of experimental science.

We thus adopted the *pragmatic criterion* (Hesse 1980, Introduction) of successful prediction and control as the criterion of "truth" or "acceptability" that operates in natural science. The pragmatic criterion assesses the "model-in-the-head" of theory against a stock of observations constrained by, but not exhausting a "space-time reality" external to the observer, and yet it is strong enough to prevent the model's being purely an intersubjective representation of social realities. But the process of theory creation is "constructive" in that it is not a matter of simply aggregating facts but may depend on the individual and social experience of the theory builders. Only recently have we come to understand how many characteristics of science previously thought to be "objective" ("read off" the real world) are, after all, products of construction, albeit constrained by the pragmatic criterion.

1. Science is in a literal sense constructive of new facts. It has no fixed body of facts passively awaiting explanation.

2. The facts themselves are *theory laden.*

3. There is a sense in which theory is empirically underdetermined by data.

Logically speaking, theories are not deducible from data, and many theories may fit the same data reasonably well.

When we engage in neuroscience or cognitive science, and especially when we seek to relate them to issues in the human sciences and theology, talk of prediction may be acceptable, but talk of control may seem unethical. However, when we consider medicine or education, or even the relative efficacy of different techniques of meditation, then we are considering forms of control which do appear morally acceptable when subject to certain limitations and here, as in the wheel example, we expect there to be cases in which the details of neurophysiology move to centre stage.

In summary, then, my shaping concern is to help create a coherent multilevel theory integrating the data of personal experience with the data of psychophysics, neuropsychology, brain imaging, neurophysiology, neurochemistry and more into a multilevel theory with explanatory connections up the levels and constraints down the levels so that we may, ultimately, understand how our brains support the reality of personal experience (Arbib 1999b), and why at times the dynamics of the micro level supports a macrodynamics that appears autonomous, and yet at other times the lower level proves disruptive of our macro level expectations. I cannot hope to offer a complete theory in this chapter (nor, I regret, in my lifetime), but I will try to set forth a number of research results that advance the general quest to place the apparent plausibility of top-down causation within a framework of the brain's micro-dynamics via multiple levels of analysis.

3.3. MULTILEVEL ANALYSIS OF THE ROLE OF THE CEREBELLUM IN PRISM ADAPTATION

To focus our thinking, let us consider a topic for which we have both empirical data and computational models: adaptation to prisms that displace the view of a target horizontally, when throwing a ball at the target. In terms of downward causation, we may say that by consciously forming the intention to throw the ball, we cause the arm muscles to contract in a temporal trajectory that propels the ball toward the target. But, because we have just donned prisms, we throw wide of the mark. We might then say, quite reasonably, that downward causation is still in play — but with the displacing effect of the prisms part of the top-down account of what causes the throw to be in error. Again, at this high-level of description, we might say that we decide to throw the ball again and again until we get on target. And, indeed, over time our error decreases, until our successive throws are centred on the target with some modest standard deviation.

Observation 1: It seems aberrant to claim that this residual error is an effect of downward causation. There are many nuances of intention that correspond to the same form of words for that intention, and there are many variations in neural state that correspond to a given form of that intention — just as thermodynamic variables express averages over astronomically many different states of the underlying molecules. Thus, the residual errors express fluctuations in the state of the brain and muscles, and are more consistent with an intention expressible as "I will throw this ball at that target." A possible save here might be to insist on downward causation, while accepting that it is stochastic rather than deterministic.

Observation 2: The compensation for the error depends on the intention to keep throwing the ball till the error is corrected. However, error correction is not itself a top-down process. While it is possible to throw once, note the magnitude of the error, and make a conscious effort to throw the ball next time at a "virtual target" at the same distance to the other side of the actual target, this is not the usual process of prism adaptation. Rather, by simply making repeated attempts to throw on target, we activate circuitry in our brain for error-based learning which yields a downward trend in the size of the error over repeated trials. The size of the trial-by-trial reduction of the error, and the mechanisms which support it, are totally outside our conscious control. Thus, rather than speaking of downward causation, we might want to speak of downward constraints: the top-level dynamics (deciding to throw the ball again) activates and constrains lower-level learning processes whose internal dynamics are opaque to consciousness.

Observation 3: A patient with a focal olivocerebellar lesion cannot adapt for the prism displacement (Martin *et al.* 1996a). Having donned the prism, he throws where the target appears to be (indeed missing it) and never adapts to the prism. But as soon as the prism comes off he is back on target again precisely because he never adapted. The goal of getting the dart to hit the target may be at the mental level, but the behavior reflects that the high level goal has no direct causal efficacy — all it can do is constrain a learning process that cannot proceed without the machinery of appropriate parts of the cerebellum being intact (Spoelstra and Arbib 2001).

Observation 4: With a different lesion to the cerebellum, a patient may be unable to coordinate the timing of arm and hand movements. A normal person can preshape and then enclose their hand even as they control their arm to move toward an object they intend to grasp, doing so in such a way that the aperture of the hand matches the width of the object when and where it contacts the object. The necessary patterns of coordination are subconscious and the fruits of learning. The cerebellum provides the necessary coordination of timing. Lacking this, the cerebellar patient — to avoid bumping or missing the object instead of grasping it — rapidly acquires a new strategy of *decomposition of movement*: open the hand, then move it near the object, and only then enclose the object with the fingers. Thus, graceful coordination of movements is replaced by serial execution of movements, each of which is less skilful than that which a normal subject would execute. The cerebellar patient might describe this as involving downward causation — "I decided to shift strategy to make up for my clumsiness" — but the shift in strategy can indeed be explained in terms of subconscious mechanisms which use reinforcement learning to adapt neural correlates of the executability and desirability of various actions (Bonaiuto and Arbib 2010).

Observation 5: A patient who has adapted to the prisms and is again throwing on target might initially believe that he has indeed learned a new skill of "seeing the world where it really is" in compensation for the prisms. However, the situation is more complex (Martin *et al.* 1996b). Surprisingly, it turns out that one has *not* learned to see "where the world really is." Many subjects (not all) will adapt to the prisms after repeatedly throwing the ball overarm, yet show little or no compensation when, in the same session of wearing the prisms, they first throw the ball overarm. Thus, contrary to introspection, one has not shifted one's view of the target but has instead learned a private visuomotor transformation for the skill of underarm throwing. Of course, once the subject has experienced the full range of conditions, they can put the results into words — but this is a case of upward causation where the conscious mind comes to observe those effects of low-level processes that can be captured in higher-level terms, just as was the case for the cerebellar patient with decomposition of movement.

While I will not burden this chapter with details, I will just mention that our models of these phenomena (Arbib *et al.* 1995; Spoelstra and Arbib 2001) take us well below the detail of different brain regions to look at the neural circuitry connecting cerebellar cortex, cerebellar nuclei, parietal cortex, the inferior olive and more; then down to the microcomplexes, each linking a subregion of cerebellar cortex and nuclei to support adaptation serving a specific motor pattern generator; then to the level of the synapse to assess what rules of synaptic plasticity yield that adaptation; and finally to a new model of the neurochemical mechanisms supporting "eligibility" of synapses to respond to training signals from the inferior olive. It might be argued that many of the subneural details are needed to make the system function but are bound by downward causation from the level of neural firing. However, the level of the microcomplex does seem relevant to "upward causation," at least according to our current theory: We posit that the extent to which there is transfer of adaptation between different skills — e.g., underarm versus overarm throwing — is explained by the degree of overlap between the circuitry of the microcomplexes which support them.

In summary, I reject Sperry's view that "neuron-talk" ceases to be relevant once it has yielded the emergent properties of conscious activity. But this in no way excludes the importance of a neuroscience couched at "the level of conscious activity." I say "at the level" to emphasize that an account of mental processes at this high level cannot be restricted to conscious processes. In the next section, I introduce "pure" schema theory as a language for describing how the mind operates at this level; "neural" schema theory then assesses how an initial schema model of some aspect of the mind must be reconfigured if contact is to be made with data from neuroscience. Such a reconfigured model will position us to understand where the schema level operations can be viewed as having a relatively autonomous dynamics, and where we must probe into the neural details (as in our assessment of the role of cerebellum in prism adaptation) to refine our initial understanding. The following section will then move up a level — our understanding of a person in terms of the schemas "in the head" must be complemented by an assessment of social interactions and of the "social schemas" which emerge within society.

3.4. OF NEURONS AND SCHEMAS

Figure 3.1 presents the distinctive neural architecture of key regions of the human brain. A banal but crucial hypothesis is that these details (and even

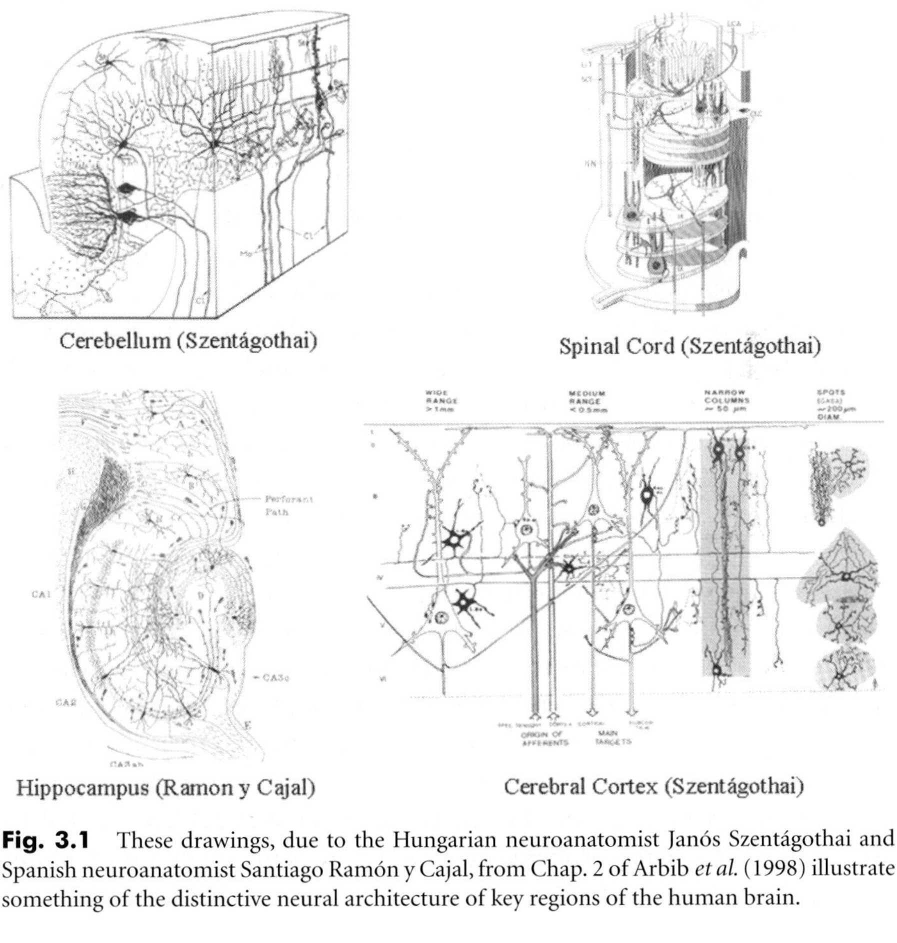

Fig. 3.1 These drawings, due to the Hungarian neuroanatomist Janós Szentágothai and Spanish neuroanatomist Santiago Ramón y Cajal, from Chap. 2 of Arbib *et al.* (1998) illustrate something of the distinctive neural architecture of key regions of the human brain.

finer details of neuron morphology, neurochemistry and gene expression) will continue to be important no matter how far "mind-level" psychology progresses. In seeking to bridge between the conscious level or the mental level and the neural level, I have found it useful to think about computational entities called schemas (Arbib 1975; Arbib 1981; Arbib 1992; Arbib *et al.* 1987; *et seq.*). Your ability to recognize an apple will, for example, be a perceptual schema; something in between the level of what the whole brain does and what individual neurons do. Similarly, most schemas that are able to deal with that apple involve very large neural networks to carry out this basic skill. And then, of course, in perceiving a complex scene, we get an assemblage of schemas when we put together our perception of a complex scene with a variety of actions to yield successful performance; then (whether consciously or not) we may invoke other coordinated schemas, abstract schemas, to make everything work. Thus, I seek to understand how multiple hypotheses about the state of the world and the possible course of action are activated as schemas that, through patterns of neural activity and through patterns of competition, cooperation, excitation and inhibition, finally converge to a state which commits the organism to a particular course of action.

"*Pure*" *schema theory* studies schemas for action, perception and more as dynamical, interacting systems underlying mental and overt behavior (in contrast to, e.g., characterizing the brain at the mental level by a "language of thought" given by a set of propositions which can be characterized by something akin to propositions of predicate logic). A schema is both a store of knowledge and the description of a process for applying that knowledge. A schema may be instantiated to form multiple schema instances as active copies of the process to apply that knowledge. This is mediated through the activity of a network of interacting computing agents, the schema instances, which among them provide processes for going from a particular situation and a particular structure of goals and tasks to a suitable course of action. A perceptual schema embodies the process whereby the system determines whether a given domain of interaction is present in the environment. A schema assemblage combines an estimate of environmental state with a representation of goals and needs. The internal state is also updated by knowledge of the state of execution of current plans made up of motor schemas, which are akin to control systems but distinguished by the fact that they can be combined to form coordinated control programs that interweave their activations in accordance with the current task and sensory environment to mediate more complex behaviors. Other schemas — coordinating schemas and abstract schemas — may be added atop these to create a cooperation/competition model of the mind to explain mental function. In this way, new schemas can be defined hierarchically in terms of other schemas. Much psychological theorizing can be done at this level.

An expanded perspective on the assemblage of perceptual schemas came about at the University of Massachusetts at Amherst, where my colleagues Ed Riseman and Allen Hanson were developing a scene understanding system that could move from a color photograph to a recognition of the various objects contained in the scene. Their VISIONS system (Arbib 1989, Sec. 5.3; Draper *et al.* 1989; Hanson and Riseman 1978) was implemented on a serial computer but the underlying computational architecture offered useful insights into the way I believe the brain operates, via *cooperative computation* — the competition and cooperation of a multitude of schema instances. Before the schemas can operate, low-level processes take an image of an outdoor visual scene and extract an *intermediate representation* — including contours and surfaces tagged with features such as color, texture, shape, size and location (Fig. 3.2). Perceptual schemas process different features of the intermediate representation to form confidence values for the presence of objects like houses, walls and trees. The knowledge required for interpretation is stored in long-term memory as a network of schemas, while the state of interpretation of the particular scene unfolds in working memory as a network of schema instances. Note that this working memory is not defined in terms of recency (as in very short-term memory) but rather in terms of continuing relevance.

Interpretation of a novel scene starts with the data-driven instantiation of several schemas (e.g., a certain range of color and texture might cue an instance of the foliage schema for a certain region of the image). When a schema instance is activated, it is linked with an associated area of the image and an associated set of local variables. Each schema instance in working memory has an associated confidence level that changes on the basis of interactions with other units in working memory. The working memory network makes context explicit: each object represents a context for further processing. Thus, once several schema instances are active, they may instantiate others

Fig. 3.2 Segmentation of a scene into candidate regions provides the bridge between the original image and the interpretation of a scene in VISIONS by associating regions of the image with schema instances. In this example, VISIONS classifies regions of the scene as sky, roof, wall, shutter foliage and grass, but leaves other areas uninterpreted.

in a "hypothesis-driven" way (e.g., recognizing what appears to be a roof will activate an instance of the house schema to seek confirming evidence such as the presence of walls in the region below that of the putative roof). Ensuing computation is based on the competition and cooperation of concurrently active schema instances. Once a number of schema instances have been activated, the schema network is invoked to formulate hypotheses, set goals, and then iterate the process of adjusting the activity level of schemas linked to the image until a coherent scene interpretation of (part of) the scene is obtained. Cooperation yields a pattern of "strengthened alliances" between mutually consistent schema instances, which allows them to achieve high activity levels to constitute the overall solution of a problem. As a result of competition, instances that do not meet the evolving consensus lose activity, and thus are not part of this solution (though their continuing subthreshold activity may well affect later behavior). Successful instances of perceptual schemas become part of the current short-term model of the environment.

"*Neural*" *schema theory* then analyzes data from neurophysiology, lesion studies and brain imaging to see how schemas are related to distributed neural mechanisms. In this way, neural schema theory extends basic schemas downward to neural mechanisms, possibly changing the definition of constituent schemas from those initially posited on the basis of introspection or psychological and behavioral data alone.

An example will clarify this distinction. Consider a very basic phenomenon: frogs or toads will snap at small moving objects and will jump away from large moving objects. A preliminary model (Fig. 3.3) has two perceptual schemas, one to recognize small moving objects and another for large moving objects. Activation of the former will trigger a motor schema for snapping while the latter will trigger a motor schema for avoidance. Now, for reasons that seemed clear already in the 1960s, it made sense to hypothesize that the tectum (a visual region in the midbrain, homologous to the mammalian superior colliculus) recognizes small moving objects and triggers the snapping while the pretectum (a region just in front of the tectum)

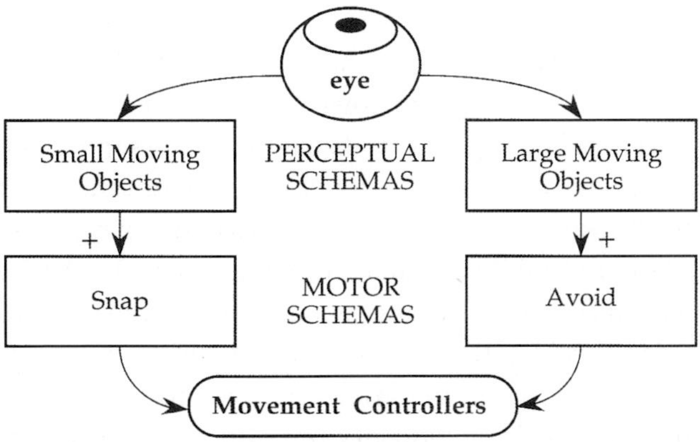

Fig. 3.3 A preliminary model of interacting schemas serving the ability of toads to snap at prey and avoid predators.

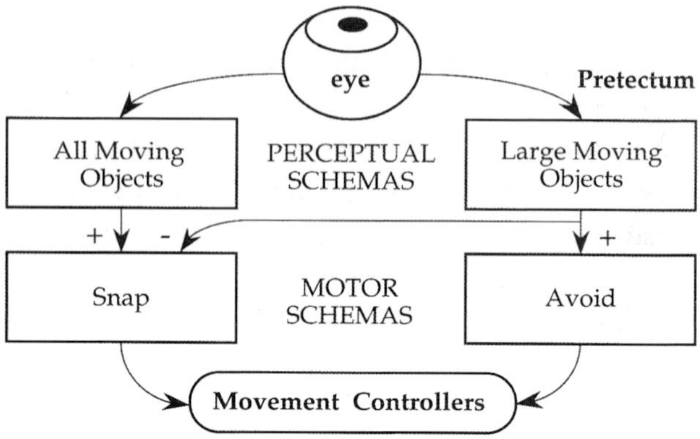

Fig. 3.4 An update of the model of Fig. 3.3 reflecting data on the effect of pretectal lesions.

recognizes large moving objects and triggers jumping away from them. If this hypothesis were true, the prediction is that if an experimentalist were to lesion the toad's pretectum then it would no longer recognize large moving objects. Thus, the model of Fig. 3.3 predicts that a pretectum-lesioned toad would snap at small moving objects but would not respond at all to large moving objects. But, as a matter of fact, the animal, after such a lesion would snap at *any* moving object. This led to a refinement (Fig. 3.4) in which we have the tectum recognizing all moving objects and triggering a snap. The pretectum, in contrast, recognizes if a moving object is large and in that case will inhibit the snapping and triggers jumping (Ewert and von Seelen 1974).

This example shows that one can have neural models even at the level of schemas without necessarily going down to the level of the neuron. A given schema, defined functionally, may be distributed across more than one brain region. A given brain region may be involved in many schemas. Hypotheses about the localization of (sub)schemas in the brain may be tested against data from lesion studies, brain imaging, and so on. To the extent that a schema model survives such testing, we can (if appropriate with regard to the issues under consideration) proceed to give more detail of how interactions between the neurons in various parts of the brain could not only implement the suggested functionality but also explain more detailed data on behavior. For example, in seeking to understand toad visuomotor coordination we can seek to understand in more detail why the shape of an object, and not just its size, affects the toad's response. We can also look at how shape and direction of motion affect both the direction and speed of the animal's response. In doing so, we may refine the description of the schemas to the point where a verbal level of description is no longer useful, and we must instead assess peculiarities of parametric description that make most sense when linked to at least some of the details of the underlying network. [Ewert and Abib (1989) provide a relevant collection of articles on mechanisms of visuomotor coordination in amphibians, models and robots.]

We have now to come back to the idea of competition and cooperation of perceptual schemas. Our schemas for visual perception embody patterns of competition which in general ensure that activation of one schema to describe a region of the image will inhibit others. (Of course, there are subtleties — we can see a region as a wall and as part of a house.) Our schemas embody patterns of cooperation which build on the fact that, on the bases of our past interactions, we have come to expect (but not guarantee) certain relations between objects in the environment. For example, if we see a tree trunk then we expect foliage above it, and when we see a certain type of cone with the point facing downward, we expect ice cream above it. As a result, if we were to see a somewhat ill-defined picture whose lower part appears to be either a tree trunk or an ice cream cone, and whose greenish upper part appears to be green foliage or ice cream, our schema network will tend to settle into a state where we will see either ice cream above a cone or foliage above a tree trunk. However, we are not forced to always see that which custom has given us. In general, we will find ourselves in situations where we can make global sense on the basis of the local constraints. However, to see the contrary, consider how René Magritte employs realism in the service of Surrealism. If we look solely at the upper third of his *Le Château des Pyrénées* of 1959, we see a castle on a mountain top against a background of clouds and blue sky. However, if we follow the mountain top downward, we find ourselves looking not at a mountain but at a large boulder suspended above the waves of the sea. In the complete picture, local realism battles global constraints to allow us to exhibit the breaking with reality that Magritte intended.

3.5. SOCIAL SCHEMAS

Both "pure" and "neural" schema theory as sketched above are theories of "computations in the head." But if we really want to move from downward causation to a neuroscience of the person, we need to understand how the perception, action and memory of the individual become embedded within a social matrix, both defining the matrix and being constrained by it. In *The Construction of Reality*, Mary Hesse and I (Arbib and Hesse 1986), sought a shared epistemology linking my schema-based, action-oriented neuroscience to Hesse's analysis of the "social construction" of scientific theories. We thus extended schema theory to include not only schemas as processes internal to the heads of individuals (*internal schemas*) but also *social schemas* which express the social reality that each individual experiences through interaction with individuals behaving according to limited variations on a set of internal schemas. For example, the English language is a social schema defined by a community whose members can communicate using words and grammatical constructions like those I use to write this chapter — and yet each person will have a different vocabulary from mine, will understand some shared words in discordant ways, and may regard as grammatical some sentences that seem ill-structured to me. Nonetheless, the young child can extract statistical regularities from a variety of speakers in her community to herself become a speaker of English. In this way, social schema theory extends basic schemas upwards to society. We may thus consider the hierarchy:

- A society made up of interacting persons and itself defining social schemas
- Persons defined by their schemas "in the head"
- Schemas in each head, defining both conscious experience and diverse subconscious processes which support experience and behavior more generally
- Neurons and circuitry whose function can be approximated by schema interaction
- Membranes
- Molecules

In seeking scientific explanations of mind and brain, we will find certain phenomena that can be explained at a single level. However, we will learn much more if we explore the interactions of each level with the level above and that below. Here what we see is not downward causation but a multiplicity of two-way reductions. As a computational neuroscientist seeking to chart these multiple levels, I must emphasize that the map is not the territory. Just as a full-scale map would not be helpful (Borges 1975) — and just as we need maps that highlight different details for driving a car versus hiking cross-country — so we, as scientists, are always trying to isolate the details which are essential to meeting the pragmatic criterion in some domain of discourse. The success of a model or theory depends as much on what is omitted as on what is included, and we may have to take an incremental approach to the understanding of what details are important and how to address them.

Where "pure" schema theory studies schemas as dynamical, interacting systems that underlie mental and overt behavior (in contrast to classical "symbolic" AI or "Language of Thought") and "neural" schema theory analyzes data from neurophysiology, lesion studies and brain imaging to see how schemas are related to distributed neural mechanisms, "social" schema theory attempts to relate "schemas in the head" to the collective representation of a society — whether a language, religion, ideology, myth or scientific society (Arbib and Hesse 1986). The crucial point is that our experience in interacting with others exposes us to patterns of behavior which provide our personal samples of a whole range of social schemas, and that the schemas forming in our own heads then interiorize aspects of these patterns in ways which may lead us to conform with current social norms or to depart from them. From this perspective, the self may be understood as a "schema encyclopaedia." One's sense of self depends in part on the ability to fill diverse social roles (son, husband, father, professor, tourist, visitor at the Vatican) in varied contexts (at home, university, shop, restaurant) in a way that meets one's own expectations — and to learn when one fails to do so. These roles are played out through the ability to deploy appropriate schemas, respecting current affordances to set expectations and meet goals. Filling these roles is a function not only of one's own action patterns but also internal models of the behavior of others (whether people known to me as individuals or people who I know only by their roles, such as airline stewardess or post-office clerk). For humans, language plays a crucial role in defining and embodying many of these roles. It is hard to imagine how we could come to have so many roles without the support that language gives us in terms of understanding our social situation. Thus, language acquisition provides a key example of how individuals become enculturated. Rather than give an account of language acquisition here (Arbib *et al.* 1987; Clark 2003; Hill 1983; Tomasello 2003), let me turn to a brief account of how language may have evolved (Arbib 2005; Rizzolatti and Arbib 1998).

3.6. THE MIRROR SYSTEM HYPOTHESIS FOR THE EVOLUTION OF THE LANGUAGE-READY BRAIN

A 1978 conference presentation by Marc Jeannerod [later published as Jeannerod and Biguer (1982)] initiated my continuing interest in relating schema theory and computational neuroscience to the visual control of hand movements [starting with Arbib (1981); Arbib *et al.* (1985); Iberall *et al.* (1986)], and this in turn led to a collaboration with Jeannerod, Giacomo Rizzolatti and Hideo Sakata [whose first fruits were summarized by Jeannerod *et al.* (1995)]. During the course of this collaboration Giacomo Rizzolatti and his colleagues in Parma discovered mirror neurons (di Pellegrino *et al.* 1992; Gallese *et al.* 1996) in area F5 of the premotor cortex of the monkey. Activation of these neurons in the brain of a macaque monkey correlates both with execution of a particular type of action by the monkey and observation of a somewhat related action by another monkey. Although we could not look at single neurons in the human brain, we could — thanks in part to brain imaging done at USC under the leadership of Scott Grafton — see whether some part of the brain is more highly active as compared to just looking at objects both when the subject observes grasping actions and when s/he executes such actions. Several such areas were found in the human brain. In particular, activation in the frontal lobe was found in Broca's area, an area traditionally considered a speech production area. What does grasping have to do with speech? Fortunately, I had spent some time visiting the laboratory of Ursula Bellugi at the Salk Institute in La Jolla, where they studied sign language and found that damage to a large region including Broca's area would yield aphasia in signers analogous to that seen in speakers (Poizner *et al.* 1987). In other words, Broca's area is not so much a speech area as a language area. Such considerations provided the foundation for the paper "Language within our grasp" (Rizzolatti and Arbib 1998), which presented the *mirror system hypothesis* for the evolution of the language-ready brain: the mechanisms supporting language in the human brain and yielding language parity (the ability of the hearer to recognize the words uttered by the speaker, and similarly for a signed language) evolved "atop" mechanisms for the control and recognition of manual actions.

The hypothesis has been extended and subjected to extensive commentaries (Arbib 2005), with many further developments since then [e.g., Arbib (2010; 2012)]. Since the details have been well reported elsewhere, let me just summarize the stages posited in the current form of the theory, referring the reader elsewhere for arguments supporting each stage.

We postulate an evolutionary progression of mirror systems and supporting circuitry "beyond the mirror":

- Circuitry for grasping and manual praxic actions (present in the last common ancestor of humans and monkeys).
- Circuitry for simple imitation of grasping and manual praxic actions (present in the last common ancestor of humans and great apes).
- Circuitry for complex imitation of grasping and manual praxic actions (unique to the human line since that last common ancestor).
- Pantomime of grasping and manual praxic actions: allowing one to communicate about objects and actions even in the absence of the objects. Such pantomime provides the breakthrough towards a freely extensible open-ended semantics. A further step was the pantomime of actions outside the pantomimic's own behavioral repertoire (e.g., flapping the arms to mime a flying bird).
- Protosign: the use of conventionalized gestures to formalize and disambiguate pantomime (e.g., to distinguish "bird" from "flying").
- Protosign and protospeech then developed in an expanding spiral — developing a system of conventionalized manual, facial and vocal communicative gestures ("protowords") separate from pantomime.
- While the preceding stages all involve changes in the brain based on natural selection operating on the genome, it is postulated that the transition from protolanguage to language depends not on biological evolution but rather historical evolution, the cumulative effect of human inventions made in the past 100,000 years.

Two famous studies have led us to distinguish the contributions of dorsal (via parietal cortex) and ventral (via inferotemporal cortex) streams processing the outflow of primary visual cortex. A lesion of the ventral path yields a patient unable to indicate the size or orientation of an object either verbally or by pantomime and yet able to use visual information to preshape and orient the hand appropriately when reaching to grasp the object [patient DF, Goodale *et al.* (1991)]. By contrast, a lesion to the dorsal path leaves the patient able to communicate about the size and orientation of the object yet unable to use visual information to preshape the hand in anticipation of the size or orientation of, say, a cylinder [patient AT, Jeannerod *et al.* (1994)]. The point here is that even schemas that seem to be normally under conscious control can in fact proceed without the subject being conscious of their activity. This dorsal-ventral distinction also plays an important role in accounts of the human language-ready brain within the framework of the mirror system hypothesis (Arbib 2010). Here, mirror systems for action continue to be important but have also provided the evolutionary basis for mirror systems for words considered not as meaningful entities but as phonological actions. A word in this sense, without any semantic correlate, is an articulatory or manual action that can be heard (or seen) and repeated. The hypothesis is that the perceptual and motor schemas (recall Fig. 3.2 for their role in visual perception) are located in the ventral brain. The division of labor, then, is between a dorsal stream supporting the generation as well as the recognition of praxic and communicative actions and a ventral stream which can support the formation of schema assemblages that support the planning of action and the recognition of the meaning of actions in relation to agents and objects in the world (Fig. 3.5). The claim is that biological and cultural evolution together elevate the ventral mechanisms to handling not only the ancestral abilities for the domain of action within the world but also the specifically human ability to talk about the world (Arbib 2008; Arbib and Lee 2008; Gallese and Lakoff 2005). In this sense, we may see the dorsal and ventral stream as providing the neural mechanisms to support Saussure's signifiers and signifieds, respectively (Hurford 2004; Saussure 1916).

The implication of all this is that we will better understand specifically human abilities if we develop an evolutionary perspective on comparative neurobiology than we will if we just treat them as "miracles of the human brain." My group seeks to develop computational models of neural circuitry based on the data of people studying the macaque's brain regions, anatomists assessing homologies between parts of the human brain and the

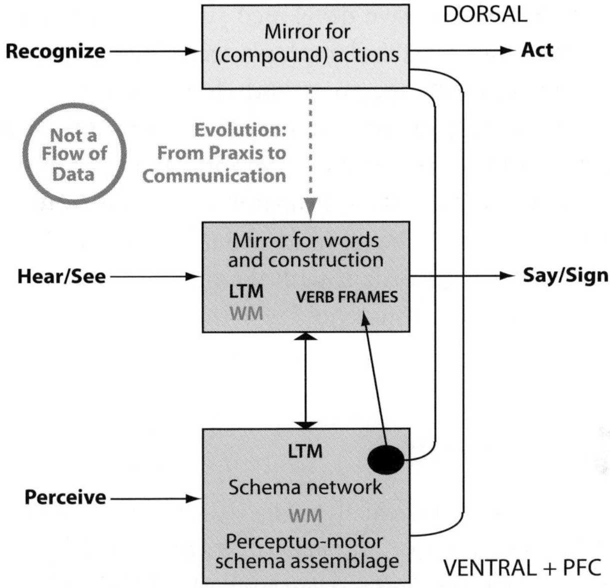

Fig. 3.5 A mirror system for practical actions (which involves diverse types of neurons involved in the control of action, not just mirror neurons) provides the evolutionary basis for a mirror system for communicative actions. Moreover, the circuitry (including regions beyond the mirror system) that supports complex imitation and compound actions has evolved to support neural mechanisms that implement the grammatical constructions that combine words, morphemes and phrases hierarchically and recursively to yield utterances, and the inverse mechanisms for parsing such utterances. However, the dorsal system for generating and recognizing words and utterances as vocal or manual actions does not encode the semantics of these words and utterances. Rather, this is achieved by the integration of dorsal codes with schema assemblages encoded within the ventral stream. (PFC, prefrontal cortex; WM, working memory; LTM, long-term memory.) (Adapted from Arbib 2010).

macaque brain, and human brain imaging and neurology. Monkey studies have the advantage of relating detailed neurophysiology to details of circuitry, correlating the activity of particular cells with different roles in supporting action and perception, whereas human data have the advantage of giving us a large-scale view of the relative importance of different brain regions in different tasks, including cognitive tasks of which non-humans are incapable. My strategy is thus in some sense to computationally "insert" the monkey brain within the human brain and then make plausible hypotheses in terms of this evolutionary framework as to how other portions of the human brain may be seen as elaborations or variations upon the macaque brain. Once I have developed such a model, guided by comparative neurobiology, I can then employ a technique called *synthetic brain imaging* (Arbib *et al.* 2000; Tagamets and Horwitz 2000), which provides systematic ways to average over the activity in different regions of a detailed neural network model to make predictions about what brain imaging will show. More recently, we have added a capability for synthesis of event-related potentials from neural and schema model (Barrès *et al.* 2013). We may then update models that are concordant with the data of monkey neurophysiology and those of human neuropsychology in the light of critical evolutionary hypotheses.

3.7. CONSCIOUSNESS, BRIEFLY

As regards consciousness I want to make essentially the following claims. Consciousness is not a unitary property of the brain. It is quite possible for neural activity to effectively link perception and action without any possible intervention of consciousness. For example, in "blindsight," a person with a lesion of primary visual cortex may be unable to consciously perceive even the location of objects let alone their identity, and yet may be able to navigate on the basis of variations of luminance or movement corresponding to objects in the ground plane (Stoerig and Cowey 1997; Weiskrantz 1974). Moreover, as the data on AT and DF in the previous section make clear, consciousness is not a direct property of having neurons of a particular structure or complexity because the same data can be represented in two networks of comparable neural complexity, yet be accessible to consciousness only when one of the networks rather than the other is intact.

Human consciousness, as we normally experience it, is a property of the brain, rather than some separate "mind stuff." It is possible that portions of our brain can support forms of "animal awareness" that may enrich our consciousness. However, human consciousness seems qualitatively different in nature. What makes human consciousness so different is that it incorporates the functions of the brain that express language. I would complement the discussion of the mirror system hypothesis by stressing the consequences of having a "précis" — a gesturable representation — of intended future movements as a means to coordinate behavior with other members of the group (Arbib 2001; Arbib and Hesse 1986). The brain of each group member must be able not only to generate such signals but also to integrate signals from other members of the group into its own ongoing motor planning. The "communication plexus," as we call the neural circuitry that provides this précis, enriches the "information environment" for the rest of the brain. A crucial element of our analysis is that the evolution of this new system provided an environment for further evolution of older neural systems. As a result, the précis comes to serve not only as a basis for communication between the members of a group but also as a resource for planning and coordination within the brain itself and, we claim, supports the emergence of a specifically human form of consciousness. Moreover, this account makes clear why at times consciousness may seem to be an observer and at other times a controller of our behavior. In any case, consciousness is a distributed property that has little access to many brain regions, and provides a précis (only partially verbalizable) based on the state of other brain regions.

In conclusion, a fundamental methodological prescription is not to try to take any one of the levels concerned (neurochemistry, synapses, neurons, neural circuitry, schemas, persons and societies) and say that it is causally complete and determinative of all that happens at the mental level. Rather, we should try to understand how each level serves higher levels, as well as how it is served by what is known about levels further down. A multiplicity of two-way reductions, or — if you dislike the word reduction, two-way conversations — will provide a more appropriate framework for developing a neuroscience of the person (Arbib 1999a; Arbib 1999b) than an over-emphasis on downward causation.

DISCUSSION

Pierre Jacob: I have a first question about schemas. Are you a realist about schemas? A realist, in the following sense: how much do you think you want to constrain the notion of a perceptual schema by conscious experience and conceptual knowledge (your example of the concept of a house and our perceiving a house)? So, is our concept of a house a constraint that determines what is a perceptual schema? Are the words in our vocabulary (in English or French and other languages) constraints on what counts as a perceptual schema?

Michael Arbib: Am I a realist on schemas? Yes, in the following sense: as a concrete example, consider that, when a toad looks at moving objects in the environment, we can basically distinguish those which trigger approach behavior and those which trigger the escape behavior; as such, this defines true perceptual schemas. Then, one can ask what are the brain mechanisms that support them. In terms of the behavior, we can say there are two motor schemas: one would be the motion towards the object followed by jaw opening and the extension of the tongue, and the other, dramatically different, mode of behavior, would be turning and jumping away. As I have said, by lesioning the pretectum, you could then discover that, in fact, the perceptual schema for preying is not localized in the tectum (as was initially hypothesized) but that this schema involves at least pretectum and tectum working together (based on input from the retina, of course). I think this is a level of analysis that is necessary if we are to bridge between behavior and brain regions and neural mechanisms. So, what realism means in this sense is not that there is a little black box in the brain labeled "perceptual schema" and a box labeled "motor schema." This analysis, on the contrary, provides an appropriate bridge, necessary to update psychological vocabulary in a way that is suited to understanding the brain. As to the role of words, I think there is an interaction between experience and words, but it is not a very direct interaction. We make use of the word not because there is a well-defined perceptual schema for it but because we use it in many social situations in which uttering that word seems appropriate. It may be that this needs to be supported by a certain pattern of schemas, but there is no one-to-one correspondence between words and schemas. Conversely, we have perceptual experiences that do not depend on words.

Pierre Jacob: One basic issue is that when you actually undertake verbal communication using language, at least three things can and ought to be distinguished: a lexicon, a syntax with certain specific characters through which we form structured sequences of words, and another aspect that I think is quite important: the ability to represent another individual's communicative intentions. Many linguists and philosophers have thought about the peculiar properties of both syntax and communicative intentions. Arguably, given that the lexicon of a language is finite, but the set of grammatical sentences of a language is not, it follows that syntax must include recursive rules. Furthermore, as stressed by linguists and philosophers since Paul Grice, a speaker's communicative intention is reflexive in the following sense: suppose the speaker's communicative intention is that her addressee forms a new belief as a result of the communicative act. Then the speaker's communicative intention can only succeed if and when the addressee recognizes it. So, it is part and parcel of the content of a speaker's communicative intention that the addressee should recognize it. Now, my question is: in your view, which do you think mirror neuron activity can explain — the recursivity of the syntax of natural languages or the reflexivity of a speaker's communicative intention? These are two different issues.

Michael Arbib: I would answer very briefly by referring to my distinction between a dorsal system which includes mirror neurons for actions (praxic actions) and mirror neurons for words (phonological actions), on the one hand, and a ventral system for assembling perceptual and motor schemas in planning decisions, on the other. The suggestion was that it is the linkage of the mirror system into this larger system that imbues words with meaning. I would say that recursion is not unique to language, but it is a property that language inherits from perception and action once it becomes able to express the richness of our embodied experience of the world. Thus, recognizing another individual's communicative intention may use the mirror neurons but only as part of a larger process beyond the mirror system.

Jean-Pierre Changeux: You mentioned schema competition. Well, does competition imply selection? How do you define this? I think you have to explain to us what is the meaning of the word "competition." This is also related to the role of mirror systems in emotions.

Michael Arbib: To "answer" your question in regard to emotions: this is a whole other subject I deliberately left out of this presentation. Learning depends on the reward systems and the motivations to act. If I am hungry, how I will act would be different than if I were thirsty, and so on and so forth. This is a huge topic, which I have begun to address elsewhere (Arbib and Fellous 2004; Fellous and Arbib 2005).

Jean-Pierre Changeux: Yes, but is the discourse of emotions relevant for your thinking on these matters?

Michael Arbib: Oh sure, it is! I gave you a 50-minute sample of my thinking and left that topic out without any sense of downplaying its importance. Let me now turn to schema competition. My focus was not in terms of learning except in my discussion of prism adaptation. To find out the appropriate actions, in certain circumstances, one will have to learn to pay attention to features that may prescribe action. It is here that the theory of reward and emotions has to come into play.

Jean-Pierre Changeux: Ok! Allow me to go to the question of consciousness. When you say that consciousness is not the property of some particular neural systems, I would say that we have to specify these systems. For example, there are drugs that can cause you to lose consciousness.

Michael Arbib: Yes, I think that it would be nice to discuss all those things but I simply took the time I had available to give you a functional analysis of what was involved.

Jean-Pierre Changeux: I think you should provide a definition and specify your conclusion. I am just trying to deal with your conclusions. One of them is about communication; and this seems to say that consciousness emerges from a sequence of steps without having specific kinds of system being involved in this process. I just want to know if these have a definite position within the brain.

Michael Arbib: Well, I think you are putting words in my mouth. I gave you the data on patients AT and DF as a case where two types of visual processing appearing at first glance to be similar seem to be such that one had a conscious component and the other did not. That is what I have as data for this discussion. Yes, there are drugs that cause one to lose consciousness; I have nothing to add to that, so I did not talk about it.

Jean-Pierre Changeux: You have said that you are a specialist of specific systems in the brain.

Michael Arbib: If by "specific" you mean that I believe I can point to a specific brain region and say this is the locus of consciousness (as distinct from an activating system for consciousness), I say no! If by your question you mean, "Do I believe that an analysis of distributed neural networks can eventually allow us to understand consciousness?" then I say yes!

Gennaro Auletta: I found your contribution very interesting and helpful. Well, there are two ways of causation: bottom-up and top-down; and allow me to say that if I have stressed top-down causation, it is only because I am sure that bottom-up causation is very useful and has a very solid tradition. I would only like to add that there are situations in which, probably, old-fashioned explanations centred on bottom-up causation may be no longer sufficient or no longer interesting and must be integrated with other ones. I have found especially interesting your point about global constraints and local interactions. In this sense, I am probably more reductionist than other people here because I started from quantum mechanics and this is exactly what quantum mechanics suggests to me. At a very basic level the world works exactly in that way. If I am allowed to somehow reformulate an old Kantian way of speaking, I would say that global constraints are unable to determine local interactions but local interactions are blind without global constraints. That is, starting from local interactions scholars will never acknowledge global constraints. On the other hand, global constraints cannot alone determine those interactions. I have also very much appreciated the issue of brain stratification, and especially your point that reciprocal connections can modify the older schemas. This was exactly my point before: you can have genetically something emerging that in a certain way "back-reacts" on the previous situation. I think this is very important, because, quite likely, this is the way in which many layers in our brain are constituted.

Michael Arbib: I am not quite sure I understand why local interactions are blind to global constraints. [Note added later by MAA: But G Auletta said "without," not "to," so this answer is misleading.] Take the example of a magnet. Then given the local spins we get a situation which throws a preponderance of co-aligned spins, and this is what you mean by global constraint. . .

Gennaro Auletta: I mean something very, very easy. When you have entanglement in quantum mechanics, you cannot find entanglement starting from a single measurement on a single system. You need a statistics of many measurements and perform a comparison. Therefore, even a lot of local measurements will never allow the understanding of a global constraint. There are some general physical lessons here that could be interesting for us.

Michael Arbib: So, as I said, the issue is whether local interactions are blind to global constraints. You have just given us an example for which that applies, but I think for our discussion a better example is that where we have a macro system such that the population effects of the components of that system do affect local interactions as in the case of magnetism, where the global magnetism affects the local spins. My point number two returns to a comment you made in your talk. I think the notion of genetics and its evolution is relevant because, without getting into questions of the mind and top-down causation, what is at the heart of our evolutionary theory of natural selection is the interaction between the genotype and the phenotype, where we have the genotype as a very nice linear code and the phenotype as this complex thing interacting with the environment. Now, its interactions with the very complex environment and their differential effect on survival determine which of those genetic codes survives. Here we have another case where this general issue of two very different levels of analysis to be put into conversation again comes into play.

REFERENCES

Arbib MA. (1975) Artificial intelligence and brain theory: Unities and diversities. *Ann Biomed Eng* **3**: 238–274.

Arbib MA. (1981) Perceptual structures and distributed motor control. In: VB Brooks (ed.), *Handbook of Physiology — The Nervous System II. Motor Control.* American Physiological Society, Bethesda, MD, pp. 1449–1480.

Arbib MA. (1989) *The Metaphorical Brain 2: Neural Networks and Beyond.* Wiley-Interscience, New York.

Arbib MA. (1991) Neurons, schemas, persons, and society. In: JE Earley (ed.), *Individuality and Cooperative Action.* Georgetown University Press, Georgetown, pp. 63–86 (see also General Discussion, pp. 179–180).

Arbib MA. (1992) Schema theory. In: S Shapiro (ed.), *The Encyclopedia of Artificial Intelligence.* Wiley-Interscience, New York, pp. 1427–1443.

Arbib MA. (1999a) Crusoe's brain: Of solitude and society. In: RJ Russell, N Murphy, TC Meyering and MA Arbib (eds.), *Neuroscience and the Person: Scientific Perspectives on Divine Action.* Vatican Observatory Publications/Center for Theology and the Natural Sciences, Vatican City State/Berkeley, CA, pp. 419–448.

Arbib MA. (1999b) Towards a neuroscience of the person. In: RJ Russell, N Murphy, TC Meyering and MA Arbib (eds.), *Neuroscience and the Person:*

Scientific Perspectives on Divine Action. Vatican Observatory Publications/Center for Theology and the Natural Sciences, Vatican City State/Berkeley, CA, pp. 77–100.

Arbib MA. (2001) Co-evolution of human consciousness and language. In: PC Marijuan (ed.), *Cajal and Consciousness: Scientific Approaches to Consciousness on the Centennial of Ramon y Cajal's Textura. Ann N Y Acad Sci* **929**: 195–220.

Arbib MA. (2005) From monkey-like action recognition to human language: An evolutionary framework for neurolinguistics. *Behav Brain Sci* **28**(2): 105–124 (see also Discussion, pp. 125–167).

Arbib MA. (2008) From grasp to language: Embodied concepts and the challenge of abstraction. *J Physiol Paris* **102**(1–3): 4–20.

Arbib MA. (2010) Mirror system activity for action and language is embedded in the integration of dorsal and ventral pathways. *Brain Lang* **112**(1): 12–24.

Arbib MA. (2012) *How the Brain Got Language.* Oxford University Press, New York, Oxford.

Arbib MA, Billard A, Iacoboni M, Oztop E. (2000) Synthetic brain imaging: Grasping, mirror neurons and imitation. *Neural Networks* **13**: 975–997.

Arbib MA, Conklin EJ, Hill JC. (1987) *From Schema Theory to Language.* Oxford University Press, New York.

Arbib MA, Érdi P, Szentágothai J. (1998) *Neural Organization: Structure, Function, and Dynamics.* MIT Press, Cambridge, MA.

Arbib MA, Fellous JM. (2004) Emotions: From brain to robot. *Trends Cogn Sci* **8**(12): 554–561.

Arbib MA, Hesse MB. (1986) *The Construction of Reality.* Cambridge University Press, Cambridge.

Arbib MA, Iberall T, Lyons D. (1985) Coordinated control programs for control of the hands. In: AW Goodwin and I Darian-Smith (eds.), *Hand Function and the Neocortex.* Springer-Verlag, Berlin, pp. 111–129.

Arbib MA, Lee J. (2008) Describing visual scenes: Towards a neurolinguistics based on construction grammar. *Brain Res* **1225**: 146–162.

Arbib MA, Schweighofer N, Thach WT. (1995) Modeling the cerebellum: From adaptation to coordination. In: DJ Glencross and JP Piek (eds.), *Motor Control and Sensory-Motor Integration: Issues and Directions.* North-Holland Elsevier Science, Amsterdam, pp. 11–36.

Barrès V, Simons A, Arbib MA. (2013) Synthetic event-related potentials: A computational bridge between neurolinguistic models and experiments. *Neural Networks*: in press.

Bonaiuto J, Arbib MA. (2010) Extending the mirror neuron system model, II: What did I just do? A new role for mirror neurons. *Biol Cybern* **102**: 341–359.

Borges JL. (1975) Of exactitude in science. In: *A Universal History of Infamy*. Penguin Books, Harmondsworth, pp. 131.

Clark E. (2003) *First Language Acquisition.* Cambridge University Press, Cambridge.

di Pellegrino G, Fadiga L, Fogassi L, *et al.* (1992). Understanding motor events: A neurophysiological study. *Exp Brain Res* **91**(1): 176–180.

Draper BA, Collins RT, Brolio J, *et al.* (1989) The schema system. *Int J Comput Vision* **2**: 209–250.

Ewert J-P, Arbib MA (eds.). (1989) *Visuomotor Coordination: Amphibians, Comparisons, Models and Robots.* Plenum Press, New York.

Ewert J-P, von Seelen W. (1974) Neurobiology and system theory of a visual pattern recognition mechanism in the toad. *Kybernetik* **14**: 167–183.

Fellous J-M, Arbib MA (eds.). (2005) *Who Needs Emotions: The Brain Meets the Robot.* Oxford University Press, Oxford, New York.

Gallese V, Fadiga L, Fogassi L, Rizzolatti G. (1996) Action recognition in the premotor cortex. *Brain* **119**: 593–609.

Gallese V, Lakoff G. (2005) The brain's concepts: The role of the sensory-motor system in conceptual knowledge. *Cogn Neuropsychol* **22**: 455–479.

Goodale MA, Milner AD, Jakobson LS, Carey DP. (1991) A neurological dissociation between perceiving objects and grasping them. *Nature* **349**(6305): 154–156.

Hanson AR, Riseman EM. (1978) VISIONS: A computer system for interpreting scenes. In: AR Hanson and EM Riseman (eds.), *Computer Vision Systems.* Academic Press, New York, pp. 129–163.

Hesse MB. (1980) *Revolutions and Reconstructions in the Philosophy of Science.* Indiana University Press, Bloomington, IN.

Hill JC. (1983) A computational model of language acquisition in the two-year-old. *Cogn Brain Theory* **6**: 287–317.

Hurford JR. (2004) Language beyond our grasp: What mirror neurons can, and cannot, do for language evolution. In: DK Oller and U Griebel (eds.), *Evolution of Communication Systems: A Comparative Approach.* MIT Press, Cambridge, MA, pp. 297–313.

Iberall T, Bingham G, Arbib MA. (1986) Opposition space as a structuring concept for the analysis of skilled hand movements. In: H Heuer and C Fromm (eds.), *Generation and Modulation of Action Patterns.* Springer-Verlag, Berlin, pp. 158–173.

Jeannerod M, Arbib MA, Rizzolatti G, Sakata H. (1995) Grasping objects: The cortical mechanisms of visuomotor transformation. *Trends Neurosci* **18**(7): 314–320.

Jeannerod M, Biguer B. (1982) Visuomotor mechanisms in reaching within extrapersonal space. In: DJ Ingle, RJW Mansfield and MA Goodale (eds.), *Advances in the Analysis of Visual Behavior.* MIT Press, Cambridge, MA, pp. 387–409.

Jeannerod M, Decety J, Michel F. (1994) Impairment of grasping movements following a bilateral posterior parietal lesion. *Neuropsychologia* **32**(4): 369–380.

Martin TA, Keating JG, Goodkin HP, *et al.* (1996a) Throwing while looking through prisms. I. Focal olivocerebellar lesions impair adaptation. *Brain* **119**(Pt 4): 1183–1198.

Martin TA, Keating JG, Goodkin HP, *et al.* (1996b) Throwing while looking through prisms. II. Specificity and storage of multiple gaze-throw calibrations. *Brain* **119** (Pt 4): 1199–1211.

Poizner H, Klima ES, Bellugi U. (1987) *What the Hands Reveal about the Brain.* MIT Press, Cambridge, MA.

Rizzolatti G, Arbib MA. (1998) Language within our grasp. *Trends Neurosci* **21**(5): 188–194.

Saussure F. (1916) *Cours de linguistique générale* (eds. C. Bally and A. Sechehaye, with the collaboration of A. Riedlinger). Payot, Lausanne and Paris. [English translation by W. Baskin. (1977) *Course in General Linguistics.* Fontana/Collins, Glasgow.]

Sperry RW. (1964) Problems outstanding in the evolution of brain function. The James Arthur Lecture, American Museum of Natural History, New York. Reprinted in 1977. In: R Duncan and M Weston-Smith (eds.), *The Encyclopaedia of Ignorance* (eds.). Pergamon, Oxford, pp. 423–433.

Sperry RW. (1980) Mind–brain interaction: Mentalism, yes; dualism, no. *Neuroscience* **5**: 195–206.

Spoelstra J, Arbib MA. (2001) Cerebellar microcomplexes and the modulation of motor pattern generators. *Autonomous Robots* **11**: 273–278.

Stoerig P, Cowey A. (1997) Blindsight in man and monkey. *Brain* **120**(3): 535–559.

Tagamets MA, Horwitz B. (2000) A model of working memory: Bridging the gap between electrophysiology and human brain imaging. *Neural Networks* **13**(8–9): 941–952.

Tomasello M. (2003) *Constructing a Language: A Usage-Based Theory of Language Acquisition.* Harvard University Press, Cambridge, MA.

Weiskrantz L. (1974) The interaction between occipital and temporal cortex in vision: An overview. In: FO Schmitt and FG Worden (eds.), *The Neurosciences Third Study Program.* MIT Press, Cambridge, MA, pp. 189–204.

CHAPTER 4

The Neural Basis of Consciousness and the Issue of Downward Causation

Notes by Gennaro Auletta and Ivan Colagè on Antonio Damasio's contribution[1]

How can the mind exert downward causality? Specifically, how can mental processes exert an influence on the physical matter of the body? Of course, the answer should take the brain into consideration, but such questions have been made barely tractable by a distorted view of what the brain is. Consciousness endowed with a kind of self displays an oddity: the mind is interpreted by, and accessible to, no other entity but the organism that has that mind. Often, mind and consciousness are not properly distinguished but they should: since traditionally there has been no satisfactory biological solution for the issue of consciousness, biological accounts of mind are opposed, and the same also applies to mental downward causality. The terms "mind" as well as "consciousness" and "brain" have the advantage of being in accordance with our basic intuitions of ourselves. However, the mind is an entirely natural biological process that can only be understood from an evolutionary perspective and the point of view of the organism's self-regulation.

[1]Two of the Editors (A.G. and I.C.) have summarized Damasio's contribution on the basis of the transcripts because the Author was unavailable to revise the text. It is clear that the Editors remain the only responsible of any misunderstanding.

In other words, any account of the mind should begin with simple organisms — in fact with simple eukaryotic cells — and their survival needs. Once we take an upward road rather than a downward one, we can e.g. discover that "un-brained" and "un-minded" forms of regulation came long before brained ones, and that the relationship between brain and mind appeared only later in evolution. Minds that are still non-conscious developed what can be called biological selves. In fact, we can suspect that even for humans, conscious states constitute a very small part of the brain's intrinsic activity, which is very important for the general life management but is actually never known to us except through the small window of consciousness. Finally, it can be said that the mind, biologically speaking, is the product of very complex brain maps deeply interconnected among one another. Because such brain maps are physical states of neural circuitries, they can also exert downward causality over other neural circuitries as well as non-neural components of the body.

Sometimes, when we look at the literature on consciousness, we find a variety of phrasings like "the mind in general", "perception", "conscious states" and so forth. Another term often used is "awareness" of the self and its surroundings, and there is also a standard dictionary definition of it. In fact, we could agree with a variation of such a definition, namely: consciousness in its strict sense refers to the knowledge of one's own existence and one's surroundings. A conscious mind is one whose contents are attributed to a knower/protagonist of the events going on in the mind (in the self). We could agree with concepts like those used by Michael Arbib (Chap. 3), e.g. a broad self with its whole social interactivity. The essence of consciousness is therefore basically constituted by the mind plus the sense of subjectivity. Consciousness is a mind process (including *qualia*) together with a self-process (characterized by a sense of ownership). The mind's contents are constructed in the organism's perspective.

A problem we face in neurobiology of consciousness is the nature of the mind itself. The possibility of having cortical maps of both sensory and motor data is fundamental for the definition of the mind. Organisms that do not have cortical maps are supposed to have limited mind processes, although they can still have extremely sophisticated forms of behavior. Any definition of consciousness centered on mental events alone without taking into account the basic sense of a possessor (protagonist) of that mind and the perspective that is unique to that organism should be considered as unsatisfactory.

Let us now focus on two problems related to consciousness: its function and its antecedents. About its function, we can say that consciousness provides an optimization of the management of life in complex brain-governed organisms that obviously live in very complex and challenging environments. About the antecedents, what we usually call consciousness has antecessors in a variety of organisms up to single eukaryotic cells (which are obviously not provided with a brain) and organisms that are multicellular but still un-brained; and then, climbing all the steps up to organisms with primitive brains and those that, ultimately, having cortical maps, display the possibility of what can be called true representations. Therefore, the antecedents of consciousness should be found in the homeostasis of single eukaryotic cells, the mechanisms for the management of life in multicellular organisms (pain and punishment devices, pleasure and reward mechanisms, predictional devices and so on), and the emotions, which are "the crown achievements" of a long history of evolutionary developments of life-regulation. The distinction between feelings and knowing such feelings is absolutely critical here. Indeed, we can conceive of organisms with sufficient representation of mental states so to have feelings (which in turn may help for managing life-regulation) without implying that those organisms *know* that they have such feelings (which therefore remain entirely non-conscious).

In addressing the issue of consciousness, some considerations about the possible characterization of the notion of "self" are needed. The self-process is grounded first on the fact that any organism is singular. Another crucial element is the fact that there is a continuity and a critical stability of a substantial part of the life of any organism. This leads to the notion of proto-self, which can be conceived as the collection of neural maps regulating and representing the state of the organism and its overall relationship with objects in the surroundings. The proto-self consists of non-verbal relationships stemming from the interaction of object and organism and the changes that such an interactions brings about on the state of the latter (Fig. 4.1a). All that also suggests the general hypothesis that the change induced in the proto-self when the organism deals with an object results in the "attentional" focusing on the object. Such an engagement happens at two different levels: one is mostly muscular-skeletal and is concerned with externally-directed sensory devices. The other is due to the fact that such an engagement also comes from internal recognition mechanisms and it is therefore generally related

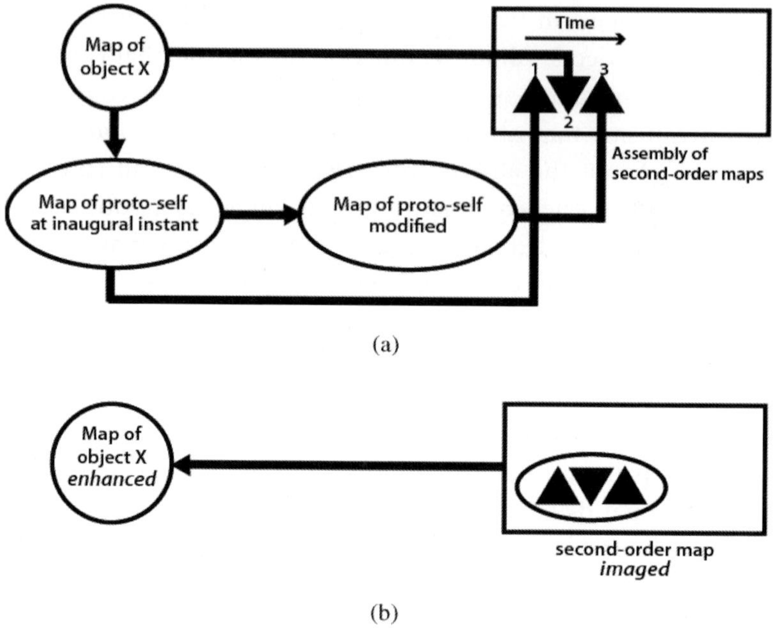

Fig. 4.1 First-order (a) and second order (b) mapping of the object as constitutive of the proto-self.

to the emotional aspects. The issue of reward raised earlier by Jean-Pierre Changeux is related to this problem.

An intriguing possibility is the assembly of a second-order map "narrating" these modifications non-verbally (Fig. 4.1b). The second-order map then has the possibility of acting on the original map of the object creating a momentary link between the state of the organism and the particular object that has caused the original change. There is no implication whatsoever that these second-order maps assemble in one specific site. In fact, it should be assumed that there is no center for consciousness as such. Another suggestion is that such second-order maps produce something that could be truly described as "a feeling of knowing", although still in non-verbal terms.

Let us now consider the tripartite structure shown in Fig. 4.2. The first level is the proto-self: it is entirely non-conscious and can be considered as a collection of representations of the multiple dimensions of the organism's current state; it may be regarded as the pre-conscious biological antecedent of both the core self and the autobiographical self. The second level is the core self, an intermediate level accompanying the transition to the conscious

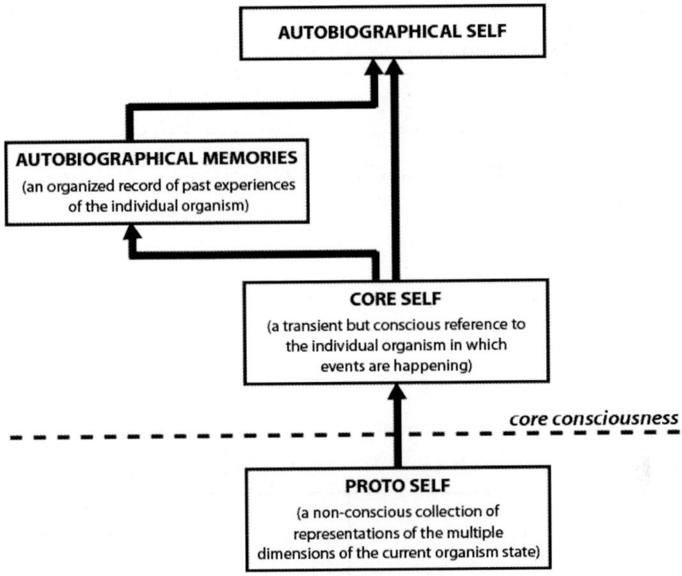

Fig. 4.2 Core self as bridging from proto-self to autobiographical self.

self-reference of the organisms. The core self can be regarded as the collection of non-verbal, second-order maps (built on the basis of the first-order maps characterizing the proto-self) that conjoin in time the neural pattern related to the object, the neural pattern related to the state of the organism, and the relationship between the organism and the object. Finally, we have a far more complex kind of self organization that can be called autobiographical self. The autobiographical self is built upon the core self thanks to the integration of memorized past experiences, plans for the future, and truly linguistic elements.

The self needs to rely on important structures in the brain stem, the tegmentum and related structures in the hypothalamus (Fig. 4.3). The highest level of this organization in all likelihood includes the insular cortex and a variety of other sensory cortices. The point is that, although the particular information processed in those nuclei is very important, wakefulness is a crucial ingredient of consciousness.

It is important to look for structures that would be able to assemble a second-order map and create some kind of mapped relationship between the organism's current state and the object that originated the change in

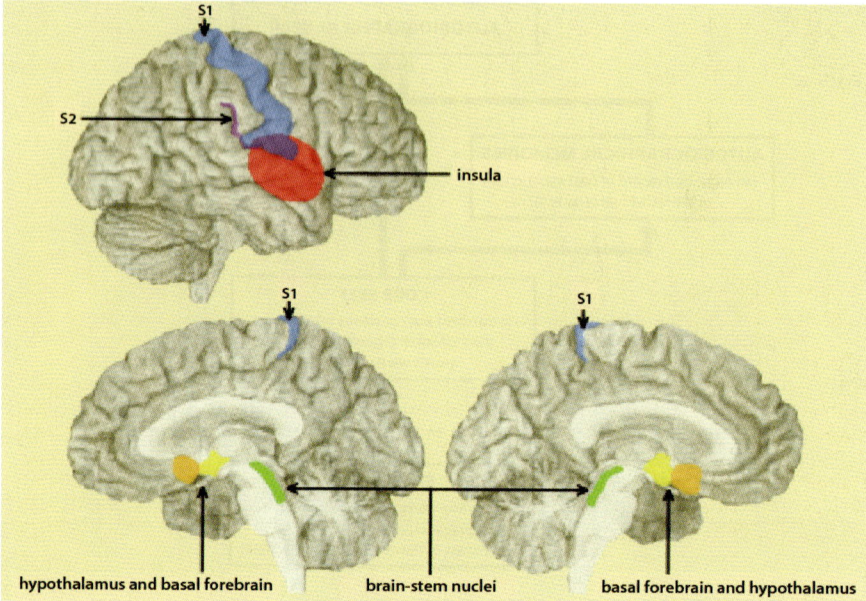

Fig. 4.3

the latter. One of the candidates, though not the only one, is a collection of structures in the posterior region of the brain. Superficially they might be called the parietal cortices, but in fact they are not just parietal since also include area 23 and the retro-splenial area. A reasonable term for this region is postero medial cortices (PMC). We know that patients in an advanced stage of the Alzheimer's disease have major changes in this area. They are in a state of enigmatic mutism showing a separation between a mind process and a self-process: they may recognize an object and even use it purposefully in a very narrow context, but never in wider contexts. They have no way of accessing their own identity. Their mental processes are still present but their consciousness is not. A lesion in the anterior part of the cingulate cortex can cause this kind of condition but only transiently. With a lesion of the PMC the process is more sustained and more like it appears in Alzheimer's disease. In the early stages of this disease, patients are ordinarily conscious in the complete sense of the term. They have autobiographical memory difficulties,

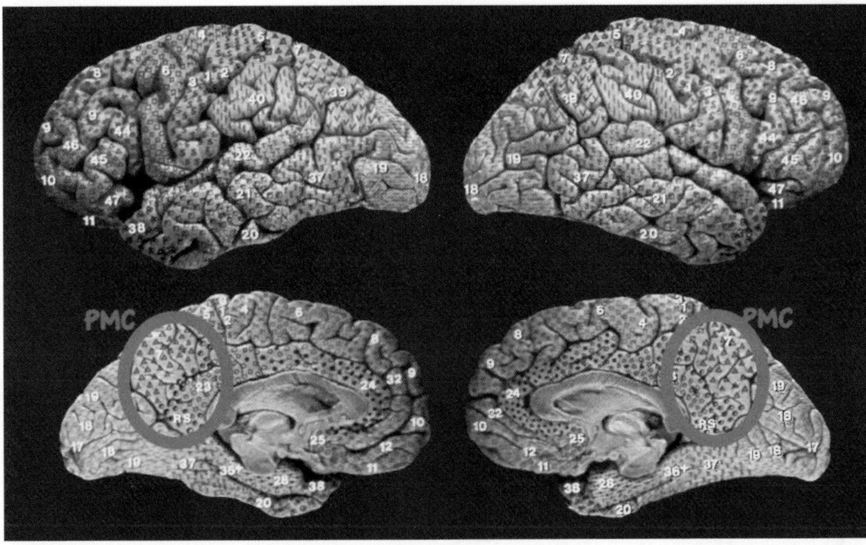

Fig. 4.4

do not learn facts, but they are still conscious, and in fact preserve their social relationships.

Additional evidence in favor of the relevance of those brain regions for consciousness is provided by anesthetics that can target selectively structures that are in the brain stem, in the thalamus and in the posterior medial cortices (PMC; see Fig. 4.4). We know, for example, that looking at the posterior medial cortices during propofol anesthesia, we can remark that the regional blood flow is most reduced when we reach level three anesthesia, in which there is true unconsciousness allowing surgery. Propofol acts very selectively upon posteromedial cortices, the thalamus, and a portion of the brain stem. When we look at decrease in metabolism during vegetative state, the PMC is one of the areas that is most involved, and the really fortunate patients that recover from such states display recovery first of this region.

Let us now consider the neuroanatomy of PMC. Such territory is huge and includes areas 31, 29, 30, 23, and 7m (Fig. 4.5). A crucial question is: do these areas work together? When looking at the connectivity, they are in fact distinct regions but nevertheless they work together as a unit. (This

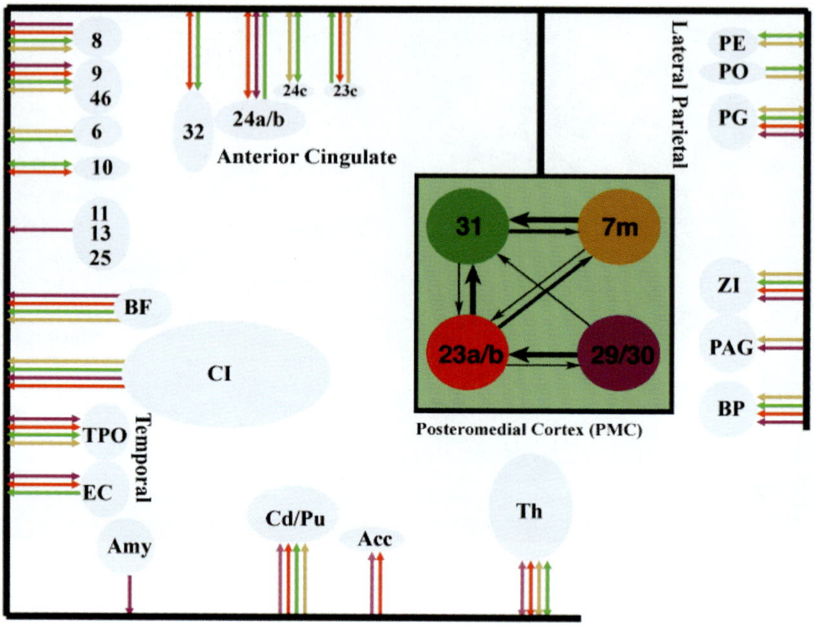

Fig. 4.5 Adapted from Parvizi *et al.* (2006).

does not mean that there are no subunits that have considerable functions independently, of course.) The PMC projects to the dorsal thalamus in a very specific band but not to the intra-laminar nuclei. Moreover, it projects quite massively to the pons as well as to the caudate and the putamen. When we put together connections to and from the PMC (as shown in Fig. 4.6) something quite intriguing is revealed: this structure is operating fundamentally as a *convergence-divergence* set of territories that can serve as a pivot among a very large number of neural influences and responses.

The PMC can receive massive inputs across a hierarchy of small territories of cortical regions, with the possibility that the PMC reciprocates the inputs all the way down to their sources. An interesting study of Hagmann and colleagues (see Fig. 4.7) shows that there is a sort of subdivision between what they call provincial hubs and central hubs. These two kinds of hubs are distinguished by the connectivity that goes in the two directions, but the main result of the study is that there is a specific territory working as the highest order of hubs available in the human brain, and such a territory largely

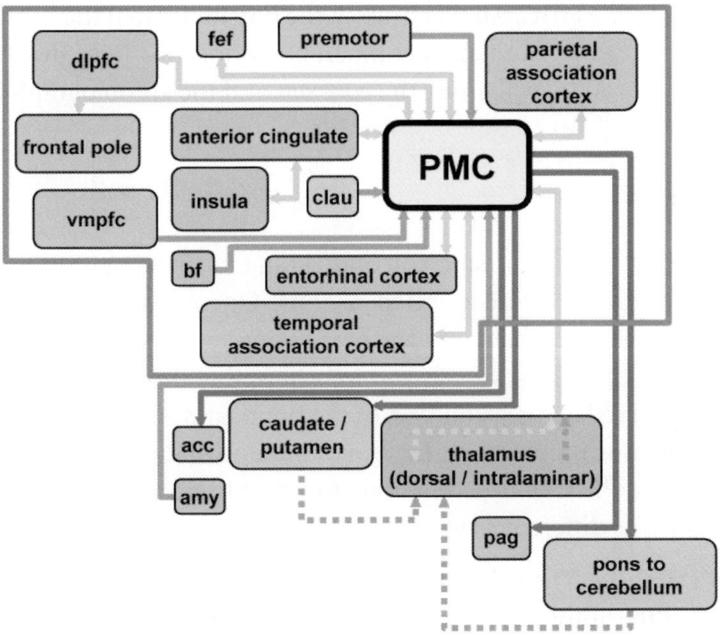

Fig. 4.6 Overview of efferent and afferent connectivity of posteromedial cortices (PMC).

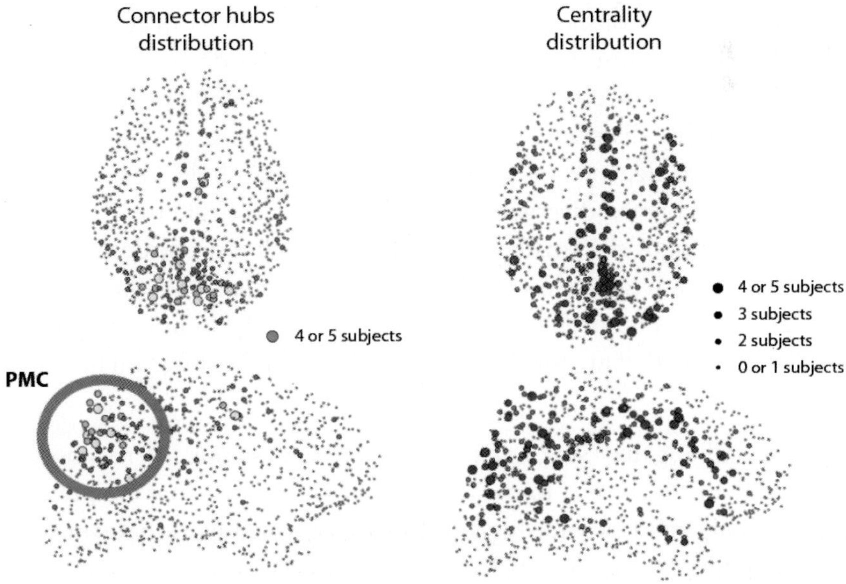

Fig. 4.7 Adapted from Hagmann *et al.* (2008).

corresponds to the PMC. All that hints at the possibility that the connectivity displayed by the PMC (and other comparable connecting hubs that feed into it) could be crucial to explaining the core self and the autobiographical self.

DISCUSSION

Michael Arbib: I want to thank you and Hanna for making us aware of this complex of strongly interconnected regions. One aspect of your talk that disturbs me is the fact that you have based your analysis so much on saying that data from the monkey leads you to believe that this structure (PMC) is strongly conserved across evolution of the brain, and so I wonder if you can complement your presentation by remarking on what you think must have changed in the PMC and its relations to other brain regions to make our consciousness as different as we believe it to be from that of the macaque.

Antonio Damasio: The role that the PMC could have in constructing what I have called the autobiographical self is to allow the relatively simultaneous reactivation of information that is not contained within the PMC but within cortices such as those of the temporal region or the prefrontal territory. This operation will be different in humans by virtue of what difference there is in kind and amount of memories. You have the possibility of reactivating in time, within a short time window, regions that are very diverse, activities that could not possibly be recorded within one simple structure or center but are likely to be highly distributed. This structure is serving, in fact, as a pivot, and what it does depends on what the other cortices hold. It simply functions as a convergence-divergence system. You can activate the PMC and it will, under the right circumstances, produce a much more widespread reactivation of many areas within largely the same time unit. So, the continuity across evolution is no problem whatsoever. Obviously, in humans you have temporal cortices that are much more complex than the ones you have in the monkey.

Jean-Pierre Changeux: Now, my question is: is such strong interconnectivity unique to the PMC or is it found also in other territories that are present in the prefrontal cortex? My feeling, but this is just a feeling, is that you should have also other systems with such characteristics. My second question, which is related to that of Michael Arbib, is: when you look at the differential development of the different territories in the course of evolution, how do some of them develop faster than others?

Antonio Damasio: Excellent questions. I think that the "unsaid" here is the fact that what the PMC is probably providing is the highest level of interconnectivity and integration, over a variety of other areas that are organized the same way and do exactly the same thing. Let me be specific. The temporal parietal junction, for example, is an area with exactly the same organization principle and it works probably just like the PMC. Also some territories in the parietal temporal cortex do exactly the same thing, as well as territories of the lateral, ventromedial and polar frontal cortex. I like the term convergence-divergence (to describe what happens here), as it is very transparent about what is involved physiologically. But I like also the terminology of hubs and spokes (used also by Olaf Sporns). There are multiple hubs and multiple spokes; they are all inter-connected. Then, we have something that is a slightly higher-order hub that does have more centrality and that is the PMC. In graph theory, we can talk about a higher level of centrality, which means that there are more things that are being interconnected by that territory than in other such hubs. We can distinguish between provincial and non-provincial hubs. What is the difference between Bill Clinton and a minor president of the United States? Bill Clinton has a centrality that is very high. These are people who have connected many more people than others. When you insert more modules, you just get more sub-hubs in the system. In the monkey where autobiographical memory is certainly more limited than in humans, I think that the structure will be doing a much lower level of binding than it does in humans. Historically, going back to the seventies, the hubs that we referred to were one or two, in the prefrontal cortex and the temporal parietal junction. Now, the connectivity inferred from diffusion imaging and from functional imaging suggests that we really have a larger set of substations.

Wolf Singer: You have said that this territory in the medial side of the hemispheres is probably the best-conserved hub that we have. The big changes have occurred on the lateral side (like for example the language areas that exchange also exclusive paths with many different spokes). Would there be room for another interpretation pointing out that what is required for core consciousness (which monkeys do not have, of course) is these medial structures that are conserved, and that what makes the big difference is what came on the more lateral side, the language area, for example, or the dorsal lateral prefrontal cortex, that is, areas that the monkeys do not have?

Antonio Damasio: Those areas make a contribution, of course, but you still need pivots to get them all together. For example, the monkey does not have an area 37; one of the greatest glories of the human temporal cortex is this large chunk of area 37, which in the monkey roughly corresponds to the areas TF and TH of Bonin. The monkey's structural core is nothing compared to ours; the subdivisions of the ventral prefrontal cortex are much richer, the dorsal level is also much richer. But still you need a connector (a super-connector) that can bind all this together. So, yes, the subunits will provide us with information that enables us to have an autobiographical self; and yet you need some kind of connector. What I imagine is a hierarchy that still preserves that super-binder, which is the PMC.

Wolf Singer: I am afraid that if you maintain this super-binder, then you trace back to the Cartesian dualist view.

Antonio Damasio: Not at all! This is *not* Cartesian at all. I am not saying, as I pointed out earlier on, that when you construct this large-scale map of different things that are connected in time, they are contained in one particular cortex. Not at all. This cortex functions as one of the binders. It's absence leads to a loss of consciousness. The PMC is not the "theater" of the mind; it is one of the makers of the "theater" of the mind.

SUGGESTED READING

Damasio A. (2010) *Self Comes to Mind: Constructing the Conscious Brain.* Pantheon, New York.

Damasio AR. (1999) *The Feeling of What Happens: Body and Emotion in the Making of Consciousness.* Harcourt, New York (2nd edition 2000.)

Hagmann P, Cammoun L, Gigandet X, *et al.* (2008) Mapping the structural core of human cerebral cortex. *PLoS Biology* **6**: e159.

Meyer K, Damasio A. (2009) Convergence and divergence in a neural architecture for recognition and memory. *Trends in Neurosciences* **32**: 376–82.

Parvizi J, Damasio A. (2001) Consciousness and the brainstem. *Cognition* **79**: 135–159.

Parvizi J, Van Hoesen G, Buckwalter J, Damasio A. (2006). Neural connections of the posteromedial cortex in the macaque: Implications for the understanding of the neural basis of consciousness. *Proc Natl Acad Sci USA* **103**: 1563–68.

CHAPTER 5

The Functional Role of Conscious Will in Voluntary Action: Cause or Consequence? A Position Paper

Marc Jeannerod

5.1. INTRODUCTION

The main feature of voluntarily controlled actions is that they are generated from within, and that their generation requires the pre-existence of an internal state (a representation) whereby they can be encoded, stored and ultimately performed independently from external influences. Thus, they cannot be accounted for by automatic sensorimotor mechanisms where a causal relationship between an immediate (sensory) cause and a visible behavior can be identified. Instead, voluntary actions are to be considered, not as mere responses to a stimulus, but as the self-generated expression of conscious cognitive states. Obviously, these definitions raise the question of how these actions are produced and how direct is the role of the conscious agent in their generation.

The present debate, a classical one in cognitive neuroscience [see Shallice (1988)], is thus about whether mechanisms responsible for consciously willing an action, and those for producing that action, are causally related, or can be dissociated from each other. Here, I will present arguments as to the existence of such a dissociation. I will draw from these arguments the conclusion that the conscious knowledge that the agent gets about his will has the function, not to cause behavior, but to provide cues for authorship and self-consciousness. I will propose that it is the comparison of the cognitive states prior to the action ("I want X") and after the action ("I did X") that provides the feeling of authorship and, ultimately, that this feeling relies on neural computation that follows the action, not on neural states that precede it.

The proposed description, which takes apart a "cognitive" level and an "executive" level for the generation of voluntary action, superficially resembles substance dualism between mind and brain, although it radically differs from it. The main difference between the present position and classical dualism is twofold. Firstly, as opposed to classical dualism, the present position postulates that the two levels are of the same nature and share the same category of mechanisms, namely, neural mechanisms; thus, the proposed duality is not a matter of substance, it is a matter of function. Secondly, the present position proposes that the two levels have distinct functions, which can be described separately, as we will see below.

5.2. WILLING AND ACTING

Let us start with the hypothesis that willing an action and doing it are two separate things. Daily experience shows us that actions can be performed automatically in the absence of awareness. Conversely, we may consciously plan an action, think about it, imagine its consequences, and so on, without executing it. A separation between these states of willing and acting had been suspected for a long time from indirect clinical and experimental evidence (Spence 2009). For example, deep brain stimulation produces mental states detached from the behavior to which they are normally associated; this is notably the case for the strong feeling of "willing to move" triggered by electrical stimulation of anterior cingular areas (Fried *et al.* 1991) or areas in the posterior parietal lobe (Desmurget *et al.* 2009). Conscious states of volition can also be observed to be separated from overt behavior in psychotic patients who claim that their own actions escape their volition, or are produced by external influences, or that their own volition produces actions in other persons (see Jeannerod 2009a).

Here, I will focus on experimental situations specifically created for the purpose of demonstrating this dissociation in healthy people, namely situations where agents attribute to themselves actions that they have not produced or that they have produced differently. This has been the subject of extensive research since the pioneering experiments of Torsten Nielsen more than 50 years ago (Nielsen 1963). In a more recent version of Nielsen's experiment performed with P. Fourneret, we quantified the subjective reports of participants about what they had just done in trying to reach for a visual target (Fourneret and Jeannerod 1998). Participants were instructed to draw straight lines between a starting position and a target, using a stylus on a digital tablet (Fig. 5.1). The output of the stylus and the target were displayed on a computer screen. The participants saw the computer screen in a mirror placed so as to hide their hand. In some trials, the line seen in the mirror was electronically made to deviate from the line actually drawn by the subject. Thus, in order to reach the target, the subject had to deviate his/her movement in the direction opposite to that of the line seen in the mirror. At the end of each trial, the subject was asked to indicate verbally in which direction he/she thought his/her hand had actually moved. The results were twofold: first, three subjects were consistently able to trace lines that reached the target, that is, they accurately corrected for the deviation. Second, they gave verbal responses indicating that they thought their hand had moved in the direction of the target, hence ignoring the actual movements they had performed. Thus, although they correctly performed the task, they were unable to consciously monitor the discordance between the different signals generated by their own movements, and falsely believed that the direction of their hand corresponded to that of the line: they tended to adhere to the visible aspect of their performance, and to ignore the way it had been achieved. In other words, according to this result, not only do we ignore what our hand has done, but in addition we report as ours a movement that departs from that we have actually done [see also Blanke *et al.* (2002)].

In fact, the access to motor awareness seems to be a matter of threshold. When, in the above experiment, the deviation of the line was progressively increased, the subjects' behavior suddenly changed. They became able to report that the movement of their hand erred in a direction different from that seen on the screen and that, in order to fulfil the instruction of

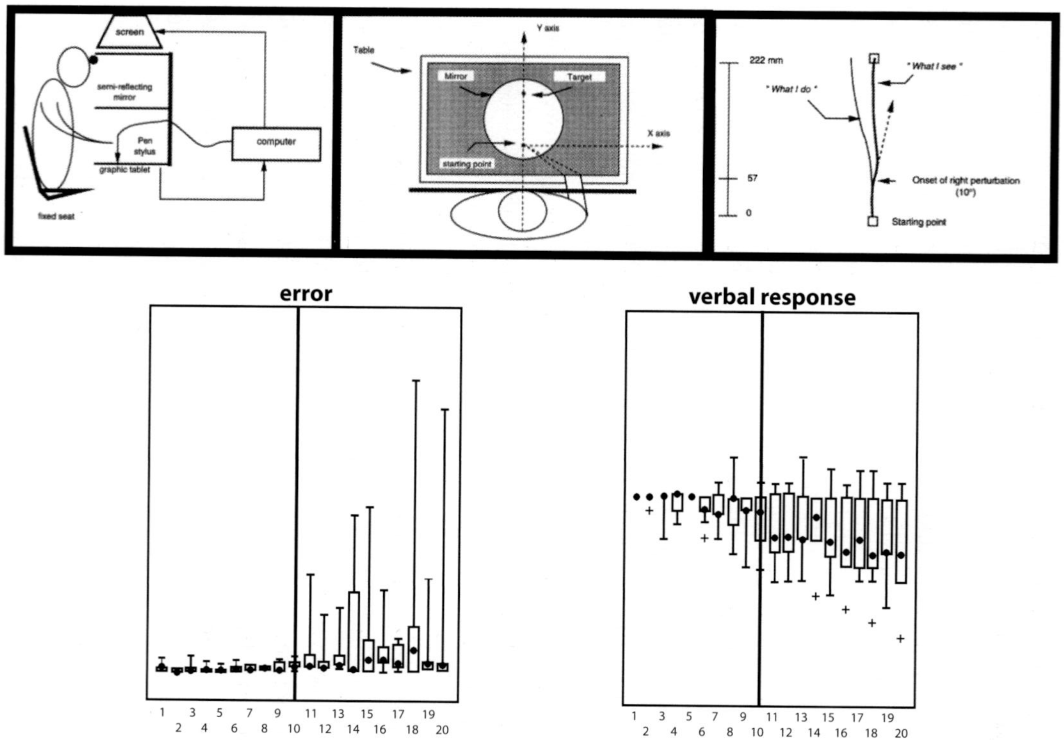

Fig. 5.1 Effect of a visuomotor conflict on action awareness. Upper part: An electronic device creates a conflict between what the subjects do (trace a line straight to a target) and what they see from their motor performance (here deviated by 10° to the right). Lower left: the subjects compensate for the deviation and reach the target with no error for deviations up to about 12° (thick vertical line). Their verbal responses (lower right) indicate that subjects believe that their hand has moved in the direction of the target, hence ignoring the deviation and the compensation effected by the automatic visuomotor system. When the deviation is increased, large errors appear and the subjects become aware of their attempts to reach the target [see Jeannerod (2006)].

reaching the target, they had to deliberately orient their hand movement in a direction different from that of the target. Simultaneously, their accuracy in reaching the target severely deteriorated (Fig. 5.1). This laboratory situation has a counterpart in everyday life, e.g., when an action which was unfolding automatically and outside the subject's awareness fails because of some unexpected event. Think, for example, of starting to drink tea: if the tea is too hot, this automatic action is stopped and remains incompletely executed. As a consequence of the mismatch between the desired motor output and its observed result, the action is de-automatized, delayed, and will continue under conscious control.

5.3. THE TIMING OF ACTION AWARENESS

In the above situation of drinking tea, knowing about the action came as a consequence of a breakdown of the automatic mechanism. The order of appearance of the two events (action precedes consciousness, not the reverse) obviously does not fit the idea of a causal role of consciousness on action generation. The role of consciousness, like that of any other phenomenon suspected to play a causal role, is constrained by time: it should be present before the event that it is supposed to cause. The fact is that execution of most actions starts automatically and is not preceded by a clear insight about what is going to happen. This was the core finding of Libet in his classical experiment. He considered that his own results "lead to the conclusion that cerebral initiation . . . of a spontaneous voluntary act . . . can and usually does begin *unconsciously*." He said, "The brain 'decides' to initiate or, at least, to prepare to initiate the act before there is any reportable subjective awareness that such a decision has taken place" [Libet *et al.* 1983; see also Libet (2011)]. In daily experience, the implicit relation that is thought to exist between conscious will (when present) and acting would thus only be based on the approximate synchronization between the two, leading to the appearance of a causal relationship. We are therefore faced with a paradox: whereas we feel and strongly believe that our thoughts determine our behavior, we realize that conscious will is a consequence, not a cause, of the brain activity that itself causes the action to appear.

There are other situations, however — those that fit into the above definition of voluntary actions — where a clear insight of what has to be done precedes the occurrence of the action. In voluntary action, by opposition with automatically generated action, the conscious mental activity of willing refers to the future, not to the past. It anticipates several steps, including the achievement of a goal, to which the final result will be compared when it is obtained. The problem is thus to describe the precise relationship of this insight — the conscious sense of willing to act — to the forthcoming action. We have to consider first that conscious knowledge of what happens is unevenly distributed throughout this cycle. It is present during the phase that precedes the action itself, with the role of representing the goal and anticipating the realization. It is also present during the phase that follows execution, with the role of checking what has just happened. In between, during the phase that immediately precedes execution and during execution itself, consciousness is scarce and incomplete, as clearly suggested by Libet's results: changes are too fast for a conscious experience to occur (Fig. 5.2). We also have to consider the fact that voluntary actions start far ahead of their manifestation and have consequences that extend long after their termination. There is a clear separation between the two stages: before the action, the

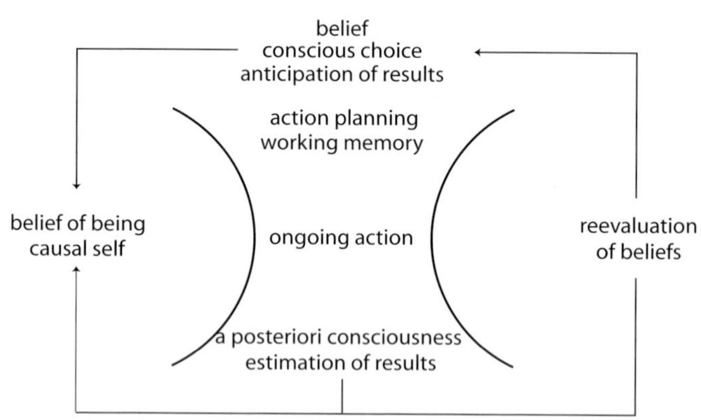

Fig. 5.2 Repartition of the degree of awareness through the action cycle. Anticipatory beliefs related to willing are fully conscious. When the action is about to begin, it proceeds through a narrow channel where the processing gets less and less conscious. Consciousness returns when the action is completed. The knowledge of the results allows strengthening or invalidating the anticipatory beliefs.

subject has an abstract representation of the intended goal; after the action, the results are fully accessible to the subject, as concrete changes of the body and its relation to the external world.

From these considerations about the timing of events throughout the action cycle, it appears that conscious knowledge about willing an action should essentially be a *post-hoc* phenomenon: it is bound to the signals resulting *a posteriori* from the completion of the action itself, not to the central signals that arise prior to the action. It is therefore critical to define the role of the conscious events that precede the execution; this is indeed the problem of matching visible facts that pertain to the external world, with an invisible mental representation that pertains to the inner world.

5.4. THE PHENOMENOLOGY OF CONSCIOUS WILL

In keeping with our initial working hypothesis of a separation of willing and acting as two different processes, we postulate further the existence of two distinct neural networks corresponding to consciously willing and acting, respectively. These two networks have in common that they represent specific aspects of the action, in other words, they "simulate" the forthcoming events related to that action. What distinguishes them from one another is that the representation produced by the simulation process in the cognitive "willing" network is consciously accessible through introspection, whereas the motor representation produced by simulation in the acting network is not. Thus, conscious mental states that precede action have distinct phenomenal properties. They manifest themselves as a particular feeling: we know what it is to want, to desire, to hesitate, etc. We are able to distinguish between these feelings, which we can report verbally. All of us may occasionally experience feelings of concentration, fatigue or even exhaustion when faced with a difficult problem, a multiple-choice decision, even when no action is actually planned. These feelings show many similarities with physical fatigue when fighting against resistance: in both cases one feels the need for rest and recovery. Mental fatigue can indeed be measured experimentally. Subjects who have been previously tired by a task requiring mental effort are less proficient than control (non-fatigued) subjects in a secondary effortful mental task (Bayne and Levy 2006). By contrast, the content of the motor representation is poorly (if at all) accessible to consciousness. This content refers to the detailed aspects of the action (kinematics, force, biomechanical constraints, inter-limb coordination, etc.) and corresponds to somatotopic activation of the motor system (premotor and motor cortex, cerebellum). Research in this field has shown that the motor representation is formed implicitly and remains outside the agent's awareness. Imagining actions, hearing action words, observing actions made by other people are among the cases where this mechanism of motor simulation has been studied [see Jeannerod (2006)].

The above hypothesis has the advantage of referring to the identifiable brain mechanism of neural simulation: according to this hypothesis, we would access consciously our intentions because simulation within the cognitive network provides them with detectable intrinsic properties. This view is criticized on the basis that it should require a second-order simulation heuristic to determine the output of the first-order simulation, such that the neural processing would continue endlessly without succeeding in the task of giving the expected result. This critique actually merges with the question of how the two networks interact. Assuming that they do interact, one has to consider the fact that the information transferred, say, from the cognitive network to the motor network undergoes transformations, such that the output of the motor network will not be a simple copy of the input received from the cognitive network. It will represent new information where the action will have acquired new characteristics as a result of this top-down control. In turn, after completion of the action, these new characteristics will become observable and will affect the content of the cognitive network.

5.5. WHY CONSCIOUSNESS? THE VALIDATION PROCESS

Assuming that conscious representations that precede an action have no direct causal influence on the appearance of that action raises the issue of understanding the function of these conscious representations. Here, I will discuss the idea that their function cannot be understood, unless they are compared with the other conscious events that arise after the action is completed [see Jeannerod (2009b)]. When this comparison becomes possible, the author of the action becomes able to establish a relationship between the "I want X" prior to the action and the "I did X" after the action.

This comparison is effected at several levels in the action system. At the lowest level, according to the well-known forward model, the neural signals generated by the movements are compared with the anticipated outcome formed prior to the action: whenever the two coincide, which signals that the anticipation is confirmed, the action proceeds automatically, and no conscious experience arises; this is the requirement for fast and precise behavior (Wolpert *et al.* 1995). Consciousness may arise (within limits) if the anticipation is disconfirmed, as we saw in the experiment described in Sec. 5.2. The hypothesis I will propose is that the same sort of comparison arises at the upper level, where the "I want X" is compared to the "I did X"; the important point is that confirmation of the anticipation, although it happens after the action is completed, will become the cue for the subject to feel like the author of the action. According to this hypothesis, the feeling of being the cause of an action would rely on mental states that follow that action, not on mental states that preceded it (Fig. 5.3).

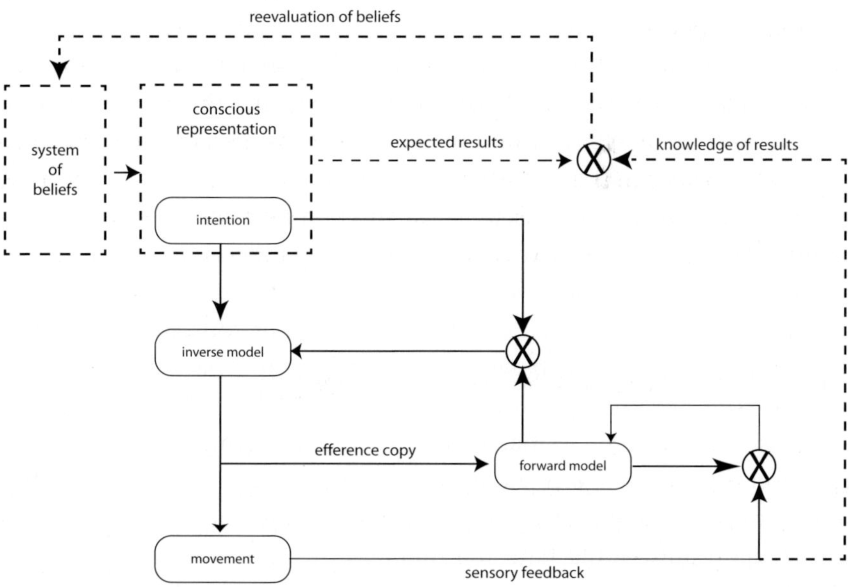

Fig. 5.3 The classical diagram of the forward model for automatic non-conscious comparison of an anticipated action with its effects (solid lines) has been expanded to account for the cognitive conscious process of validation of the results of an action when concordant with the expectation (dashed lines).

This paradox (where an effect turns into the appearance of a cause) has been claimed by some people to be the origin of an "illusion of conscious will."As in other types of illusions (e.g., optical illusions), the conscious report is contradicted by reality and by experimental evidence. According to Wegner, "The intentions we have to act may or may not *be* causes, but this does not matter, as it is only critical that we *perceive* them as causes if we are to experience conscious will. . . . Will is experienced as a result of an interpretation of the *apparent* link between the conscious thoughts that appear in association with action and the nature of the observed action" (Wegner 2002). What is critical here is not reality; it is indeed the appearance, and the appearance, in order to be compatible with a causal interpretation, must be that of a sequence where the (conscious) feeling of willing should precede the (conscious) feeling of having done. When this is achieved, the sequence of events feels like a causal relation. This is the case in situations in daily life where the output of the two networks is quasi-synchronous. It is only in laboratory experiments that the illusory nature of this feeling can be revealed, by distorting or even reverting the sequence.

The above considerations might suggest a rather negative view of the role of consciousness in voluntary action, namely that it would be a source of attribution errors generating illusions and false interpretations of reality (Bongers and Dijksterhuis 2009). On the contrary, what I am developing here is the idea that consciousness in voluntary action has a genuine cognitive function, which expands far beyond motor cognition. It has the function of providing the subject with authorship and self-consciousness, a role that automatic action, i.e., performed without consciousness, cannot afford to play. This idea is based on the mechanism I have outlined above for comparing the anticipated effect and the observed effect of the action. As we saw, if the degree of match between the two is high, the outcome of the action is registered as a success. In other words, in a voluntary action, the concordance of the conscious anticipation and the conscious knowledge of the results would have the function of validating the belief of the author to be a cause of his action. Thus, the author would become able to endorse the consequences of the action as resulting from his own choice and decision.

5.6. HOW DO WE GET THE SENSE OF BEING THE AUTHOR OF OUR ACTIONS?

This validation function is the unique property of voluntary actions, i.e., actions involving a conscious cognitive counterpart. To account for the specificity of this type of action with respect to automatically executed actions, I will use a metaphor in the domain of theater. Consider a new play appearing in a theater program. This play has been conceived, imagined and written by an author; in the theater, however, it will be played by an actor. My point is that the author and the actor have completely different relations to the play. The author knows about the reasons, the choices he has made, the purpose of the play he has written; he has a historical or narrative perspective on the play. However, although he has an idea about how it should be interpreted, he has no direct influence on the actor. He may be absent or even dead when the play is shown to the public. By contrast, the actor must translate the text into actions, speech and movements, by following (or not) the instructions given by the author. In this metaphor, the actor owns the body through which the action of the play is performed, whereas the author owns the cognitive system that generates the narrative of the play. In human actions, however, at variance with a play in a theater, the author and the actor are one and the same person: by pushing the metaphor, the actor can be seen as the body of the author, he is the medium through which the play will be made visible to the public.

The merit of this metaphor is to illustrate the different degrees of self-consciousness that can be experienced during each type of action. In automatic actions, those that are not preceded by a deliberative phase, the self is mostly "embodied," in the sense that it is the product of body signals generated during the action. These signals provide a minimal sense of self, limited to the ongoing event. In voluntary actions, by contrast, the sense of self is disconnected from body signals. It relies on cognitive feelings produced by changes in activity in cognitive networks with a pure mental output. By way of consequence, this conscious sense of self is not limited by the timing of the action itself: its time span spreads over a large period which may include events that have occurred prior to the action as well as those that follow it. This is a narrative self, which establishes a temporal continuity between these events.

In conclusion, the voluntariness of voluntary action is a factor of self-consciousness. The sense of being the author of the action (the sense of will) has the function, not to cause actions, but of self-attributing one's own actions. The sense of will arises from the belief that our thoughts have a causal influence on our behavior and, because the conscious thought and the observed action are consistently associated, the narrative self tends to build a cause-and-effect story.

DISCUSSION

Pierre Jacob: Your model of action is quite clear when you have both the motor instructions or motor intentions (that is, intentions in action), that constitute a model of the action, and the sensory feedback following the executed action, so that you can compare the expected outcome with the observed one. Now, I think that you want to extend this model as far as you can also for prior intentions, in order to be able to provide a model for prior intentions and not merely for the motor intentions. In this case, you want a uniform account for both a prior intention that is followed by an overt action and also for actions that are not executed, that is, for actions that you are planning or imagining but not executing. But in the case where the prior intention is not followed by an explicit action, you do not have sensory feedback so that you cannot actually use the feed-forward model (see Fig. 2.3 in Auletta's contribution, Chap. 2). The dilemma is that if you do not have feedback, the only way in which you could represent your prior intention is by actually providing a meta-representation. So, for the part of the self relative to the awareness of the prior intention that is not followed by an execution and that, therefore, cannot provide a sensory feedback, the forward inverse model cannot supply ground for awareness of the actor.

Marc Jeannerod: Well, this is not entirely true, because you have the simulation process.

Pierre Jacob: My question is: Does the simulation provide a prediction of the outcome of the (non-executed) action?

Marc Jeannerod: Yes, it does. The simulation gives you some insights on the content of the representation. Of course, if you have a representation and a pure simulation they do not have the same value as the action itself. I agree on this. The process of validation is fully effective when the action is known; specifically, when the action corresponds to the anticipation. So, there is an added value of actually doing an action. A person that only simulates but never executes actions would not cross the gap of self-transition.

Gennaro Auletta: As you know, I rely very much on your work, and I have tried to frame the issues of free will, consciousness and intentions in actions in more general, also philosophical, terms. Now, I have a question, I mean, a genuine question. Do you think that in certain situations the model of Libet can be useful, especially as far as, for very specific problems, it is a way to connect prior intentions and intentions in action? This is perhaps not the general model, as you have shown, but in very specific situations it could still provide a connection, otherwise the risk is that the two remain always unconnected from each other.

Marc Jeannerod: I see that Libet's data give some hints about what is going on at the level of the minimal self. But I do not think it is useful in terms of the building of representations and of self-consciousness. In any case, Libet's experiment has so many limitations. You were speaking this morning of the veto because within Libet's protocol, according to him, there is a little time interval that can be exploited to block the action, to stop it. But this may give rise to misleading lessons. So, there are people who see this in terms of brain activity that is needed to generate a concrete activity. But this view might be misleading because you do not have the time to generate a new conscious activity able to possibly interfere with the previous one.

Jean-Pierre Changeux: Just trying to look for the objective measurement of conscious processing itself, there seems to be a time window for the conscious process that lasts much longer than non-conscious ones. Possibly there are criteria for differentiating consciousness and non-conscious processes. Now, referring to the scheme that you have shown (see Fig. 5.3), it could mean that if our views are correct (on which a question mark should always be put), the timing of the processes on the top would be different from the timing of the processes behind it. Do you accept this?

Marc Jeannerod: In the general view of consciousness, conscious process is a slow system. If we were to rely on conscious processes to stop our car when we are going to strike an obstacle, we would have accidents. Consciousness, in fact, comes later, too late to act effectively. This is also testified by ordinary experience of driving a car: sometimes you first react to a danger and only after realize what actually happened.

Gennaro Auletta: I think that there is still a problem. Of course, I see your point and all of us agree that we have some unconscious motor processes that go on automatically and spontaneously. Now, my point is how consciousness can react on this level, provided that you have different timings. Would you agree, for instance, that consciousness is able to jump from one context to another? I mean: suppose that you have two different contexts that are distant from a pure unconscious brain's information-processing point of view. Perhaps, consciousness may allow for jumping from one to the other, that is, to switch from a course of action to another one, thus setting off another automatic and unconscious motor action that could not spontaneously arise without such a bridge.

Jean-Pierre Changeux: Ok, but the real question is about the timing of this switch.

Gennaro Auletta: This is the reason why I think that the model of Libet could be interesting: it would allow for blocking an action during its execution and then for switching on another one.

Michael Arbib: I have to say two things about this. First of all, the diagram which we are discussing has a meaning that does not necessarily invoke consciousness. There is a vital distinction between feedback correction of our own movements and learning correction that adjusts the action's automatic trial. The prism adaptation process which I spoke about does not involve a conscious re-evaluation of beliefs; it is simply the comparison of the expected result with the target; if the action deviates from the target, this adjusts the cerebellar circuitries and their relationships. In this case, the actual adjustment involves an awareness perhaps of the error but no conscious representations and no change in the system of beliefs. This is a subconscious process. The second point is a general problem with diagrams like this. They just lack the richness of a real-life story where you are consciously aware, as a theorist, that there are many schemas involved. In many cases this gives the impression that there is only one inverse model, only one forward model and we just figure out the right intention. Instead, in fact, the intention may not simply be passing information to the inverse model, but it may require activating the appropriate set of motor schemas. So, it may be that consciousness enters more in a situation where you have to change strategy rather than when you have simply to adjust some sensorimotor programs.

Marc Jeannerod: I fully agree.

Ivan Colagè: My question concerns the very last part of your presentation. Namely, it concerns the difference between the first, original part of the scheme (the one characterized by the forward and inverse model but without considering the system of beliefs and its re-evaluation) and the enlarged model that also includes the knowledge of the result of the action, the conscious representation and the system of beliefs (see Fig. 5.3). Now, we have seen several times in our discussions that one becomes conscious of something especially and mainly when something is going wrong. My question is the following: already concerning the first part of the scheme we can have error correction, and so we could in principle become aware of the fact that there is some failure or some problem. At this level, however, we are not yet concerned with beliefs. Indeed, as far as I have understood it, this first part of the scheme is about unconscious processes, whereas the enlarged one, having to do with beliefs, concerns conscious processes. Now, the difference between the two situations is that in the second case (involving beliefs) one is aware not only of the fact that things are going wrong and that you are then correcting something, but also of the specific way in which one is correcting what is actually being corrected.

Marc Jeannerod: Ok. I think I have stressed that the self-consciousness you get after the automatic system failing is not at the same level as when in the conscious system you are comparing the state prior to the action and the state after the action. Maybe this is a good opportunity to clarify that the self-consciousness that you get through the first level is not of the same nature as that reached when the other system is at play. In fact, this is the reason why I made the distinction between actor and author. In the first case, self-consciousness is transient, discontinuous, and vanishes as soon as one moves to the next step, whereas in the second one self-consciousness has a temporal window that exceeds the action duration and is continuous over successive episodes. I think this is a different way of characterizing self-consciousness.

Maybe we use the same word (self-consciousness) but connected with different kinds of self: minimal self in one case, narrative self in the other.

Ivan Colagè: Do you think that a way to characterize, at least in part, the difference between the first scheme and the second one is that in the latter you can know exactly the way in which you are revising your beliefs whereas in the former you actually do not know what you are correcting?

Marc Jeannerod: Yes, because the former situation is automatic. If you have some failure in the automatic system you get some consciousness, or awareness in this case. But from this maybe you cannot jump directly to the higher level of self-consciousness.

Gennaro Auletta: I find this distinction between two varieties of consciousness very interesting. This was also my point earlier, because I think that it may be incorrect to say that consciousness is always slow. In many situations it is slow, indeed. But there are situations in which you have to take very quick decisions. And one can do this exactly when one can connect two contexts that are very far away from each other at the level of unconscious processing. This is a way, so to speak, to create a diagonal cross-context between two different contexts that allows for quick decisions.

Jean-Pierre Changeux: I think that here there is a problem of precisely defining the time scales involved in the arising of consciousness.

Marc Jeannerod: Let me turn to this idea of having two degrees of consciousness. I think that if we put the two systems at the same level of explanation then we may well be wrong because we are speaking of systems of a different nature. The problem of the time scale in which you get consciousness has been very well settled in experiments. If you run experiments, you get it in the range of milliseconds and seconds. If things go wrong, we immediately make a correction and go to the next step. The temporal windows of the automatic system are now well known. I mean, the subjective present is very different from the objective present, but not for the automatic system; for this the timing is constant and very well fixed.

Jean-Pierre Changeux: What is the timing for the access to consciousness? I mean, from the time of the presentation of the stimulus, how long does it take to become conscious of it?

Gennaro Auletta: I think something between 100 and 400 milliseconds. Anyway, apart from quantitative factors of timing, I have also a qualitative example from Penrose. He was amazed by the fact that human beings can have insights on problems that normally take a lot of normal computation time to solve but which we can solve quickly even if we do not have all the conditions for extracting the solution (that is, the problem is not well defined). Probably, this may be done exactly because we can put together things that are actually distant from a computational and processing point of view.

REFERENCES

Bayne, TJ, Levy N. (2006) The feeling of doing: Deconstructing the phenonenology of agency. In N Sebanz, W Prinz (eds.), *Disorders of Volition.* MIT Press, pp. 49–68.

Blanke O, Ortigue S, Landis T, Seeck M. (2002) Stimulating illusory own-body perceptions. *Nature* **419**: 269–270.

Bongers KCA, Dijksterhuis A. (2009) Consciousness as a troubleshooting device? The role of consciousness in goal pursuit. In: E Morsella, JA Bargh and PM Gollwitzer (eds.), *Oxford Handbook of Human Action.* Oxford University Press, New York, pp. 587–602.

Desmurget M, Reilly KT, Richard N, *et al.* (2009) Movement intention after parietal cortex stimulation in humans. *Science* **324**: 811–813.

Fourneret P, Jeannerod M. (1998) Limited conscious monitoring of motor performance in normal subjects. *Neuropsychologia* **36**: 1133–1140.

Fried I, Katz A, McCarthy G, *et al.* (1991) Functional organization of human supplementary motor cortex studied by electrical stimulation. *J Neurosci* **11**: 3656–3666.

Jeannerod M. (2006) *Motor Cognition. What Actions Tell the Self.* Oxford University Press, Oxford.

Jeannerod M. (2009a) The sense of agency and its disturbances in schizophrenia. A reappraisal. *Exp Brain Res* **192**: 527–532.

Jeannerod M. (2009b) *Le cerveau volontaire.* Odile Jacob, Paris.

Libet B, Gleason CA, Wright EW, Pearl DK. (1983) Time of conscious intention to act in relation to onset of cerebral activity (readiness-potential). The unconscious initiation of a freely voluntary act. *Brain* **106**: 623–642.

Libet B. (2011) Do we have free will? In: L Nadel and W Sinnott-Armstrong (eds.), *Libet, Free Will and Responsibility.* Oxford University Press, Oxford.

Nielsen TI. (1963) Volition: A new experimental approach. *Scand J Psychol* **4**: 225–230.

Shallice T. (1988) *From Neuropsychology to Mental Structure.* Cambridge University Press, Cambridge.

Spence SA. (2009) *The Actor's Brain. Exploring the Cognitive Neuroscience of Free Will.* Oxford University Press, Oxford.

Wegner D. (2002) *The Illusion of Conscious Will.* MIT Press, Cambridge, MA.

Wolpert DM, Ghahramani Z, Jordan MI. (1995) An internal model for sensorimotor integration. *Science* **269**: 1880–1882.

CHAPTER 6

Can the Self Be Considered a Cause?

Juan J. Sanguineti

6.1. METHODOLOGICAL PREMISES. A FIRST PHENOMENOLOGICAL LAYOUT

This chapter is philosophical and uses a phenomenological method (in a broad sense). It is likewise based on science and attempts to present a philosophical interpretation of scientific achievements, which means that I do not claim that the empirical methods used in natural sciences can be used to demonstrate its conclusions. In my view, neuroscience and behavioral sciences applied to men and animals are "mixed" sciences, since they use empirical and theoretical concepts typical of the physical sciences along with psychological and social concepts strictly irreducible to physical notions. Philosophy of neuroscience, of physics, of biology, together with philosophy of mind, adds to these scientific fields a particular epistemological framework (for example, in discussions about causality, truth, reality, finality). Debates in these philosophical areas, not always clearly delimited, are sometimes semantically obscure, especially those concerning reductionism (since science is methodologically reductive, but not philosophy).

In an initial phenomenological approach to my topic, the "self" (*my* self) appears to be the primary source of actions. The self includes our (*my*) own identity, it is self-conscious in its actions, and contains a power to act. We act according to physical laws and yet our actions go beyond these laws. We are inclined to operate in very specific ways, but these inclinations are, in many respects, "what we like to do," or "what we love to do." Here "love," in a broad sense, refers to ends (or values and goods, including persons), whereas reason refers to the means to attain those ends. Therefore, we act *for some love* and *according to consequent reasons.*[1] *Freedom* means that nothing is *absolutely* determinant for our personal actions, except our own decisions. A choice is a decision to act, implying the control of our body. The subjective experience of having a real influence on bodily changes is a privileged source of our universal experience of causality and dominion.

This is the platform, hard to deny, though debatable, which I take as a point of reference in my following considerations (Searle 2007). Now I will delineate a sort of "natural philosophy," which I claim to be consistent both with sciences and with the phenomenological view so far presented.

6.2. CAUSALITY IN THE NATURAL WORLD

Efficient causality in nature means that new alterations are provoked in the physical world *through* previous events. In physics, natural causality is analyzed in the context of *laws* (functional laws) and *systems* (without necessarily speaking of an "agent" that causes). Systems are totalities integrated by interacting parts (elements) and situated in an interactive environment. Complex systems have different and variable levels of causality.[2] Accordingly, their outcomes are difficult to predict in the long run. Bottom-up causes are more elementary external entities or even parts acting over another part of a whole, whose action, consequently, modifies the whole (*material causes*). Although several ordered causes may be operating, a single "ultimate" action, under some circumstances, usually "triggers" a mechanism or a series of effects [*triggering causes*, different from *structuring causes* (Dretske 1993)].

A system can be a cause as a whole (many structured elements can exert an efficient causality; for example, a wheel causes alterations on the ground). It seems inadequate to say that a whole "acts" over its parts (a house does not act over its parts), unless it is more than a mere "whole," as in the case of self-organizing unities and living beings, which we shall consider in the

---

[1] I develop this point in Sanguineti (2008). For the duality love/reasons (or good/reasons, end/means, values/means, motivation/reasons), I am inspired by the Aristotelian account of human choice: *The Nicomachean Ethics*, V, 1139 a30–b15 (Aristotle 1934); *Movement of the Animals*, 701 a1–25 (Aristotle 1961); *On the Soul*, III, 433 a10–434 a20 (Aristotle 1957).

[2] For this theme, see Heylighen *et al.* (1999) and Heylighen (2008).

next section. In living beings some parts of the organism ("governing parts") control or exert an overall effect over the other parts (this is the case of the DNA and the nervous system).

6.3. SELF-ORGANIZATION[3]

Spontaneously self-shaped structures (crystals, molecular self-assemblies, dissipative structures) can be determinant for the positions and the precise ways of operating of their parts. In nature, actions result from collective or compound entities (simple entities are likely problematic). Structures are normally more than mere aggregates of parts. Self-organization means the existence of self-induced structures showing specific or very selected interrelations between their parts.

The most characteristic self-organized beings in nature are the living beings (they show a strong self-organization). Since organisms exhibit a sort of "personality" (adaptation, identity, self-repairing, individuality, self-teleology), they manifest, for the first time in nature, a kind of *self*. I think this "self" fits very well in the Aristotelian primary model of a natural *substance*. The "self" indicates in some degree the property of "immanence" (not opposed to external relationships), i.e., an autonomous or spontaneous self-regulation and a development of the living being in order to satisfy its "own" needs:

- A living being *self-organizes, self-develops, self-sustains, self-defends*. This property creates a variety of functions or a "division of tasks" (organic functions), which are assigned to specialized parts of the body. Control functions are characterized by *information control* (Ellis 2008; Auletta *et al.* 2008). Living beings display, in this sense, a "semiotic causality": arriving signals (information), once recognized, induce physical effects in the organism. Consequent "messages" are forwarded for the benefit of the whole. The living being's self-control is in a constant relation to its environment and to its own differentiated parts.
- Since the living being "self-governs for the sake of itself" (or for the benefit of the species, at least in favor of a group), every bottom-up input from outside will be useful (functional) and not harmful for the organism *only*

[3]For this topic, see Davies (2006); Maturana and Varela (1980); Kauffman (1993).

if integrated in the whole (therefore, it should be received in a very selective way). In this sense, *bottom-up causality in living beings always includes a top-down causation*, i.e., external influences normally are received and regulated by the totality.

- As a whole, the living being is "governed" by its DNA. Animals add to this a special system with the task of controlling the whole functional body: the nervous system, which in some way is the "center" of the body.
- Living beings possess an *identity* that cannot be "representationally" observed, i.e., it cannot be grasped by empirical scientific methods (Jonas 1982, pp. 64–92). This "self" is not a "thing," but it is real and displays a *praxis*, namely a differentiated array of operations accomplished for its own benefit, in continuous interactions with the environment and with other living units. This set of operations is variable (homeostasis) and it is permanently used, among the capacities of the organism, as a "network," as a "habit," and as "a procedural memory." This feature explains why living beings are able not only to correct themselves, but to *learn* as well, i.e., to improve their praxis starting from more adequate conditions (within certain limits), which is the origin of the historical dimension of life.

6.4. INTENTIONAL LIFE

Many features of life are transferred to the level of *sensations* and *feelings* in animals. Sensations are not merely "supervenient" properties, but something essential to animal life (including ourselves). The unity of the organism, inasmuch as it is *self-felt*, becomes a *subject*.

(i) Sensations, feelings, perceptions, emotions, etc., are *psychosomatic acts*. They are not two acts associated (mental and physical or neural), as hypothesized by dualism, but any act includes a psychic dimension and a neural dimension (like the Aristotelian unity between form and matter).

(ii) In virtue of sensations, raised at the level of consciousness, we have a regular *phenomenal/realistic access* to our life (the separation between appearance and reality is secondary), though we know that not everything of our personality is attested by consciousness.

(iii) In a first degree, tasting, touching, smelling and other sensitive capabilities (interoception, proprioception, sense of balance) imply the "self-feeling" of many aspects of the *organic* causal processes (lively movements, feeding, breathing, etc.). In a second degree, there are sensations/perceptions (like sight or hearing) more strictly *intentional*, in that they concern external objects *without a direct reference to vegetative functions* (Sanguineti 2009; Jonas 1982, pp. 104–105). Their "objects" (physical objects and in a very special way *other subjects*) "hide" their underlying biological structures. This kind of separation (like an "abstraction") presents things as *objects* (better than "representations"), which are often linked to motor *affordances* (a place to "rest on," etc.) (Gibson 1979). Intentionality enables human beings to overcome the causal ties to the material causes. As humans, but also as animals, we perceive *things* and *subjects*, or subjects in an environment of things (we also perceive pure representations, in a secondary way).

6.5. THE UNITY OF SENSITIVE CONSCIOUSNESS

Beyond the organic totality of the animal body, sensitive life implies an additional "intentional unity" that may be considered a very important aspect of the "animal self." Animals are endowed with a sensitive sense of unity (a non-reflective "subjectivity"), encompassing vegetative, cognitive and emotional contents. In other words, the animal self is constituted by its *sensory vegetative-emotional-motor intentional body* (not merely by its cognitive "unity of perception"). The animal perceives its own body as a whole situated in a landscape populated by meaningful objects. Animals experience their body as a dynamic totality that can be meaningfully articulated and moved in their environment (Jonas 1991, pp. 192–202). This is not "cognition," in the classical sense, but rather *Erlebnis* in the phenomenological sense (an "experience").[4]

"Voluntary actions," in humans, include not only the awareness of having intentions, but also the "bodily feeling" (phenomenological, not physiological) of the motor control of many parts of our body (the "voluntary body"). This "unitary self-perception" is dynamic and is integrated by different

[4]In Italian, *vissuto*; in Spanish, *vivencia*. This terminology comes from phenomenology. Some linked words, very useful for what we intend to say, are: *self-sentient body*; *Leib* (lively body), in German, as opposed to *Körper* (merely dimensional body).

aspects and steps. Because it is biological, it undergoes an evolving process and it can be disintegrated (partially or totally).

6.6. INTENTIONAL AND NATURAL LEVELS

The intentional level includes cognition, feelings and many practical operations. Yet, there is an underlying dynamic structure, namely a vegetative/sensitive level, not conscious, which we could call "natural" (recalling the Greek notion of *nature*). This "background level" has its roots in the genetic and epigenetic basis and *is*, in a certain sense, the whole organism as endowed with vegetative and sensitive powers to be developed in an environment. It includes structural psychological potentialities, such as animal instincts (or basic affective and behavioral inclinations) and intelligent animal skills.

The natural level, though innate, can be modified to a certain extent by experience, memories and habits (at a lower level, by conditioned reflexes; at an upper level, we can include learned associative animal languages), and it can therefore be considered as a "structuring cause" or, better, a set of structural and functional conditions and positive potentialities that enable or even push to perform operations at the intentional level. Accordingly, the latter level modifies the former, so that we have a circularity in the mutual feedback causation between both levels.[5] Downward and upward causation should be understood in the context of this duality, as I will explain later on. The underlying natural structures are like a platform over which the intentional operations can instantiate concrete ends and can exercise a limited control over some physical powers.

6.7. ANIMAL ACTIONS AND THE ANIMAL SELF

The foregoing considerations lead us to conclude that the "animal self" (in the sense of the animal identity as a unitary self-organizing living being) includes several aspects, and helps us to understand to what extent the animal "self-moves." Some authors, like Jaak Panksepp [with his notion of "affective mind"

[5] For a layered vision of the intentional and neurological complexity and causality, see Murphy and Brown (2007); Emmeche *et al.* (2000); Ellis (2006); Deacon (2006).

(Panksepp 1998)] and Antonio Damasio [with his account of the proto-self (Damasio 1999)] helpfully approach this topic, though I have no space here to make comparisons and technical distinctions to better appreciate their contributions.

(i) An *animal deep self* could be viewed as a background including a natural teleology, both in vegetative and sensitive domains, which emerges from genetic and epigenetic structures and operates mainly through subcortical controls. This structural self has an endogenous dynamics and it is the primary source of the various "activations" of the animal life and behavioral outputs (to see, to feel, to move the limbs, to look for food, to build a nest, etc.). It is the source of the intentional animal life (for example, instincts), opening a space where the animal performs intentionally controlled activities in a certain domain. But these activities cannot control the deep self (Fig. 6.1).

(ii) *The intentional animal self* is situated in the *perceptual-emotional-motor* space. This is the space of sensitive consciousness, which guides the animal intentional behavior. It is primarily activated by sensations, feelings and emotions, normally associated with cognitive operations (perception, imagination, memory), with effects in the "physiological body" (via the viscerosomatic nervous system and the endocrine system) and with behavioral results (via the brain motor circuits). Consequently,

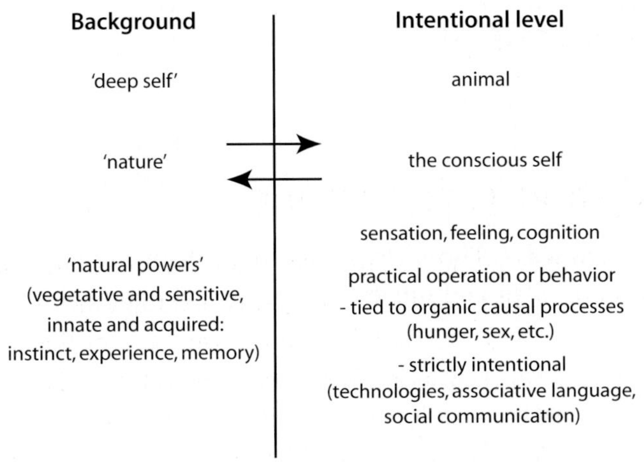

Fig. 6.1

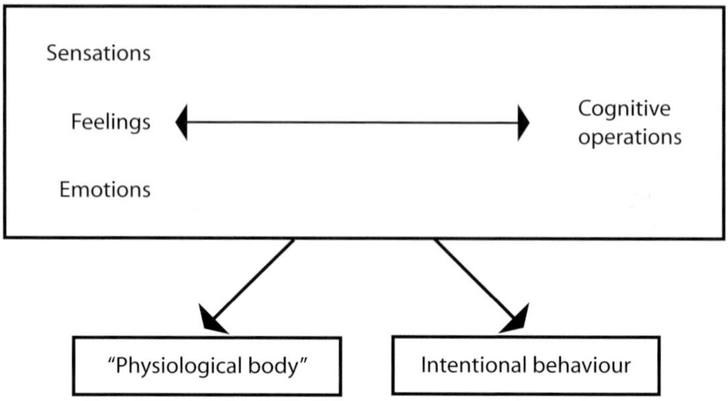

Fig. 6.2

animals as "subjects" can self-regulate their displacements in their territory and their reactions, facing meaningful stimuli in a world of objects and "affordances." They can move their body intentionally (head, eyes, limbs) in their "horizon of objectivity," and they can communicate their intentions to other animals. Animals as subjects pursue intentional ends, which are instantiations of their natural biological ends. All the elements so far indicated are active, of course, only through interactions with the environment and with other living subjects (Fig. 6.2).

6.8. HUMAN HIGHER LEVELS

Humans "amplify" animal powers to the level of universal reason, free will and self-consciousness.[6] Accordingly, our subjectivity deserves the special name of *personhood.* There is no space here to develop these fundamental aspects, which allow the delineation of the significant difference between humans and animals.[7] Those elements are intentionally related to notions such as universal truth and universal good, and are applied to the world as

[6]Regarding the notion of "universal reason" as the root of freedom, based upon Aristotelian non-dualist premises, yet endorsing the human soul's spirituality, I take some basic assumptions by Thomas Aquinas as tenable in modern contexts: *Summa Theologiae*, I, q. 82, a. 1; I–II, q. 6, a. 1; I–II, q. 9, a. 1, a. 3 and a. 4 (Aquinas 1998).

[7]The difference between humans and animals is discussed in MacIntyre (1999).

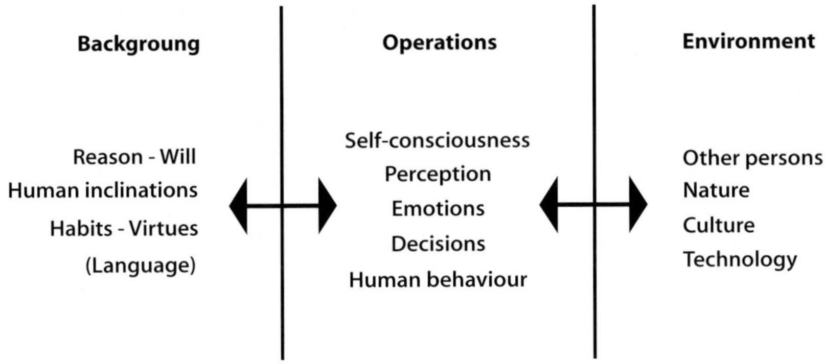

Fig. 6.3

such and to the acknowledgment of other persons (including ourselves). This higher level or layer (whatever name we can use to designate it: intellect, will, reason, spirit) is deeply rooted in the other psychobiological layers, creating the capacity of intelligent perception and human-specific feelings and desires (Fig. 6.3).

Persons act and evolve, taking account of these levels, while they interact with other persons and with cultural creations (mainly through language and symbols) and create in themselves habits and virtues.[8] Decisions emerge from this complex background. Decisions are free (self-determined, self-conscious, and related to the whole of the person), provided all the other elements are adequately activated (reason, consciousness, inclinations, emotions, intentional spaces, habits, neurophysiologic aspects, etc.). This complexity explains the limitations and constraints (physical, psychological, sociological, etc.) of human freedom.

6.9. THE SELF AS A CAUSE

In general terms, the self (*autós, seipsum*) is the "subjective totality" displayed in the different degrees of intentional life, showing among them a progressive increase in the immanence and reflexivity typical of life. In human beings the self is the human person, whether or not it is self-conscious,

[8] I develop some of these points in Sanguineti (2007).

although it is always endowed with the potentiality of intellectual and existential self-consciousness ("I am, I exist, I will, I think").[9]

I have distinguished between a *deep self* and an *intentional* and *self-conscious self*. Although there is only one self, one person for "each one" (tautologically), who is born and dies and is indicated by a personal name, when we say *I*, we usually indicate that most intimate "part" of ourselves subjected to our intentional control (self-consciousness, will, responsible behavior), including our very self-controlling power, i.e., our will. The "intentional personal self" encompasses the sense of our personal identity, with his history, as far as we are able to remember it. In this regard, the intentional self is a narrative self. In grammar it is expressed by personal names. Of course, through this denomination (the person inasmuch as it is self-conscious) we do *refer* to the totality which "we are" (our cells, our brain), though we know nothing about those aspects of our nature that are not accessible to our consciousness.

From what I have just said, the notion of *self* seems to be semantically difficult. I have no space here to propose a philosophical analysis of the concepts of body, soul and mind. I focus my attention on the causal problem: Can the Self be considered a cause (a top-down cause)? In what sense? To what extent?

The deep human self (our psychobiological constitution) is multilayered. Regarding its highest level (intelligence, the power of the will), historically situated, the deep self is a specific source of our intentional operations. Accordingly, the deep self operates as a *structuring cause* from which the power to act for the intentional self emerges. Impairments or positive capacities of the deep self diminish or increase the operational power of the intentional self.

We, as intentional agents and persons, have the capacity to do many things, through the activation of many powers of our multilayered nature. We cannot perfectly control these capacities, and many of them are totally out of our intentional control. Those actions under our power of decision, with many external or internal foreseen consequences, are called *free* or *voluntary* actions. Only in relation to these actions we can say that we are a cause (an agent), and since these actions imply a self-control of the whole

[9]For these topics, see Bennet and Hacker (2007, pp. 331–334); Brock (1998); Fabro (2006); Gallagher and Shear (1999); Kenny (1979); Wojtyla (2003).

of our "intentional body," we can say that, inasmuch as we freely act, we are self-causing, "causes of ourselves." This does not mean that our free actions are the total causes of what we do. They are a partial cause, but decisive. Many causal dimensions intervene in the production of an act of will (background, memories, intellectual competence, social contexts, habits). A free action, a free choice, selects a possibility offered by nature, culture or many circumstances (I do not consider here the problem of free will as such).

6.10. LEVELS OF CAUSALITY

There are many senses in which something is said to be a "cause," and regarding life it is too simple and rather rare to have linear sequences of causes going on from earlier events to later outcomes. Living organisms normally present multidirectional causes forming networks, sometimes hierarchical, whenever new qualitative emergent levels are built upon more elementary structures. It is in this sense that we can speak about *bottom-up* and *top-down* causation. Bottom-up causation, in the following considerations, is the causation that comes from "lower levels" (more elementary) and has an influence upon "upper levels" (more complex, with new properties and "installed" as a structure of the lower level), whereas downward causation will be understood the other way around. Downward causation is not "pure," since it "uses" the bottom-up causal powers, with their "autonomous" dynamism, in a selective way for the benefit of the whole.

Causation along the same level could be called "horizontal." Moreover, we should always take into account endogenous causation or causation coming from outside within the threefold directions of causality (for example, if a friend tells me an idea, he is an external cause acting at the "horizontal level" of "thoughts causing new thoughts").

(i) According to the proposed distinction between a deep self and an intentional self, we can consider, firstly, in what sense the former is the cause of the latter. The deep self, as we have seen, is the *source* of the intentional self, a "structuring cause," neither downward nor upward. The downward/upward dilemma just appears whenever we consider the "layers" of those dimensions and the causality in-between. The intentional human self, conversely, can "shape" the deep self at the level of habits, virtues, culture, mentality and education. Whether humans can change their nature operating in a bottom-up way upon their genetic structure or through neuro-enhancement ("trans-humanism") is a question to be discussed elsewhere.

(ii) To place the downward/upward issue in terms of mental/physical causation would be dualistic and misleading. Intentional operations are always psychosomatic, but not in the same sense. Hence, the downward/upward issue is to be projected onto the interactions between the different levels. Running the risk of simplifying, I propose to consider at least four levels in human beings: (a) the *physical* level specifically studied by biology; (b) the *organic sensitive* level (organic-*felt* events: hunger, etc.); (c) the *psychological sensitive* level, organically accomplished but not dealing with organic events (perceptions, emotions); and (d) the *intellectual* or *volitional* level, organically realized through the psychological level, but dealing with metaphysical realities (the dignity of the human person, moral values, contemplative knowledge of the universe, abstract sciences like mathematics, arts and esthetic values, religion).

6.10.1. Bottom-Up Causation

(i) *Emergence of novelty from the bottom.*[10] This topic concerns evolution, but can be considered even without reference to evolutionary processes. A very specific material organization "induces the emergence" of new properties, which could not be predicted *a priori* from the mere knowledge of the lower levels[11] (the emergence can be necessary without entailing reductionism, provided we acknowledge the irreducible novelty of what "emerges"). A special organization of cells "produces" the eyes and the phenomenon of vision; brains endowed with certain conditions "induce" the appearance of some cognitive powers (without totally "explaining" the latter). This problem is too difficult to be dealt with in this brief chapter. Since the emergence of the "new" is a form of self-organization, it cannot be explained by a pure bottom-up process. It seems that both bottom-up and downward causal directions, combined together are necessary to explain the emergence of more and more self-organized organisms.

[10] See, for this topic, Clayton and Davies (2006).

[11] For an account of this topic in a Thomistic view, see Stump (2003, pp. 194–197); O'Rourke (2004).

(ii) *Bottom-up events as material causation.* Once we have stable living beings, specific bottom-up events are normally helpful to provide the optimal material conditions for the realization or maintenance of specific "higher operations," provided they are well integrated in the whole structure they are supposed to serve. Otherwise, the bottom-up events would become ineffective or even harmful. For example, physical or neural factors (external stimuli, neurotransmitters, drugs, metabolism), though displaying a purely electric or chemical causality, have effects in the production or modulation of sensations, perceptions and emotions (this does not mean that psychic acts are purely "mental," since they are psychosomatic). This kind of causation corresponds to the Aristotelian notion of "material cause." Notice that, according to the sense just explained, a bottom-up causation in the living being does exist only *subordinated* to some downward causation.

In a similar but not exactly coincident way, some neural conditions can "facilitate," in many different senses, the realization of intelligent insights or volitional acts. Which is the cause, for instance, of a brilliant insight coming to the mind of an artist or a scientist? In my view, the act of insight is a real "creation" (in the sense of "human creativity"), presupposing an adequate background. It is not the mere result of a particular assembly of neurons, though this assembling may be a necessary condition for the possibility of the intellectual or volitional operation.

The bottom-up causes of organic and psychosomatic operations and states are normally considered sufficient causes (a drug can cure my headache), since in this case the width and quality of the correlative downward causation (pain sensibility) is strictly tied to the material conditions of the bottom-up cause. But if we go further and further up, the relevance of bottom-up causes, even if they are necessary, becomes weaker. The more intentional are the cognitive and affective operations, the less relevance is assigned to the necessary bottom-up material causes. It would be odd to say that a drug "caused" me to understand a problem, though the drug could have been a more "proper cause" of my psychological attention. For the same reason, it is quite inappropriate to say that falling in love is a mere chemical problem.

Accordingly, intelligent and volitional acts, if they are really *original*, can be said to be facilitated, in my view, but not properly caused by bottom-up material causes. However, the "origin" of insights is different from the origin of acts of free will (and only the latter strictly pertain to the intentional Self as a cause). We control our voluntary acts, but we do not fully control our insights or intellectual inspirations.

6.10.2. Downward Causality[12]

We speak of downward causation whenever a higher-level process or event influences or selectively guides processes or events pertaining to a lower level, such as emotions like anger or sympathy guiding the behavior of an animal, with innumerable effects on the organism (physiological and behavioral), or just as attention guides many details of the perceptive field. In these cases, the problem is not to be seen in dualistic terms (the efficacy of a mental act over its "neural correlate"), since acts like perceptions or emotions are psychosomatic. Downward causation in the intentional life is rather the action of "higher" psychosomatic acts over "lower" ones (for example, perception over sensations) or over external behavior.

In this sense we can understand how the intentional self of an animal is the "moving force" of its behavior (another example: thirst moves the animal to look for water). Of course, there are "negative" modalities of these causalities, as when psychologically negative events can produce depression in animals, with hormonal consequences. Note, however, that a psychosomatic act can have an external cause, assumed by the individual at its own level: depression can be "caused" by the death of a mate.

The same can be said, but not exactly in the same sense, of human intellectual and volitional operations. These operations are psychosomatic, since they involve sensitive higher operations, but in my view they contain an immaterial (spiritual) power (will and universal intelligence) that cannot operate separated from the sensitive level (ordinarily, but not exclusively, will and intelligence operate associated to language). Through their embodiment in perceptions, emotions, sensitive awareness, language, etc., human rational and volitional acts can exercise a downward causation, very specific in the different cases, over the organism and intentional voluntary behavior. For example, when we read, we activate brain areas in a very specific way, which

[12]See Andersen *et al.* (2000); Auletta *et al.* (2008); Heil and Mele (1993); Murphy and Brown (2007); Stephan (2002); Stevenson *et al.* (1986); Jeannerod (1999; 2006).

would not have been activated if we had decided to read different things or not to read at all. What, then, is the cause of our act of reading? There is a convergent plurality of ramified causes of this act: the meanings contained in the book, as far as they are comprehended by the reader; the intellectual skills of the reader to understand and to read; his or her personal will to read; his or her attention, along with his or her optical system and his or her whole activated brain. Damage to any of this enormous number of causal factors in the act of reading can compromise the efficacy of the act.

6.10.3. Horizontal Causality

The "proper cause" of a thought is usually another thought, as when we entail consequences from some premises, which is a case of "logical causality." A thought emerges in our mind because somebody told us an idea or because we glean the idea in a book. Mathematical ideas usually come from other mathematical ideas, and the same can be said for all sciences, arts and any other human activities. In another sense, usually considered by analytic philosophers, perceptions, beliefs and desires, well coordinated and always in an appropriate context, lead us to a specific behavior (I see an ice cream, I feel the desire to eat it, I know that nothing prevents me from doing this in my context, so I eat the ice cream). All of these are examples of "horizontal causality," i.e., causality between psychosomatic operations belonging to the same level (often to the same epistemic level).

Again, dualist, parallelist and sometimes even functionalist authors analyze these processes delineating two parallel chains, one psychological and the other one neural. I think that there is rather a unitary complex dynamism made up by several converging networks pertaining to different levels, each one displaying a specific and sometimes "autonomous" dynamics and a partial contribution. The horizontal causalities should be seen, with all their links to bottom-up and downward lines of causation, always contextualized and within adequate backgrounds.

6.10.4. Do Objects Cause?

Intentional acts are open to the world and to other persons. In some way, these "objects" could be considered as "transcendent" causes of those actions. A judgment, we could say in some way, is "caused" by the truth of a fact.

Perceptions are "caused" by a spatial framework grasped in the act of seeing. The same can be said of the objects of desire and love (even persons), or of other realities deserving contemplation (for example, a beautiful landscape).

But this is not real causality, since it needs the action of the subject (contemplation, insight, love, desire) in order to be accomplished. There must be an adequate proportion between the objects and the special capacities of subjects: we can appreciate the beauty of a landscape if we have eyes and an adequate contemplative attitude.

6.11. CONCLUSION

Psychosomatic causality is systemic, with many feedback mechanisms and loops. This complex causality is not linear, but multidirectional and with many lines and ramifications. It is a historical complexity, operating in ever-changing contexts and with many levels and several aspects. The distinction between background (the deep self) and operations (the intentional self) is relevant in these analyses. It follows that neuroimaging, although useful for the study of mental processes, is in any case a partial view and always needs careful interpretation.

The human person is a powerful agent in the world, a self-conscious agent having the natural capacity to move him- or herself as a whole and to manipulate events in the physical world. The power of the human self is not something extrinsic or just added, but it is quite comprehensible in the broad world of intentional life, showing itself as the highest degree of self-organization and teleological control that we know in the universe.

DISCUSSION

Wolf Singer: It seems to me that it is difficult when speaking of self-organizing systems (like the brain, of course) to distinguish between cause and effect. And it may sometimes be not very useful even to try to do that. If we measure a self-organizing living system like the brain, we need to analyze this system in terms of its many subsystems. For example, when a part of the system attains a certain state and becomes activated, the reward system sends a reward signal with the effect that the whole system tries to maintain such a state. It is difficult for me to see what is top-down and what is bottom-up here and the way in which we speak of cause and effect. Within the system, something that looks intentional or causal is also physical, as there is a chain of reactions; this makes it difficult to distinguish cause and effect and top-down from bottom-up.

Juan José Sanguineti: Actually, I agree with you; it is very difficult to understand all that in self-organizing systems like the brain. In my contribution, I have assumed that causation is *systemic*, and in the context of a system it is difficult to distinguish between cause and effect as in some sense everything is cause and effect together. We just try to create schemes of causality. In science, there are models. In philosophy, we try to distinguish the different senses of the notion of cause, and we have to recognize all these different levels. I do not think that everything can be understood in terms of top-down or downward causation. There are degrees, however; for example, in the worm the reward system is different from that of other animals or even of man. In the latter case, reward includes understanding of the concept. A punishment in a legal system is somewhat different from punishing a dog. The difference is that a dog does not have the notion of what is just or unjust. Finally, when I have spoken about Aristotle, it is worth mentioning that his model of a substance sounds a little inappropriate if referred to non-living beings, since his model of what an entity or an individual object is, is centred on the living being. And of course it is difficult to pass from one level to a different one. He uses an analogy but this is a difficult matter to deal with adequately.

Jean-Pierre Changeux: Thank you very much for your presentation. I would like the stress the fact that you said that we need a natural philosophy. Is that what you have said?

Juan José Sanguineti: Yes.

Jean-Pierre Changeux: This is interesting. Then, I am a little bit worried about the way in which you are using the notion of self, as it seems to me that you place this notion at a different level from that of the neurosciences. For example, what I mean is that your use is not the same use as that of Damasio. We should distinguish this notion as used in expressions like "self-organization," on the one hand, and like "consciousness or self-awareness," on the other. "Self" is, of course, a key word, and I think one should be aware of the different meanings you gave this notion throughout your presentation. Moreover, there is something that I would like you to clarify and that may be linked to the issue of natural philosophy. In your presentation you said, I quote, that "the self is not a thing." Well, so what is it? Do you consider that it is a kind of mental representation, for instance? One basic question in science when we start discussing something is: is this a thing or not? I think that the self *is* a thing. And a thing is a material entity that can be described in physical and chemical terms. Maybe this is a mere question of words, but I would be really naturalistic here. Finally, I would like to address your last distinction among four levels in the human being. You distinguished the physical, the organic, the psychological, and the intellectual level. Now, I think that there is confusion here, because physical means structural (i.e., the neural network) and psychological means functional, so I do not fully grasp the real utility of your distinction.

Juan José Sanguineti: Ok, I think there are three questions. First one: I agree that the "self" in the sense of a self-organizing system is quite different from the "self" in self-awareness. I also think that this parallels the distinction that I have drawn between the deep self, which is not intentional and has no awareness, and the intentional self, which properly has self-awareness. In my opinion, it is also important to try to connect and relate these different senses of the self (I have tried to show that there is some kind of interaction between them), otherwise we could run into some form of dualism. The second question has to do with "the self is not a thing." Well, I used this expression in the sense that it is not something added to life, or to the animal. In other words, I wanted to avoid the risk or the temptation (often present in philosophy) of reifying the self, that is, of assuming that the self is something different and distinct from the living being. This would bring us toward some sort of vitalism.

Gennaro Auletta: If you wouldn't mind, I have a little question. Would you agree to say that the self is the systemic organization of an organism that establishes the distinction between self and not-self, i.e., between the self and the environment?

Jean-Pierre Changeux: I tend to see the question in terms of the processing within a given network that is distributed within the brain. Moreover, there is a further dimension of complexity: we are able to represent the self of others. This is the social dimension and involves the so-called theory of mind, for example.

Juan José Sanguineti: Now, I am going to answer the third question. When I introduced the distinction between the four levels in human beings, I did not claim that the higher ones eliminate the lower ones. A person is a biological organism but with higher levels than those that we find in animals. When we see a person, of course we see a body, but the body of the person is different from the body of an animal as it has specific properties. I want to add that when I used the term "physical," I used it in the sense in which it is used by physics as a discipline, or by the physicists, if you want (or even chemistry).

Chuck Crowell: In your talk, you mentioned Dretske and his identification of structural cores. How does this relate, on the one hand, to Aristotle and his theory of causes and, on the other hand, to Professor Auletta's example of the trees able to canalize wind in a forest?

Juan José Sanguineti: From my personal point of view, it is not possible to identify triggering or structural causes with any of the Aristotelian causes (material, formal, final and efficient cause). There are in some way homologies or similarities but no more. I also think that in speaking of triggering or structural causes you are actually speaking of systems, and Aristotle did not refer to such an issue.

Pierre Jacob: Nearly at the end of your presentation, you took the logical relation between premises and conclusions as an example of horizontal causality. Well, it seems a bit strange to me to say that logical premises *cause* a conclusion.

Juan José Sanguineti: Yes, it is a bit strange to speak of logical causality. I wanted just to say that, when we think, there is something that brings our minds to another thought. When I want to produce new ideas, of course some drugs could increase my memory or my attention, but essentially I have to study. I was referring to such a kind of causality, when thoughts cause other thoughts. Perhaps you do not use the word "causality" in this context and wish to limit the scope of the notion to something essentially material (or more material). Of course, I am not denying that at the level of thoughts there is also the biological causality at play. I was trying to somehow distinguish psychology from logic or mathematics. I am indeed against a "psychologistic" approach to mathematics or to logic.

REFERENCES

Andersen PB, Christiansen PV, Emmeche C, Finnemann NO. (2000) *Downward Causation. Mind, Bodies and Matter.* Aarhus University Press, Aarhus.

Aquinas T. (1988) *Summa Theologiae.* Edizioni Paoline, Turin.

Aristotle. (1961) *Movement of Animals.* Loeb Classical Library, Harvard University Press, Cambridge, MA.

Aristotle. (1957) *On the Soul.* Loeb Classical Library, Harvard University Press, Cambridge, MA.

Aristotle. (1934) *The Nicomachean Ethics.* Loeb Classical Library, Harvard University Press, Cambridge, MA.

Auletta G, Ellis GFR, Jaeger L. (2008) Top-down causation by information control: From a philosophical problem to a scientific research programme. *J R Soc Interface* **5**: 1159–1172.

Bennett MR, Hacker PMS. (2007) *Philosophical Foundations of Neuroscience.* Blackwell, Oxford.

Brock S. (1998) *Action and Conduct: Thomas Aquinas and the Theory of Action.* Clark, Edinburgh.

Clayton P, Davies P (eds.). (2006) *The Re-emergence of Emergence.* Oxford University Press, Oxford.

Damasio AR. (1999) *The Feeling of What Happens: Body and Emotion in the Making of Consciousness.* Harcourt Brace and Co., New York.

Davies P. (2006) The physics of downward causation. In: P Clayton and P Davies (eds.), *The Re-emergence of Emergence.* Oxford University Press, Oxford, pp. 35–52.

Deacon TW. (2006) Emergence: The hole at the wheel's tube. In: P Clayton and P Davies (eds.), *The Re-emergence of Emergence.* Oxford University Press, Oxford, pp. 111–150.

Dretske F. (1993) Mental events as structuring causes of behaviour. In: J Heiland and A Mele (eds.), *Mental Causation.* Oxford Clarendon Press, Oxford, pp. 121–136.

Ellis GFR.(2006) On the nature of emergent reality. In: P Clayton and P Davies (eds.), *The Re-emergence of Emergence.* Oxford University Press, Oxford, pp. 79–107.

Ellis GFR.(2008) On the nature of causation in complex systems. *Trans Roy Soc South Africa* **63**: 69–84.

Emmeche C, Koppe S, Stjernfelt F. (2000) Levels, emergence, and three versions of downward causation. In: PB Andersen, PV Christiansen, C Emmeche and NO Finnemann (eds.), *Downward Causation. Mind, Bodies and Matter.* Aarhus University Press, Aarhus, pp. 13–34.

Fabro C. (2006) *L'io e l'esistenza.* Edusc, Rome.

Gallagher S, Shear J (eds.). (1999) *Models of the Self.* Imprint Academic, Thorverton.

Gibson JJ. (1979) *The Ecological Approach to Visual Perception.* Houghton Mifflin, Boston.

Heil J, Mele A (eds.). (1993) *Mental Causation.* Clarendon Press, Oxford.

Heylighen F, Bollen J, Riegler A (eds.). (1999) *The Evolution of Complexity.* Kluwer Academic, Dordrecht.

Heylighen F. (2008) Complexity and Self-Organization. In: MT Bates and MN Maack (eds.), *Encyclopedia of Library and Information Sciences.* Taylor and Francis, London.

Jeannerod M. (1999) The cognitive way to action. In: RJ Russell, N Murphy, T Meyering and A Arbib (eds.), *The Neurosciences and the Person: Scientific Perspectives on Divine Action.* Vatican Observatory Publications and CTNS, Vatican City and Berkeley, CA, pp. 57–66.

Jeannerod M. (2006) *Motor Cognition: What Actions Tell to the Self.* Oxford University Press, Oxford.

Jonas H. (1982) *The Phenomenon of Life.* The University of Chicago Press, Chicago and London.

Jonas H. (1991) *Dalla fede antica all'uomo tecnologico.* Il Mulino, Bologna.

Kauffman S. (1993) *The Origins of Order: Self-Organisation and Selection in the Evolution.* Oxford University Press, Oxford.

Kenny A. (1979) *Action, Emotion and Will.* Routledge, London.

MacIntyre A. (1999) *Dependent Rational Animals: Why Human Beings Need Virtues.* Open Court, Chicago.

Maturana H, Varela F. (1980) *Autopoiesis and Cognition. The Realization of the Living.* Reidel, Dordrecht.

Murphy N, Brown W. (2007) *Did My Neurons Make Me Do It?*, Oxford University Press, Oxford.

O'Rourke F. (2004) Aristotle and the metaphysics of evolution. *Rev Metaphys* **58**: 3–59.

Panksepp J. (1998) *Affective Neuroscience: The Foundations of Human and Animal Emotions.* Oxford University Press, New York.

Sanguineti JJ. (2007) *Filosofia della mente.* Edusc, Rome.

Sanguineti JJ. (2008) La scelta razionale: Un problema di filosofia della mente e della neuroscienza. *Acta Philosophica* **17**: 247–272.

Sanguineti JJ. (2009) Automovimiento y crecimiento como características de la vida según Leonardo Polo. *Studia Poliana*: **11**: 111–131.

Searle J. (2007) *Freedom and Neurobiology.* Columbia University Press, New York.

Stephan A. (2002) Emergentism, irreducibility, and downward causation. *Grazer Philosophische Studien* **65**: 77–93.

Stevenson L, Squires R, Haldane J. (1986) *Mind, Causation and Action.* Blackwell, New York.

Stump E. (2003) *Aquinas.* Routledge, New York.

Wojtyla K. (2003) *Metafisica della persona.* Bompiani, Milan.

CHAPTER 7

The Brain: A Highly Distributed Self-Organizing System. Who Has the Initiative?

Wolf Singer

7.1. INTRODUCTION

Let me begin with an epistemic caveat that is motivated by neurobiological evidence. Our perceptions and imaginations as well as our abilities to reason are constrained by the cognitive abilities of our brains — and brains, like all other organs, are the product of an evolutionary process. Hence, our brains have become adapted to the conditions of the mesoscopic world in which life has evolved. It is the world within the scale of millimeters to meters, the world where the laws of classical physics prevail. It is not the world of quantum physics and not the world of astrophysics. As a consequence, our cognitive functions have become adjusted to a very narrow segment of the world to support survival and reproduction. To cope with these challenges, pragmatic heuristics and cognitive abilities have evolved that are in all likelihood not optimized to comprehend the essence behind the perceivable phenomena and to identify the "absolute" in the Kantian sense. Evolution did not prepare us to understand processes at subatomic or cosmic scales, because they were and are completely irrelevant for our daily struggle to survive. Thus, it is very likely that our cognition is constrained, and this to an unknown extent.

reconstructing. The sparse sensory signals that our sense organs extract from the world are interpreted by our brains in a highly idiosyncratic and selective way. These interpretations are derived from a huge amount of *a priori* knowledge about the world that is stored in the functional architecture of the brain. Utilizing this knowledge, our brains permanently formulate context-dependent hypotheses, interpret sensory signals as a function of these inferences and present the result of this reconstruction to the workspace of consciousness. What do we know about the neuronal basis of the *a priori* knowledge required for the interpretation of sensory signals and ultimately the world around us? It is commonly accepted that all the knowledge a brain can have and the rules according to which this knowledge is processed reside in the functional architecture of the brain. This contradicts the often-suggested analogy between computers and brains. Computers possess specialized processors and separate memories for programs and data. In the brain, there are only neurons and connections. Both programs and stored knowledge reside in the anatomical layout of these connections and their efficacy, the latter being modifiable by learning. The question of the origin of stored information is thus reduced to the question: Which processes determine the functional architecture?

7.5. THE ORIGIN OF IMPLICIT KNOWLEDGE AND ITS ROLE IN COGNITION

The most important determinant and hence the most important source of knowledge is, of course, evolution. What makes brain architectures comparable across human subjects is evolutionary acquired information that resides in the genes. It is knowledge about the world that is expressed in the functional architecture of brains every time an organism develops. Another source of knowledge is epigenetic modifications of architecture. In humans, brain development is slow and lasts until age 20. During this period connectivity undergoes major changes because neurons form and break connections. A large number of connections that are first formed are destroyed again in order to develop the mature architecture of the brain. This making and breaking of connections is guided by activity, i.e., by experience. Thus, the brain adapts itself to the particular environment in which it evolves. Finally, there is knowledge acquisition by lifelong learning. This is based on graded changes of the coupling strength of the existing connections between neurons. In the adult brain only a few new connections are formed and under normal conditions no breaking of connections occurs, but the efficiency of the existing connections can be up- or down-regulated.

In conclusion, evolution can be considered as a cognitive process through which the evolving organisms acquire knowledge about those features of the world whose consideration facilitates survival and reproduction, but probably not the understanding of the deeper structure of the world that assures coherence across scales. Knowledge acquired through evolution is stored in the genes, expressed in the functional architecture of the brain and then complemented by developmental imprinting and learning. Together, this *a priori* knowledge determines what and how we perceive. It defines, for example, the sensory categories according to which we assign qualities to our experiences. And these assignments are rather idiosyncratic. We arbitrarily classify electromagnetic radiation with wavelengths between 400 to 700 nanometers as light, because the photoreceptors in the eye are sensitive to this wavelength range. Radiations with slightly longer wavelength stimulate our temperature receptors, and we categorize the respective sensations as temperature. A similar arbitrariness of category boundaries is observable in other sensory domains.

A priori knowledge is also required for the definition of the nature of objects. We have clear concepts of what an object is. It must be confined by borders, should be solid, and if it moves, all components should move in a coherent way. This definition is appropriate in the mesoscopic world, but it does not apply to objects at atomic or subatomic scales. If we had no *a priori* definition of the properties of objects, we would not be able to distinguish objects; we would be unable to extract object-specific features from the two-dimensional brightness distribution that complex scenes generate on the retina. The principles according to which we associate features to objects reside in the circuitry of the visual system. Similarly, the learning rules according to which we establish associations are implemented by specific molecular mechanisms that have been preserved nearly unchanged since the evolution of primitive nervous systems. A particularly interesting epistemic problem would arise if the same also holds for the way we reason, infer and assign values to certain activity patterns and not others. The respective rules must also reside in the functional architecture of the brain and as mentioned, this architecture is the result of adaptation to a narrow segment of the world.

7.6. EVOLUTIONARY HERITAGE

In this context it needs to be emphasized that evolution is a conservative process. Adaptive modifications are conserved unless there is a major change in conditions that makes these modifications obsolete or maladapted. This is the reason why our nerve cells function in exactly the same way as those of snails, why the same learning rules and the same mechanisms of signal transduction are implemented. Also the development of structures followed a very conservative path. Since the first appearance of the cerebral cortex, the six-layered sheet of nerve cells that covers the hemispheres of the brain, no new structures have emerged. There is just more of the same, and this increase of hardware makes all the difference. It is the addition of a few more, strategically inserted cortical areas that seems to make the difference between the brain of a human being and that of our nearest neighbours, the great apes. Apparently, it is only this addition of processing substrate and the associated gain of complexity that is responsible for the difference between animals and humans, between species that failed and those who succeeded to promote cultural evolution with all its far-reaching consequences. One of them is of particular relevance here. Cultural evolution created a sociocultural environment of ever-increasing complexity that in turn contributes to the epigenetic shaping of brain architectures. Thus, even if the genetically determined layout of brain architectures has changed only little since the beginning of human civilizations, those features that can be modified by epigenetic shaping are likely to have undergone major modifications.

Here are two examples that illustrate how *a priori* knowledge structures our perceptions. The mold in Fig. 7.1 is used to produce candies. On the left you see the inside with the concavities and on the right the rear side with the corresponding convex protrusions. In reality, the right mold still shows the inside but is just upside down. The reason for these very different perceptions is that the brain makes the *a priori* assumption that light comes from above. In this case contours that have the shadow above need to be interpreted as concave and those with the shadow below as convex. Thus, an assumption of which we are not aware determines what we perceive. Another really striking example is shown in Fig. 7.2. It is hard to believe, but surfaces A and B have exactly the same luminance. They appear so different because the brain sees the shadow that is caused by the cylinder on the right. Even though the amount of light reflected from surfaces A and B and impinging on the retina

Fig. 7.1

is exactly the same, the brain interprets the brightness of the two surfaces as different because it infers the following: given that there is a shadow, surface B must be brighter than surface A, which has no shadow on it, in order to reflect the same amount of light. Thus, the brain "computes" the inferred brightness of the surfaces but we are not aware of these computations. We just perceive the result and take it as real, i.e., we see B as much brighter than A. One could spend hours demonstrating examples which indicate that the brain is generating inferences that we are not aware of, that it is permanently reconstructing the world according to *a priori* knowledge and that we, as perceiving subjects, have to take for granted what the system finally offers us as conscious experience. As expected, this is not only the case with specially designed psycho-physical experiments but is an essential feature of all our perceptual processes. And, it has a very adaptive value, as perceiving the world strictly as it is presented to the peripheral sensory system would not be so efficacious in singling out dangers and useful details. Imagine a primate that has to distinguish between poisonous red berries and similar ones that have a slightly different hue. To survive, the primates have to distinguish the hues

Fig. 7.2

of those two red berries in the morning, at lunchtime and in the evening. In these different situations the spectral composition of the light illuminating the world is radically different. The only way to pick the right berries is to first determine the spectral composition of the sunlight, to take this into account when computing the "redness" of the berries and only then make the choice. In this case the spectral composition of the actual light conditions is determined by taking the context into account, by comparing the colours of the sky, the grass, the leaves, the bark of the trees, etc. So we perceive what the brain presents to us and all this happens without us being conscious of these inferences. [For more on color constancy, see, e.g., Land (1986)].

7.7. THE MIND–BODY PROBLEM

Let me now turn to the implications of brain research for the mind–body problem. I have already said that neurobiologists assume that all mental functions including phenomenal awareness and self-consciousness are based on neuronal processes. And here we have a severe explanatory gap.

We encounter extreme difficulties when we attempt to explain how exactly the *qualia* of our subjective experience emerge from neuronal interactions. We describe the two phenomena in different language systems and are lacking satisfactory bridging theories. We can show with experiments that there is only a minor difference between imagining something and perceiving it in reality. If one investigates subjects in the scanner and asks them to imagine, for example, a rotating wheel, one observes a characteristic distribution of activities in prefrontal brain regions that need to be engaged in order to activate the readout of stored information. In addition, one observes activity in the visual areas that corresponds to the readout of the memorized images of rotating wheels. If one then asks the subject to open his or her eyes and to look at the image of a rotating wheel, there is only little additional activation and it is confined to the primary visual area of the cerebral cortex where the activity from the eyes is first processed (Goebel *et al.* 1998). Thus, the difference between the activity patterns associated with imagination and real perception is minimal. This is also the reason why the brain's threshold for hallucinations is so low that sleep deprivation or fever can blur the distinction between imagery and perception. The best-examined hallucinatory states are those occurring with schizophrenia. We investigated schizophrenic patients who suffer from verbal hallucinations and registered the activity in the auditory cortex of the speech-competent hemisphere. Every time patients reported via a button press in the scanner that they heard voices that they attributed to a real speaker, there was activity in the primary auditory cortex. This activity had about the same amplitude as the responses to voices presented by earphones. When healthy subjects imagine silent speech, they activate their speech centers, but not the primary auditory cortex. In schizophrenic patients this self-generated activity in speech centers apparently propagates all the way down to the primary auditory cortex, making it impossible for the subjects to distinguish between self-generated and evoked activity (Dierks *et al.* 1999).

Let me now turn to the phenomenon of awareness. How is it possible that we not only perceive but are in addition aware of the fact that we perceive? How can we explain in neuronal terms phenomenal awareness, the ability to be aware of ones cognitive operations? There are as yet no satisfactory answers to this question. However, evolution might provide some hints. In less-evolved brains the paths from sensory to executive areas of the cerebral cortex are short and there is little crosstalk between processing streams related to the different sensory modalities. As evolution proceeds and brains become more and more complex, one observes the addition of new cortical areas, which are no longer connected to the periphery, neither on the executive nor on the receptive side. They receive their information, their input, exclusively from the output of the phylogenetically older areas. This process is iterative, with more and more areas added that communicate only with other areas. Evidence indicates that all cortical areas, the old and the more recent ones, operate according to the same principles because they have the same intrinsic organization (Douglas and Martin 2004). Thus, the more recent areas process the output of the older areas according to the same principles, as these process the signals from the sense organs. Such iteration is, in principle, capable of supporting the generation of meta-representations, i.e., representations of representations of representations. And this is perhaps the structural basis that allows us to subject to cognitive operations not only signals from the outer world but also those resulting from internal processes, to realize various levels of meta-cognition and thereby to become aware of our own perceptual functions.

7.8. THE DISTRIBUTED ORGANIZATION OF THE BRAIN AND WRONG INTUITIONS

Another conundrum and challenge to both brain research and our intuitions is the distributed organization of our brains. The neurobiological evidence accumulated over the last decades has led to radical changes in our views of the brain. In the early days, intuition and introspection were the major sources of knowledge for the formulation of hypotheses about the organization of the brain. Now we learn that these intuitions are in drastic conflict with the evidence provided by scientific investigations, raising the interesting question as to why the brain is so agnostic to its own organization. We do not feel our brain, have little intuitive insight into its processes, and are surprised to find that it works according to principles that differ substantially from what we thought. Intuition suggests to us that somewhere in the brain there ought to be a convergence center where all information comes together to be amenable to coherent interpretations of the world. This would be the site where perception takes place, where the intentional agent is active, where

Before closing, let me briefly turn to one of the limitations of our intuition with respect to assembly coding. It is difficult for us to understand the structure of distributed representations because the essential information is encoded in non-stationary, spatiotemporal relations among large numbers of neurons rather than in the responses of individual neurons. In principle, one can measure these patterns by registering the activity of many time series and neurons that participate in a dynamically configured assembly simultaneously, which is difficult and requires the analysis of high-dimensional, non-stationary time series with advanced mathematical methods. The resulting descriptions of representations will then be vectors of activity defining a high-dimensional state space. They are not tangible and are difficult to represent in two or three dimensions; they are essentially lists of numbers or systems of equations. It is not satisfactory to have such explanations for a percept, a thought or a feeling. The problem is that, in the future, descriptions of neural states will become more and more abstract, more and more mathematical, and more and more intuitively implausible.

7.12. THE NEURONAL CORRELATE OF CONSCIOUS PROCESSING

To conclude, I would like to give at least one example of an experimental approach, in this case in search of the neuronal substrate of conscious processing. We asked the question: How does subconscious processing of stimulus information differ from conscious processing? In this experiment, subjects looked at a fixation point on a computer screen. First, there appeared a noisy mask, then a word, and then another mask. At the end of the trial, with some delay, we showed a word or its figural counterpart, and subjects were asked to press buttons indicating "same" or "different." The masks were adjusted such that the subjects did not always consciously perceive the first word, and in this case subjects had to make a forced choice and were instructed to guess. Interestingly, when there was concordance, even if subjects had not seen the first word, the reaction times were shorter. This proves that subjects had processed the stimulus material even though they were not aware of having processed it. In other trials, subjects actually perceived the word consciously and reported on it. During these experiments we investigated the respective brain activities and measured the oscillatory responses from the different areas of the cerebral cortex. When we measured only the power of the oscillations in particular frequency bands, there was not much difference between the conscious and unconscious condition. However, when we considered not only the amplitude of these oscillations but also the precision of the phase-locking of these oscillations across sensors, we noticed shortly after the encoding, and only in the conscious condition, a brief burst of very precise phase synchronization across widely distributed cortical areas. This episode of high coherence was missing in the unconscious condition. If one looks at the topological distribution of this synchronization phenomenon, one sees that an extended network of cortical areas engaged transiently into very precise synchronous oscillations when what has been processed has access to consciousness (Melloni *et al.* 2007). This suggests that access to consciousness requires that the system gets into a metastable state that is characterized by very precise phase-locking of synchronized oscillatory activity in a particular frequency range across widely distributed cortical areas. This is a very elusive description of the neuronal correlate of conscious processing as we describe a particular dynamic state, not an anatomical substrate that one can localize. There is actually no specific area that could be identified as responsible for consciousness. Most of the interesting information is in the "in-between," in the relational graph. This comes very close to what Sherrington has stated about consciousness: "Pure conjunction in time without necessarily cerebral conjunction, we mean conjunction in space, lies at the root of the solution of the problem of the unity of mind." It is the intangible simultaneity that binds processors together and ensures that distributed contents are experienced as a coherent whole, not as unity somewhere in space, but in time — and this is not what our intuition would assume.

7.13. THE OUTLOOK

So what is the outlook? I tend to predict that the more we learn about the brain, the more abstract the descriptions will become, because we shall deal with increasingly large numbers of neurons that generate spatiotemporal patterns in a highly nonlinear system. These patterns will not be stationary but dynamic trajectories, probably trajectories that never repeat identically because if they did we would have no notion of time. Many of the explanations that we will come up with (and this has already started) will further be in conflict with our primary intuitions. We have no feeling for the dynamics ruling the functions of our brain because we have difficulties imagining complex nonlinear processes. Furthermore, we will have to consider more and more that the brain is a member of a sociocultural network and that some of the phenomena that seem to be so difficult to explain in pure neuronal terms will have their explanation only when considering interactions among brains, "brain-networks." This widened perspective is probably necessary to close the explanatory gap between first- and third-person perspectives. We shall have to consider the fact that our brains are the product of their embedding in a complex cultural environment and that many of the "mental" constructs with their immaterial connotations that are so difficult to relate to brain processes, such as value systems, consciousness, intentionality and so forth, have only come into the world because brains mirror themselves reciprocally in other brains and assign properties to the experiences made during these social interactions. We are just at the beginning of social brain research, but there are already initial investigations of phenomena that exist only in the realm of social interactions such as empathy, greed, jealousy, and probably also the concept of the intentional conscious self.

DISCUSSION

Gennaro Auletta: First of all, I hope nobody of us here is a dualist, nobody of us here believes that the mind comes from the outside of the brain, and nobody of us believes that quantum mechanics is the explanation of the brain. I assume that the all of us agree on that. I am even completely convinced that the brain and man, as well as humanity and culture, are a result of evolution. Notwithstanding, we should always distinguish, in my opinion, between the genetic explanations and the results of genetic processes, that is, the functions that exist and can be performed once they have been selected. These are two different issues. So my point is the following: I completely agree that our brain functionalities are contingent, and I even believe that a lot of things in humans are contingent. Nevertheless, there are characters of our intelligence that are probably not contingent at all, because I find it a little absurd that intelligent beings coming from other planets, if you allow this consideration, would disagree with us about fundamental issues like the structure of atoms, the Big Bang, the molecular structure of matter, and so on. We would probably converge; of course, we would have different needs, different evolutionary paths, but this notwithstanding, we would probably converge. To deny this for me would be absurd and ultimately antiscientific. We also have very interesting examples: Marino's studies show that dolphins and primates, even if diverging many millions years ago and having morphologically very different brains (for instance, dolphins lack much of the prefrontal areas), have deep convergences, such as a very intelligent behavior. Also from a cultural point of view, Lord Renfrew, a specialist at Cambridge, has shown (in a talk at our 2009 Evolution Conference in Rome, at the Gregorian University) that there are interesting convergences, for instance, the State has been created both in Central America and in China, although in a completely independent way. Of course, somebody could say that this is a result of our common brain organization, but this seems to me a little "mythic" and so far has proved a not-so-good explanation. We should find true mechanisms able to explain these kinds of facts, and such an explanation may turn out to be highly complicated or at least more difficult than we may assume nowadays. Therefore, I would avoid too-simplified kinds of explanation. So, my question is whether you would agree (you as well as the all of us here) with the fact that you have aspects that cannot be completely accounted for in terms of contingency even if they are the product of evolutionary history.

Wolf Singer: Certainly there is convergence: the eye in the octopus and the eye in vertebrates has developed completely independently. If the constraints are the same, and in this case they are the same, evolution produces the same outcome. Whether this is true for other cases, well, God knows, I have not the faintest idea. I think our knowledge is severely limited, and there are many things out there which we have no idea about.

Gennaro Auletta: But the environments of dolphins and primates are quite different.

Wolf Singer: Well, as long as the description systems that we develop in our constrained world are coherent, this is fine and it is all that we can do, and the rest is belief.

Gennaro Auletta: It seems to me that what you are expressing now is also a belief.

Marc Jeannerod: I was thinking about your point on brain evolution. You mentioned the idea of the addition of new neural patterns and complexification. People who are studying comparative anatomy tend to say that all the areas that will develop in humans are already there in the lower species. The architect genes are the same in all those species, and the difference in brain configuration depends on the time of expression of certain genes. I think that it is a different perspective: you have the complexity already there, but it has just not taken place, or it has not yet fully taken place in some species, and it explodes in humans.

Wolf Singer: I am not enough of a specialist on this. Obviously, what we see is an increase in volume. There are areas that we will have difficulties identifying in the monkey and that we humans do have. Mice certainly do not have much of the prefrontal cortex that we have. I mean that it is easy to get a big cortex by just increasing the cycle time of the generation of cortical neurons. It is also very interesting the way in which human brains develop in sequence. I have built models to show how the inputs extensively rearrange the topology of the functional circuitries (that is when schizophrenia starts), and you get a complete remodeling of who talks to whom preferentially; this may of course make all the difference.

Antonio Damasio: I just want to clarify one point, because this may be a matter of wording or it may not. You have said, and I entirely agree, that there is no self-center of the brain. However, it seems to me that it is almost as if you are denying a self-process or a protagonist process. Well, I think that could be a problem, because there *is* such a thing. In other words, the fact that you did not have any area of the brain that you can identify as a consciousness center in a Cartesian way (and I entirely agree with that) does not imply that there is no self-process. There is something that is very objective in our minds in that narrow band of consciousness that we achieve. And I think that by just denying the self-center and denying that process, you are losing the possibility of articulating a lot of things that we do know about regions of the brain that are surely involved in the process of consciousness, and others that are not.

Wolf Singer: OK, yes, I fully agree. I would say that as soon as we can identify a dog, we have a multitude of sensory processing areas that when they cooperate properly are able to represent the dog properly, and if some of these areas are deficient then the dog will not have any color, or it will no longer bark, and so on. And I think the same is true of the self-process: you need the parietal areas and the prefrontal areas and you need all sorts of representations in order to make this thing work. But it is a distributed system or process.

Antonio Damasio: Oh yes, in that we entirely agree. I just wanted to avoid some possible confusion. And I think that a similar risk is hidden in your points about hierarchies. You talked about hierarchies, and I entirely agree on your point about certain prefrontal cortices being vital to confer upon us the kind of human processes that we do have. But in a way you were also denying the hierarchies: I think that not all of this is alike, and there are certain parts of the brain, for example, in the prefrontal cortex within the temporal and parietal regions, that are going to be somewhat more important than others at the end of all these convergence–divergence chains. Will you agree with that?

Wolf Singer: Yes, absolutely: the speech area is something special; it sits like a spider in a web.

Antonio Damasio: And also the prefrontal cortices and the PMC, which we were talking about yesterday?

Wolf Singer: Yes, sure.

Jean-Pierre Changeux: I am sorry, but just why do you say that the language areas are special?

Wolf Singer: The intrinsic organization of these areas, of course, is the same as everything else; I do not think there are major differences. But the way in which they are connected to other areas does make all the difference. They have a special connectivity that makes these areas special, that makes them able to do things that others cannot do. The intrinsic operation must be similar, it looks the same; but the way in which they are hooked up to other areas is very specific.

Antonio Damasio: Are you throwing away the idea that there are certain cortices that by virtue of their connectivity and perhaps by their intrinsic organization can be called mapped versus non-mapped?

Wolf Singer: We had this discussion in the last few days. The point is whether they are in principle not mapped or whether we do not see the maps because we do not ask the right questions about the categories that are met there.

Michael Arbib: This is a more technical point. You ended up with somehow seeing synchronization as the secret of consciousness. Now, if I look around this room, I recognize different people, the color of the table, and so on. It is not clear to me how just talking about a basic carrier wave and some cells locking together in phase differences would have enough precision to store and allow immediate retrieval as I look around at the hundreds of different aspects that I see in this room. So, how does your phase-locking synchronization view accord with the richness of human experience?

Wolf Singer: Yes, empirically that is what it looks like. In the case where you consciously process a stimulus in imaging recordings, if you measure the phase synchrony of the oscillatory activity, there is a transitory formation of a large network amidst other senses. This then disappears and it seems to be the trigger moment for access to consciousness. Then, you have sustained theta oscillations where the contents are brought to short-term memory paths and there is recall, which is seen by an increased downward activity at the time of comparing the target to this information sample. I think that, in order to explain the unity of consciousness, one needs much more: something very much in the sense of the workspace. I wrote here that this is actually in agreement with the workspace hypothesis: you need a trigger event that distinguishes conscious from subconscious processing, and once you get across the threshold you have this representational space in which you can book in whatever you want and this would be bound together.

Michael Arbib: It seems to me to be a little bit like explaining books by saying they have a title page. So, it seems to me that the synchronization is a fascinating fact, but it tells us nothing about the richness of consciousness. Yes, if I turn off the oscillation, maybe I am not conscious anymore; but once I am conscious I am aware of so much, and it seems to me that your theory does not begin to tap the richness of that experience.

Wolf Singer: Well, there is no problem if you have only one content in these visual areas, or even if, in addition, you have oratory connotations and certain emotions; but then, what do all the others give? Much of what you have in there is from your memory.

Gennaro Auletta: I think that Michael Arbib's point is that it seems that you take only these degrees of freedom (it is actually only one), but likely you need much more to understand consciousness.

Wolf Singer: Yes, but, you know, consciousness is constrained. You do not have to have many contents there; you would be surprised to see from psychophysics how much you are actually consciously aware of.

Gennaro Auletta: The problem is how you will somehow have all the information caught simply in some oscillatory waves or in synchronization; do you see the point? The problem is the huge complexity of that, and we expect that probably many variables are relevan there.

Wolf Singer: I do not know what happens after this trigger moment. There must be a global state that has unifying properties. Now you can think to look into these rhythms, you can concatenate them and make all this as complicated as you want if you have a phase-space available as a coding space. Anyway, none of us has any reasonable explanation so far; I am not claiming I have one, but there is no other one.

Pierre Jacob: I will just recall two questions from a previous discussion. One question is: can one make sense of synchronization? I guess you just suggested something like this: you need a master clock that can be read by neurons. Well, if you think there is such a master clock that keeps track of time, are neurons able to read it? The other question is: you seem to use two kinds of tools when you think about evolution of brain functions and cognitive abilities. The first one is the complexity of brain structure, and the second one is convergent solutions based on environmental constraints (so that you find species with fairly different brain structures coming to convergent solutions). Finally, I have an additional question about domestication. It has been pointed out that dogs, for example, unlike chimps, can understand pointing gestures. My question is: What is the connection between domestication and brain study? Is it part of the brain structure or not?

Wolf Singer: Last question first, because I cannot answer everything. I do not have the faintest idea why crows can recognize themselves in a mirror and wolves cannot do that; I do not know. Whether dogs have been selected through selective breeding to be able to do this, I do not know. The second question is whether there is a master clock. I do not think there is; I think it is a concatenation of many oscillators that agree on certain rhythms. There are certain persistent frequencies that the brain likes, others that it does not like. The brain covers a frequency range from below 1 hertz all the way up, and it exploits this frequency space in a very nice way. This seems to be a general principle in all brains: if you think about lobsters, they even use exactly the same treatment in order to reconfigure the networks, and I think the same principle is used in the entire living world.

REFERENCES

Dierks T, Linden DEJ, Jandl M, *et al.* (1999) Activation of Heschl's gyrus during auditory hallucinations. *Neuron* **22**: 615–621.

Douglas RJ, Martin KAC. (2004) Neuronal circuits of the neocortex. *Annu Rev Neurosci* **27**: 419–451.

Engel AK, Singer W. (2001) Temporal binding and the neural correlates of sensory awareness. *Trends Cogn Sci* **5**(1): 16–25.

Fell J, Klaver P, Lehnertz K, *et al.* (2001) Human memory formation is accompanied by rhinal-hippocampal coupling and decoupling. *Nat Neurosci* **4**(12): 1259–1264.

Fries P. (2005) A mechanism for cognitive dynamics: Neuronal communication through neuronal coherence. *Trends Cogn Sci* **9**(10): 474–480.

Fries P, Reynolds JH, Rorie AE, Desimone R. (2001a) Modulation of oscillatory neuronal synchronization by selective visual attention. *Science* **291**: 1560–1563.

Fries P, Neuenschwander S, Engel AK, *et al.* (2001b) Rapid feature selective neuronal synchronization through correlated latency shifting. *Nat Neurosci* **4**(2): 194–200.

Gaillard R, Dehaene S, Adam C, *et al.* (2009) Converging intracranial markers of conscious access. *PLOS Biol* **7**(3): e1000061.

Goebel R, Khorram-Sefat D, Muckli L, *et al.* (1998) The constructive nature of vision: Direct evidence from functional magnetic resonance imaging studies of apparent motion and motion imagery. *Eur J Neurosci* **10**: 1563–1573.

Gray CM, Singer W. (1989) Stimulus-specific neuronal oscillations in orientation columns of cat visual cortex. *Proc Natl Acad Sci USA* **86**: 1698–1702.

Gray CM, König P, Engel AK, Singer W. (1989) Oscillatory responses in cat visual cortex exhibit inter-columnar synchronization which reflects global stimulus properties. *Nature* **338**: 334–337.

Hipp JF, Engel AK, Siegel M. (2011) Oscillatory synchronization in large-scale cortical networks predicts perception. *Neuron* **69**: 387–396.

Land EH. (1986) Recent advances in retinex theory. *Vision Res* **26**: 7–21.

Melloni L, Molina C, Pena M, *et al.* (2007) Synchronization of neural activity across cortical areas correlates with conscious perception. *J Neurosci* **27**(11): 2858–2865.

Roelfsema PR, Engel AK, König P, Singer W. (1997) Visuomotor integration is associated with zero time-lag synchronization among cortical areas. *Nature* **385**: 157–161.

Singer W. (1999) Neuronal synchrony: A versatile code for the definition of relations? *Neuron* **24**: 49–65.

Sporns O, Chialvo DR, Kaiser M, Hilgetag CC. (2004) Organization, development and function of complex brain networks. *Trends Cogn Sci* **8**(9): 418–425.

Womelsdorf T, Schoffelen J-M, Oostenveld R, *et al.* (2007) Modulation of neuronal interactions through neuronal synchronization. *Science* **316**: 1609–1612.

CHAPTER 8

Two Views of Brain Functioning

Marcus E. Raichle

Some of my colleagues have already discussed very nicely some of the things I am going to address. My purpose is to discuss the idea of two views of brain function, which is not something that you can measure so easily. These two ideas have been around for centuries.

First, however, I want to comment on the methods that I am going to refer to repeatedly. Indeed, I have lived in the world of brain imaging for many years. We have two basic techniques. One is PET, where we can measure the blood flow in the brain and its various metabolic parameters. The key feature is that one observes an increase in blood flow when a stimulus is presented to the brain. This has been known since the work of Angela Basso, who was a distinguished investigator in Italy in the 1800s (there is more to say about that, especially with regard to the cell biology related to these issues; I return to this at the end of the chapter). It is worth noting that when the brain increases its blood flow, its oxygen consumption does not increase as much; as a result, when the blood flow goes up, a local increase in the amount of oxygen present in the brain occurs. This is the basis for fMRI to work, as this local increase in oxygen generates signals that fMRI can capture. The change in oxygenation of hemoglobin has an effect on the magnetic field, so that if there is deoxygenated hemoglobin, which is an fMRI contrast,

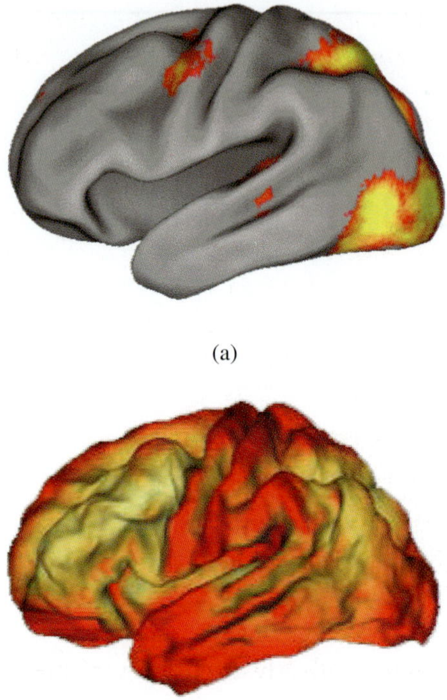

(a)

(b)

Fig. 8.1 Two views of brain function. (a) One view, the most commonly held view, posits that the brain is primarily reflexive, driven by the momentary demands of the environment. This image depicts changes in brain activity measured with functional magnetic resonance imaging during the performance of a task involving visual stimulation. Only the areas of change are shown by subtracting away the brain activity when subjects were not performing the task. (b) The other view is that the brain's operations are mainly intrinsic involving the maintenance of information for interpreting, responding to and even predicting environmental demands. This image depicts ongoing metabolism of the brain measured with positron emission tomography.

dark veins become visible, and when the oxygenation of the blood increases, such veins disappear. Now, if the blood flow of the brain increases and the oxygen consumption does not, the so-called fMRI blood oxygen dependent, or BOLD, signal is produced. So we have both the fMRI BOLD signal and the metabolic parameters identified by PET. Therefore, in this chapter I will be oscillating, if you will, between PET and fMRI.

Let us now turn to the two views of brain function (Fig. 8.1). One view, the most commonly held one, posits that the brain is primarily reflexive and driven by the momentary demands of the environment. This view governs the overwhelming number of experiments done in neuroscience; certainly in the world of brain imaging it has dominated the scene. Now, since the introduction of fMRI in 1992, there have been over 12,000 refereed publications exploring this interesting phenomenon and, as you can imagine, many brain functions have already been studied with one or the other imaging technique (not to mention the other tools that are available). There is another view of brain function that has been articulated by many scholars going all the way back to the beginning of the last century. In this view, the brain's operations are mainly intrinsic, involving the maintenance of information for interpreting, responding to and even predicting environmental demands.

Now, the issue is that if there are these two views and they are not necessarily mutually exclusive, how do we adjudicate the relative merits of these two ways of thinking about the brain? Is one more important than the other? I would like to suggest that the first thing to do is to try to take an "economic view" of the matter. We need to become "neural economists," so to speak, and ask: How much does it cost to perform each of these two different functions? The first information on this actually goes back to the 1940s, when the first quantitative measure, developed by Seymour Kety and colleagues, of the brain's oxygen consumption and energy use was performed by assessing the balance of inflow and outflow of blood, energy and oxygen. What they pointed out is that in the human being, while the brain is only two percent of the body weight, it accounts for up to 20 percent of the body's entire energy consumption. Such results have been reproduced many times. The brain is very active even during sleep. The heart supports the activity of many organs: the liver and the spleen, and all the way down to the muscles. All of these are significantly below the level of the brain. So, the brain is an extremely expensive organ to run, and one might simply ask the question: How does *evoked* activity account for all of this consumption? Another interesting finding, by the same Kety group, came in 1955. They had individuals do a difficult mental arithmetic task and measured the overall brain oxygen consumption (regional measurements were not yet possible), which is equivalent to measuring the energy consumption. They observed no changes in the blood flow or in the oxygen consumption, i.e., doing this difficult mental arithmetic had no measurable effect on the overall cost of running the brain.

we just take a neural-economist point of view, and ask where the money is being spent, the answer is: on the intrinsic activity.

Then the question is: What accounts for the high cost of this intrinsic activity? The first thing to point out is the strong evidence that most of the consumption happens at the synapses (i.e., where one neuron talks to another). There are a lot of experiments demonstrating this point quite nicely. In an early work (Kadecaro *et al.* 1985), for example, a stimulator is placed in a peripheral nerve, where the neurons under study have their origins. The change in the frequency of stimulation determines the firing rate of cell bodies in the dorsal root ganglion (Fig. 8.4), but the energy consumption that results from this activity is only seen in the spinal cord where these neurons synapse with other neurons. Just to put the things in perspective, a statistical anatomy of the cortex may point out that you have a hundred billion neurons but up to two hundred trillion synapses. Thus, synapses are a dominant factor.

In experiments where the input-output of neurotransmitters is considered, it is possible to calculate every step along the way from the release of neurotransmitters from one neuron to their interaction, at the synapse, with another neuron; their removal from the synapse into astrocytes; and then back around to neurons. Thus, we know all of the costs of every one of the steps involved, and if we add them all up, it is remarkable to note that 60 to 80 percent of the whole energy cost in the brain is consumed only by the activity of its primary excitatory neurotransmitter glutamate (Fig. 8.5). There are parallel experiments using MRI spectroscopy that, remarkably, produce the same figures as in this particular calculation and that have been verified by a number of other investigators. So, the bottom line, as I see it, is that the brain is an enormously expensive organ in which many costs are intrinsic, mostly related to a conversation among the cells of the neural system. Therefore, the functioning of the brain is essentially dominated by its intrinsic activity.

Now, the question is: For what purpose? Again this has been eloquently addressed elsewhere in this volume; I can only add a few simple anecdotal comments. When I think about this, I think about the late David H. Ingvar. He was very interested in the idea examined in his lovely essay (Ingvar 1985) where he referred to the creation of a memory of the future (in other words, the brain is in the prediction business). Well, I always like to point out this essay because I think there are some interesting ideas there.

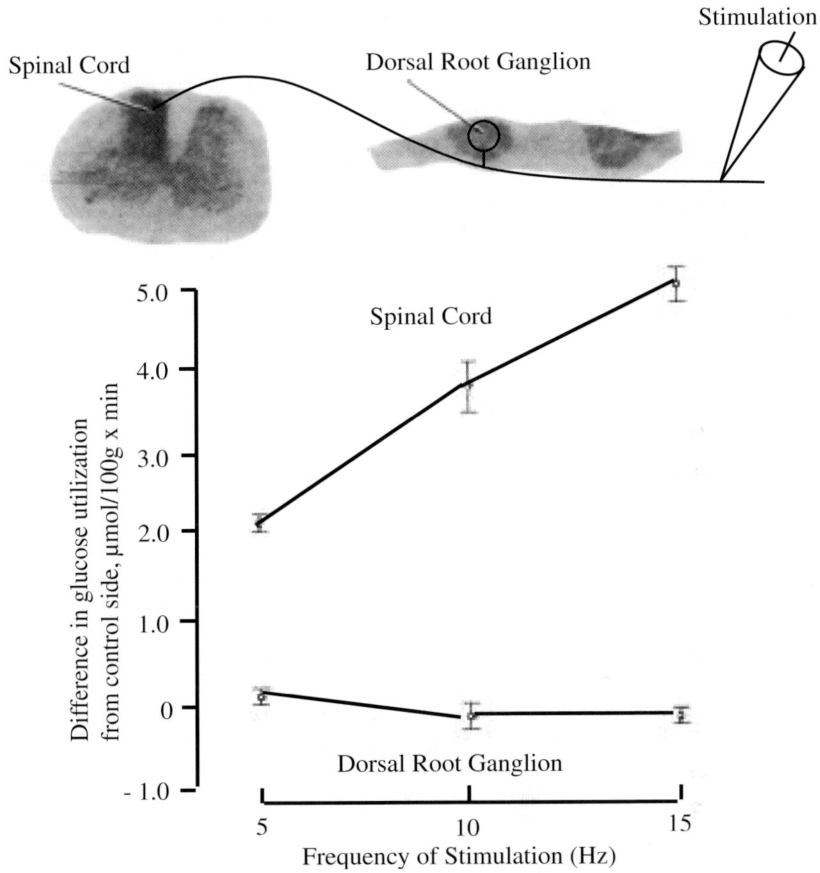

Fig. 8.4 It is important to understand what accounts for the enormous cost of brain function. This figure from the work of Sokoloff and colleagues (Kadekaro *et al.* 1985) nicely demonstrates that the increase in metabolism associated with incoming sensory information occurs at synapses, in this case in the spinal cord, and not in the cell bodies, despite the fact that spikes are recorded in the cell bodies in the dorsal root ganglion (for an extensive review of this literature, see Raichle and Mintun 2006.) (Adapted from Kadecaro *et al.*, 1985).

I am not in the position to talk about consciousness here, and having listened to the eloquent statements about consciousness, I turn to this point with some trepidation. However, I could not resist sharing with you one interesting observation because I find it extremely useful. I mentioned that conscious awareness commands a tiny fraction (about 100 bits per second) of the total amount of information being processed by the

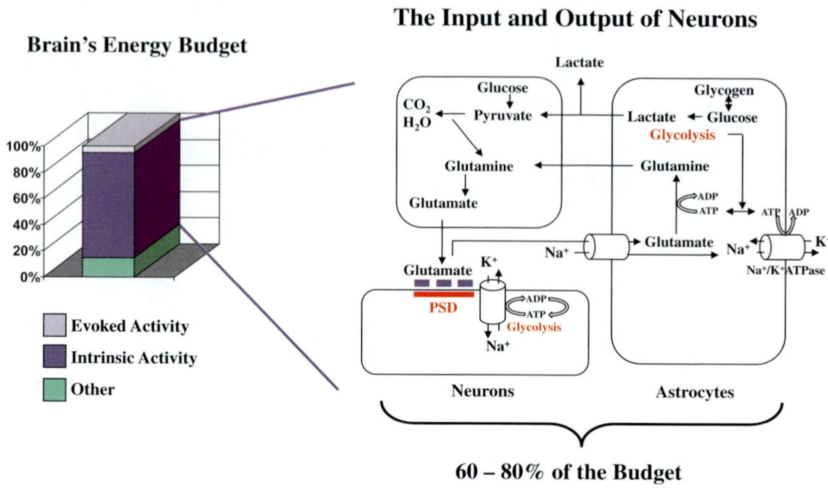

Fig. 8.5 Work by several groups has refined our understanding of the cellular events accounting for the cost of brain function. This work is summarized in this figure and in two recent reviews (Raichle and Mintun 2006; Buzsaki *et al.* 2007). Excitatory neurotransmission mediated by glutamate appears to account for a significant fraction of the cost, with the role of inhibition yet to be fully accounted for.

brain. So, among the many things that we may take into account in considering consciousness, there is one that concerns me in particular — it was provoked by a beautiful and bold observation by Michael Gazzaniga, who began his work on split brains in humans (a surgical treatment for intractable epilepsy) as a student of Roger Sperry. Thus, in the treatment of epilepsy you can split the brain in half, and you basically have two hemispheres to which you can communicate independently with visual information from the left side of the visual field going to the right hemisphere and vice versa. It is therefore possible to deliver stimuli to one hemisphere or the other but, critically, it is only the left hemisphere that is capable of giving you an account of what is going on because it is in the left hemisphere that our capacity for language resides. In this important experiment in a split-brain patient (Gazzaniga 1998; see Fig. 8.6), a picture of a house in a snowstorm on the left and a chicken claw on the right is shown to the patient, so that the right hemisphere is presented with the snowstorm and the left hemisphere with the chicken claw. The subject is then asked to select an appropriate tool from a set of objects

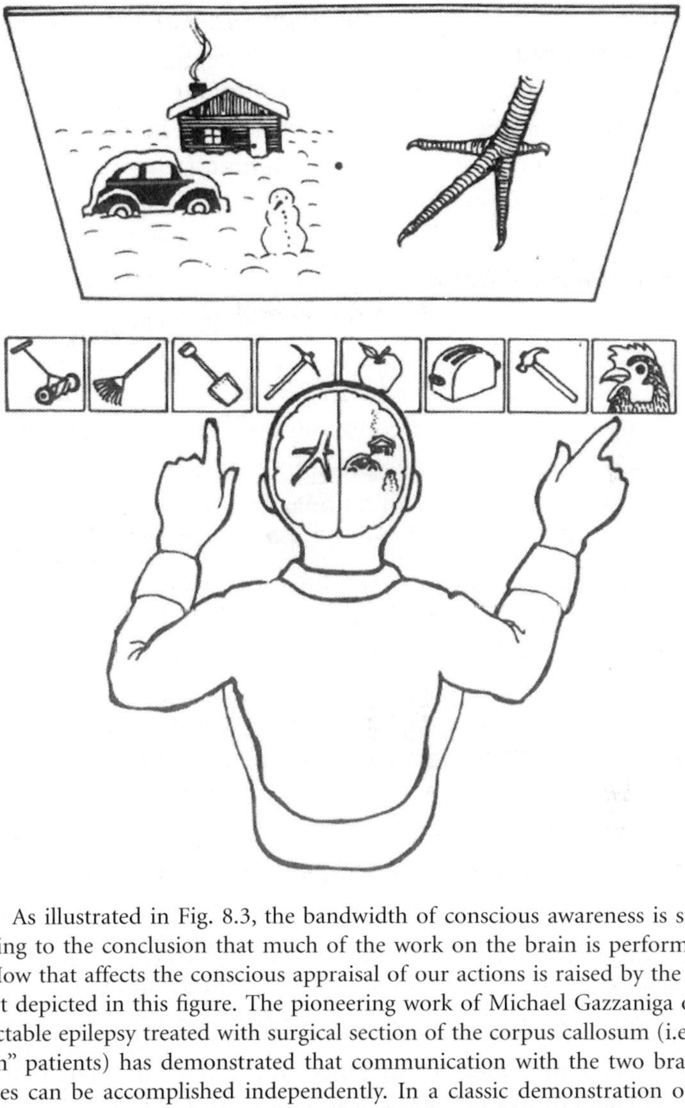

Fig. 8.6 As illustrated in Fig. 8.3, the bandwidth of conscious awareness is surprisingly small leading to the conclusion that much of the work on the brain is performed unconsciously. How that affects the conscious appraisal of our actions is raised by the interesting experiment depicted in this figure. The pioneering work of Michael Gazzaniga on patients with intractable epilepsy treated with surgical section of the corpus callosum (i.e., so-called "split brain" patients) has demonstrated that communication with the two brain cerebral hemispheres can be accomplished independently. In a classic demonstration of this phenomenon, the right (non-verbal) hemisphere is shown a winter scene and, when asked to choose from a set of pictures, the one most appropriate points with the left hand to a shovel. The left (verbal) hemisphere is shown a chicken claw and appropriately chooses the picture of a chicken. Critically, when the subject is asked why the shovel was chosen with the left hand the verbal response, which comes from the left hemisphere, is that it would be used to "clean the chicken coop." This is a very clear demonstration of post hoc reasoning arising because of the inaccessibility of a full understanding of the non-conscious activities of our brain. For a more complete description of this important work (see Gazzaniga and LeDoux 1978; Gazzaniga 2001). (Adapted from Gazzaniga *et al.* 2002).

with the right hand (associated with the left hemisphere that is presented with the chicken claw) and a tool from another set of objects with the left hand (associated with the right hemisphere that is presented with the snowstorm). Correctly, the left hand picks a shovel, and the right hand picks the chicken head. Keeping in mind that the left hemisphere (where language resides) cannot know what was presented to the right hemisphere, the subject is, nevertheless, asked to explain why he chose the shovel with his left hand. The answer can only be based on what the left hemisphere knows, which has to do with chickens: the shovel was chosen because it is appropriate to clean out the chicken shed! I find this a very important observation because it points to the use of *post hoc* reasoning by the brain to explain behaviors whose origins are outside of our conscious awareness. It is highly likely that such a phenomenon is not confined to split-brain patients and is important because of the large amount of functional activity ongoing intrinsically in the brain that resides outside of our conscious awareness.

The question now is: How did we study intrinsic activity? The difficulty, indeed, is the following: Since what the brain is doing is mainly intrinsic, we are consequently deprived of the standard experimental techniques associated with the performance of tasks. From the imaging perspective, one thing that struck us was the curious observation (done during an imaging experiment) that the intrinsic activity of the brain in the resting state is organized in detailed patterns which decrease their activity during the performance of externally directed tasks (this is something that we had not foreseen). Indeed, it is found that a set of areas (some of which Antonio Damasio talked a lot about), namely, the posterior cingulate and precuneus, as well as the medial prefrontal cortex, decrease their activity during externally directed behaviors. The fact that specific areas that are active during non-evoked intrinsic activity deactivate during externally directed behaviors clearly suggested the intrinsic activity to be somehow patterned. We referred to such intrinsic organization as a "default mode" of brain function, and the brain network most associated with this has become known as the brain's default mode network (Fig. 8.7).

How do we study intrinsic activity in even greater detail? Indeed, what I have just reported is a somewhat limited view: it indicated that there was the potential for a large-scale organization of the resting state in the brain. But, how do we take this further? What arose was that the ongoing,

Activity decreases during externally-directed behaviors

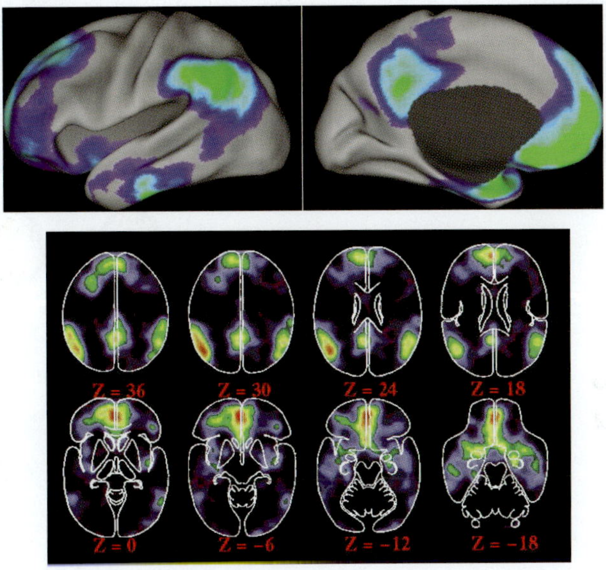

The Brain's 'Default Network'

Fig. 8.7 One of the most important demonstrations of the organization of the ongoing, intrinsic activity of the human was the occurrence of activity decreases in functional brain images during the performance of goal directed tasks as shown in this figure. The particular constellation of brain areas shown in this figure has become known as the brain's *default mode network* (for more detailed accounts, see Raichle 2010; Raichle and Snyder 2007; Raichle 2011). (Adapted from Raichle *et al.* 2001).

spontaneous fMRI BOLD signal, which appears on first examination to be very noisy, is in reality rich with information about intrinsic activity. Figure 8.8 shows the "noise" coming out of a particular area of the cortex. What is interesting is that if you look at the same area on the other side of the brain, you see a similar intrinsic noise. In 1995, a couple of MR physicists simply asked the question whether or not this noise is coordinated with anything. What you see is remarkable: these two signals in a resting state (in which no motor movement is present) are coherent. These signals and the patterns of coherence that can be observed from them have become a major focus of imaging research worldwide.

We became especially interested when it was observed that the default mode network (Fig. 8.7) also exhibited spontaneous BOLD activity that was

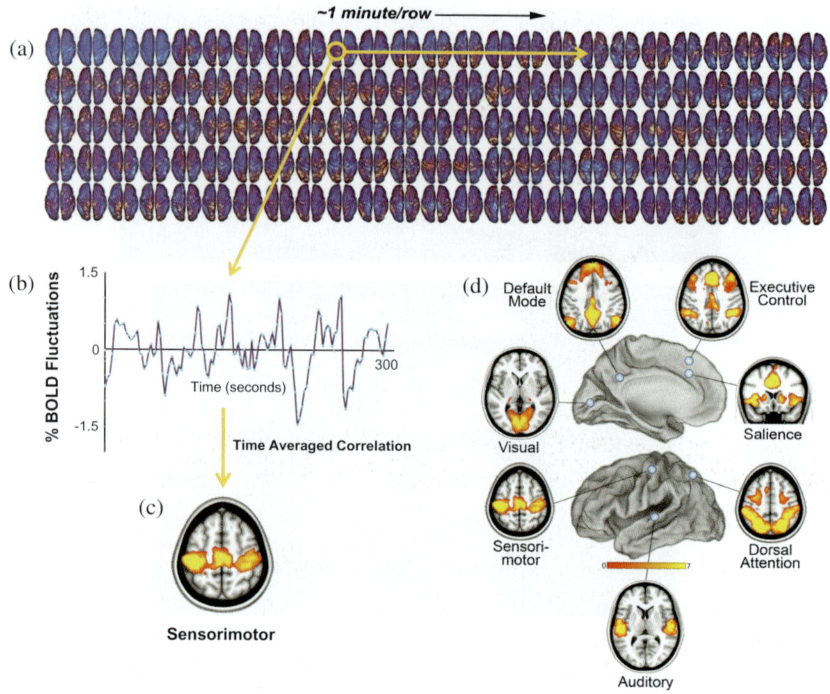

Fig. 8.8 From the fluctuating patterns of intrinsic activity seen in the human brain with fMRI BOLD imaging, striking patterns of spatial coherence within known brain systems can be extracted. A single subject example of data from which these patterns are derived is shown (a). These data were obtained continuously over a period of 5 minutes (each row is one minute, each frame is 2.3 seconds). We have found it instructive to view the data occasionally in this way as it helps one understand the slowly moving, ever-changing nature of the activity. An interpolated version of these data in a movie format may be downloaded from (ftp://imaging.wustl.edu/pub/raichlab/restless_brain). The patterns of spatial coherence shown on the bottom are obtained by placing a seed region in a single focus within a system (in this case in the sensorimotor cortex) and extracting the resulting BOLD time series (b). This time series is then used as a regressor to search the brain for correlated time series. The result are brain-network specific images of spatial coherence in the ongoing activity of the brain (c). This strategy has been applied with ever increasing sophistication to systems throughout the human brain. A more complete description of the data processing steps leading to such images is presented elsewhere along with alternate strategies (Zhang and Raichle 2010). Shown in (d) are 7 major brain networks analyzed in this way.

coherent throughout the system (Fig. 8.9). When the ongoing spontaneous signals from two regions with the default mode network (Fig. 8.9, top) are monitored with fMRI, they are strongly correlated in time (Fig. 8.9, middle). The bottom of Fig. 8.9 represents a map based simply on the spontaneous

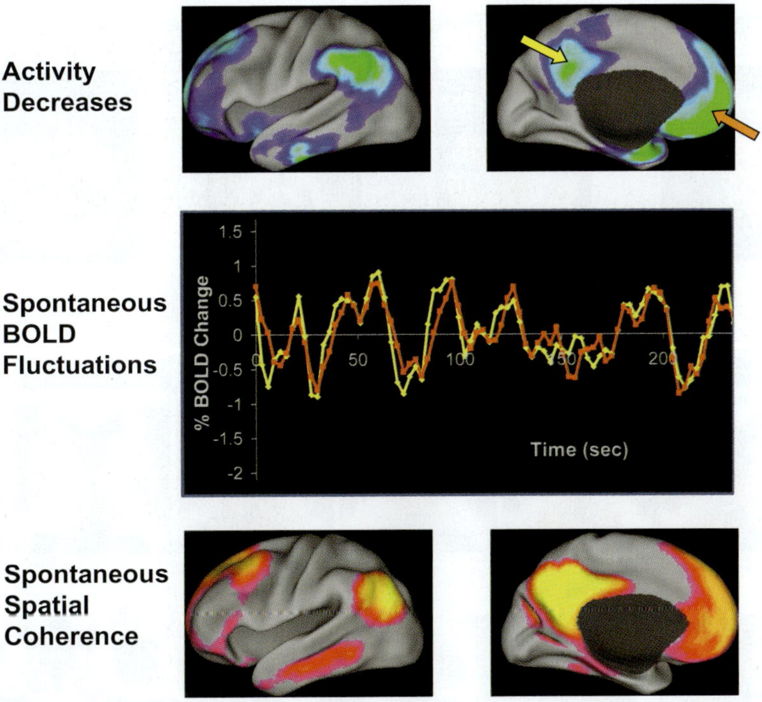

Fig. 8.9 From the anatomy of the default mode network revealed by activity decreases during the performance of a goal directed task (*top*; see also Fig. 8.7) one can select regions in which to monitor ongoing activity within this network at rest (*middle*) which reveals remarkable patterns of coherent activity within the entire network (*bottom*).

coherence oscillations in that system. Our current research is aimed at mapping most of the systems that are characterized by this coherent spontaneous activity in a resting state.

It is worth pointing out that there are, obviously, connections to the thalamus. You could partition the cortex in a way that relates to the boundaries of its large-scale organization, and inquire whether or not these patterns of coherence have different distributions within the thalamus. Furthermore, it is possible to look at the dominant inputs of the differing performance of the thalamus [these data are published in Zhang *et al.* (2008); see Fig. 8.10].

Now, one of the things that was of interest to us was the extent to which this is a matter of ongoing cognition or whether it involves something deeper.

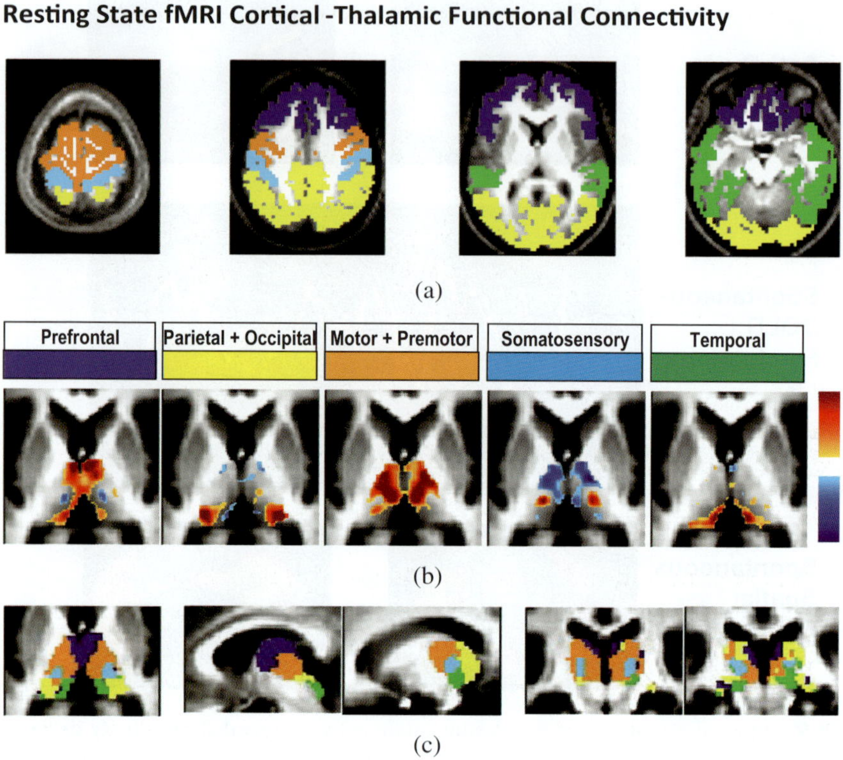

Fig. 8.10 Patterns of ongoing spontaneous activity (Figs. 8.8 and 8.9) can also be used to delineate the relationship between major anatomical subdivisions of the cerebral cortex (**a**) and the thalamus as shown in horizontal (**b**) and (**c**) sagittal and coronal views. (Adapted from Zhang *et al.* 2008).

In Vincent *et al.* 2007, we were looking at spontaneous activity in the ocular motor system (a collection of brain areas concerned with eye movements) of a group of monkeys. Figure 8.11 shows the patterns of coherence that define the classical ocular motor system in the monkey as they are manifest in ongoing spontaneous activity (in this case under general anesthesia; Fig. 8.11, left), during awake activity (i.e., when the monkey looks from side-to-side at objects; Fig. 8.11, middle), and in the underlying anatomical connections (Fig. 8.11, right). Thus, while intrinsic activity is likely very important as a context for conscious activities, its presence does not require conscious awareness.

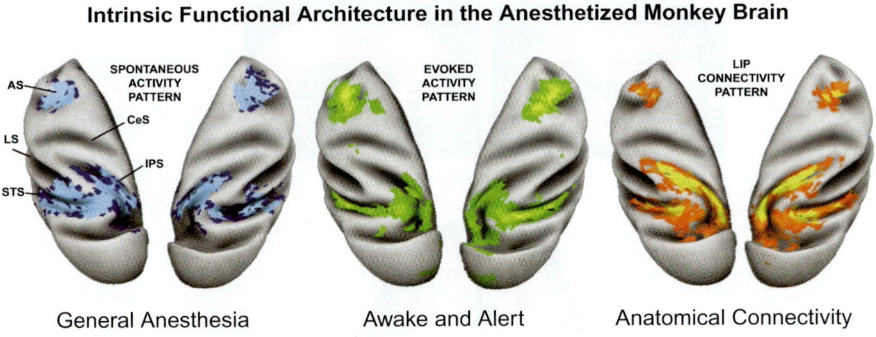

Fig. 8.11 An illustration of the preservation of patterns of spontaneous activity, in this case the monkey oculomotor system, under general anesthesia (*left*). The pattern of this spontaneous activity is remarkably similar to that evoked by monkeys performing an oculomotor task, and (*right*) the actual anatomical connectivity within the oculomotor system of the monkey. (Adapted from Vincent *et al.* 2007).

A crucial question concerns the functional importance of this spontaneous activity. Again, I come back to something that Antonio Damasio mentioned in referring to the work of Olaf Sporns (Hagmann *et al.* 2008). Figure 8.12 is particularly interesting, as it is not only based on imaging evidence for the connectivity within the brain but also combines the evidence for spontaneous activity arising from all of these areas. What can be concluded is a hierarchical organization of hubs that define not only the default networks (in the center of Fig. 8.12), but connections from these to other networks as well. This takes us beyond the notion of modularized systems within the brain. Now, leaving open the question of how these systems communicate, such communication must be based on temporal relationships in some very complex way.

Do monkeys have something corresponding to the default mode network? Indeed, as Vincent *et al.*'s (2007) *Nature* paper shows, the monkey has something analogous to this system, but it does not extend far forward in the monkey brain as it does in the human brain, where it also involves the prefrontal cortex (Fig. 8.13).

This hierarchical organization also raises the question of its implications in the case of normal human development. This is now an active topic of scientific investigation in many laboratories. In St Louis, researchers have been looking at children from the ages of six upwards, and it turns out

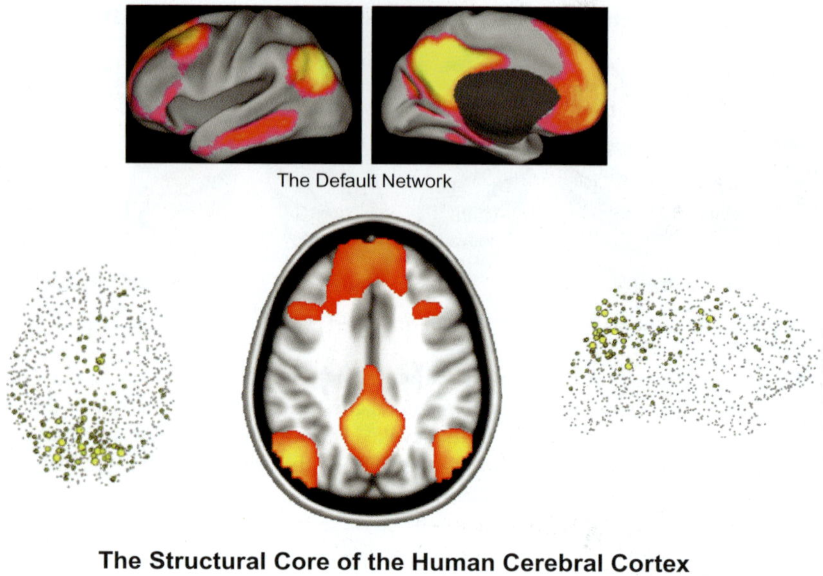

Fig. 8.12 From the work of Sporns and colleagues (Hagmann *et al.* 2008; Bullmore and Sporns 2012; Harriger *et al.* 2012), it has been increasingly clear that the brain's default mode network represents a critical structure in the overall organization of the brain as illustrated in this figure. (Taken from Hagmann *et al.* 2008).

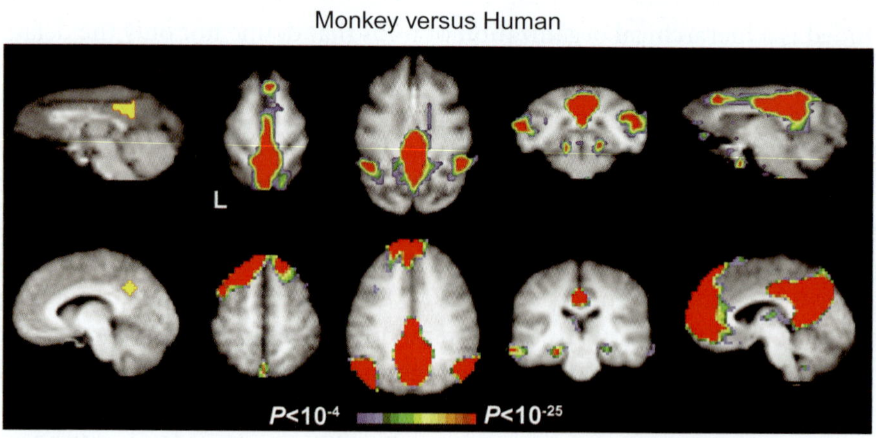

Fig. 8.13 A default mode network appears to be an evolutionarily conserved feature of mammalian brains here seen in the monkey (*top*) and human (*bottom*) brain (Vincent *et al.* 2007). It is also present in rodents (not shown; see Lu *et al.* 2012). (Adapted from Vincent *et al.* 2007).

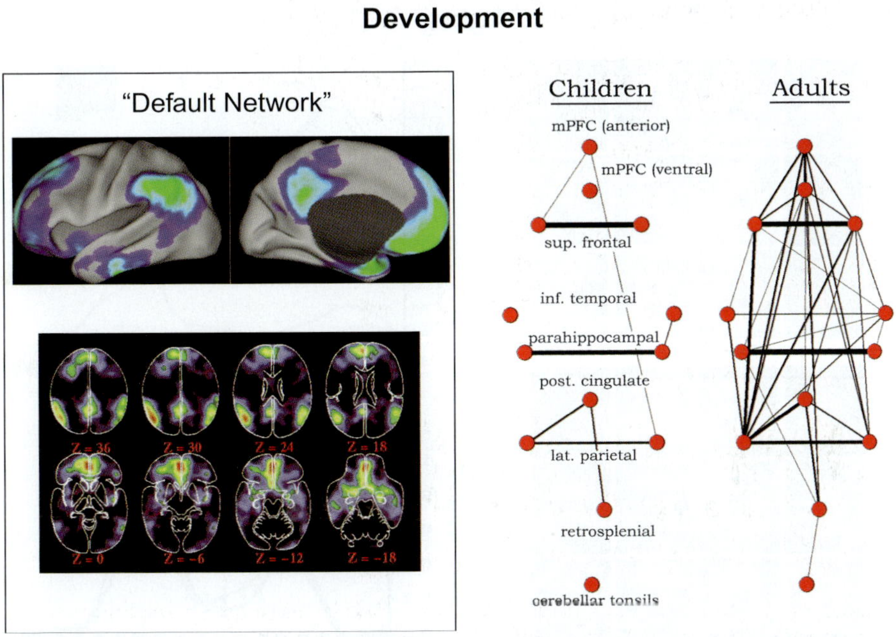

Fig. 8.14 Using measures of connection strength between nodes of the default mode network (*left*) based on patterns of spontaneous activity, it is possible to track the development of the default mode network in humans from children to adults. In the two diagrams on the right, the connection strength is denoted by the width of the lines connecting particular nodes or elements of the default mode network. (Adapted from Fair *et al.* 2008).

again that young children display the same system, but the completeness of its organization varies with age. Figure 8.14 (Fair *et al.* 2008) shows the connectivity of the default mode network. It is remarkable that while in the child the left-to-right strength of coherence is much stronger than the front-to-back one, it is the strength of coherence in the front-to-back direction that increases as we become adults.

Unfortunately, as you age the process is reversed (Andrews-Hanna *et al.* 2007). In Fig. 8.15, the front-to-back coherence is expressed in terms of the (z-transformed) correlation coefficient between the front and the back. It is also worth mentioning that these aging differences appear to affect specific systems, particularly the default mode network, while sparing other systems such as the visual one.

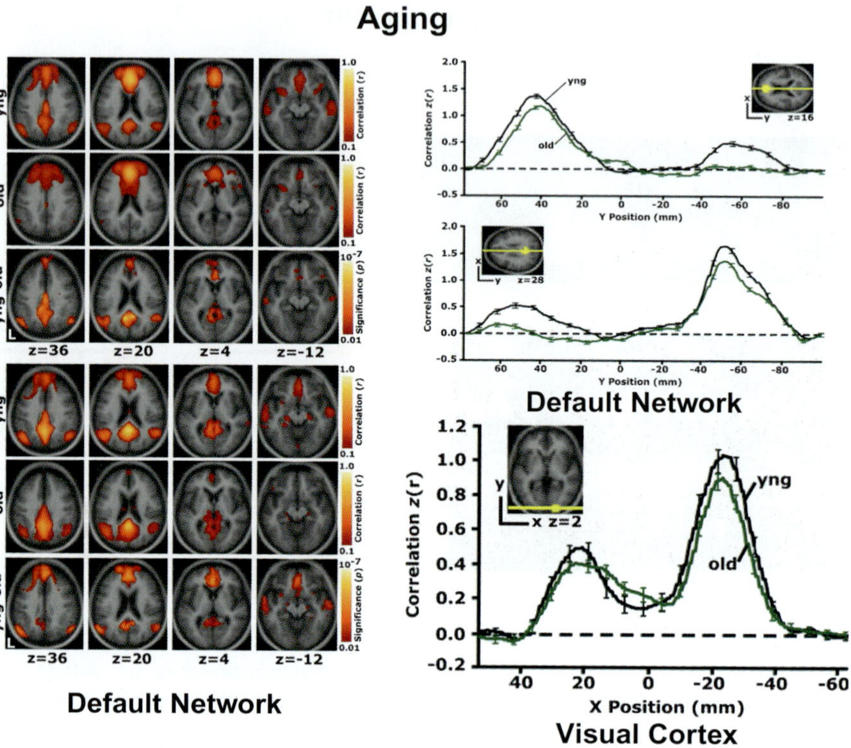

Fig. 8.15 Changes in default mode network connectivity not only increase their strength during development (Fig. 8.14) but appear to reverse that process in aging as shown in the work of Buckner and colleagues (Andrews-Hanna *et al.* 2007). This is not a characteristic of all brain systems. As shown in this figure on the lower right, resting state functional connectivity within the visual system is not affected by aging. (Adapted from Andrews-Hanna *et al.* 2007).

One of the very important questions occupying the time and attention of investigators worldwide is the nature of the brain's electrical activity represented by the spontaneous fMRI BOLD signal. As noted above (Figs. 8.4 and 8.5), given that synapses account for energy changes associated with imaging signals, it comes as no surprise that it is the electrical events at these synapses that have commanded the attention of most researchers. What has emerged from this research is that very slow fluctuations in electrical activity (as depicted in Fig. 8.16 from our own work) account in the best way for the spontaneous BOLD signal. It has been generally accepted that these slowly

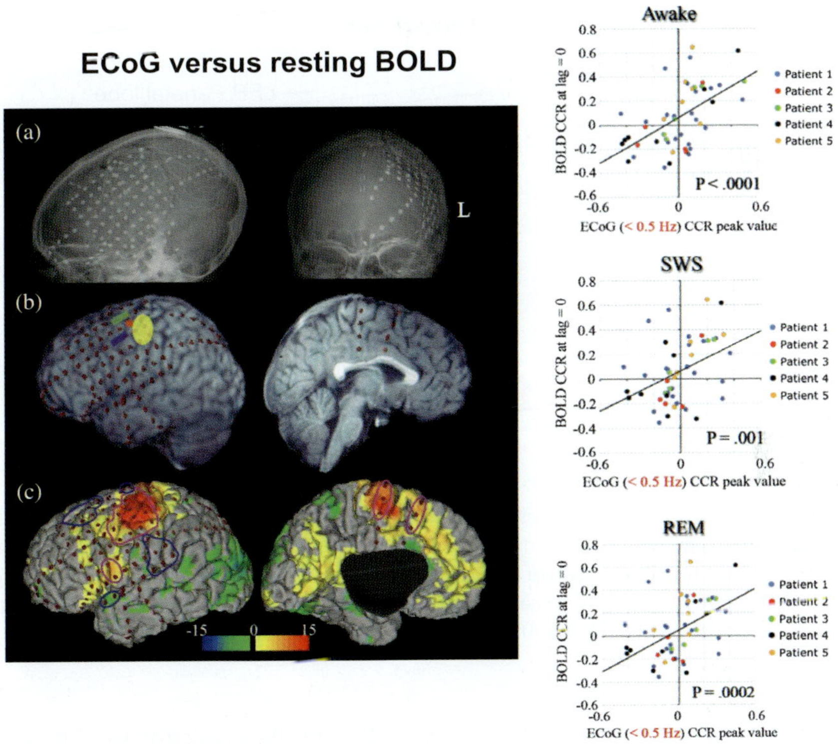

Fig. 8.16 As research on the brain's spontaneous activity has proceeded apace, it has become increasingly important to understand the underlying neurophysiology of the slow spontaneous fluctuations of the fMRI BOLD signal (see Fig. 8.8). The opportunity to record directly from large electrode grids cortex of human subjects undergoing evaluation for epilepsy surgery (*left*, a and b) and compare the data obtained with fMRI measurements of the spontaneous fluctuations of the BOLD signal in the same subjects (*left*, c) has allowed us ascribe the BOLD fluctuations to fluctuations in slow cortical potentials (i.e., field potentials) at frequencies less than 0.5 Hz (*right* graphs). This relationship appears to hold whether subjects were awake or asleep. (Adapted from He *et al.* 2008).

varying electrical potentials represent changes in the excitability of neurons, as shown in Fig. 8.17 from the work of Hahn *et al.* (2006). In this work they were able to show that fluctuations in the membrane voltage of neurons in the hippocampus (i.e., fluctuations in their excitability) were coherent with changes in synaptic electrical activity some distance away in the parietal cortex of rodents, a phenomenon remarkably similar to the patterns of coherence we observe in the fMRI BOLD signal in the resting state.

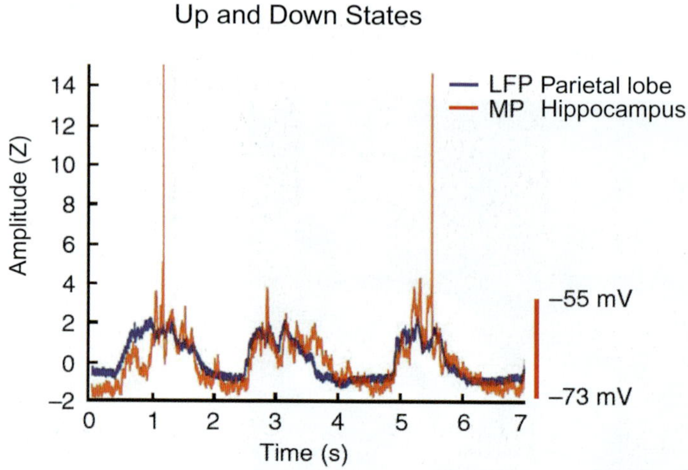

Fig. 8.17 One of the hypotheses concerning the nature of slow cortical potentials is that they represent changes in cortical excitability. For a more detailed review of this subject, see Raichle (2011). A nice illustration of how this might work is shown in this figure where changes in membrane potentials (MP), so-called up and down states, in hippocampal neurons of mice are compared with local field potentials (LFP) in the parietal lobe. (Adapted from Hahn *et al.* 2006).

Let us turn for a moment to the metabolic implications of this work. When the brain changes its activity, a parallel increase in the amount of glucose is observed even though it is not accompanied by an increase in oxygen consumption (Fig. 8.18). Most people think of glucose in the neurobiological sense as just the "fuel" for the brain; it is what the brain burns to get energy. Of course, glucose is remarkably important in this regard. In fact, it is the sole provider of the fuel for the brain's energy needs. But glucose also has the capacity to provide many other important things for the brain, as illustrated in Fig. 8.19. Remarkably these additional contributions have much to do with the health and remodeling of neurons and other cells of the brain. Why might this be important? Think, for example, about what happens at the so-called postsynaptic density. Here, you have a huge number of channels and proteins, and several patterns exist in this remarkable membrane system (Figs. 8.19 and 8.20).

Now, as Eve Marder says, "the brain of the human adult must retain the capacity to respond to novel challenges by learning while at the same time maintaining the essential structure of the networks that allow for sensation

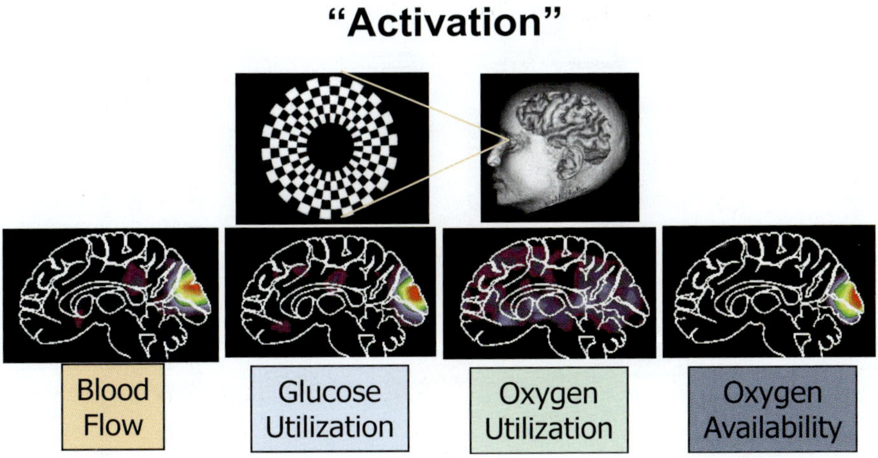

Fig. 8.18 Circulatory and metabolic changes associated with task-induced increase in brain activity. Two things are noteworthy. First, blood flow increases more than oxygen consumption leading to a decrease in deoxyhemoglobin locally in the brain vasculature (i.e., supply exceed demand). Because deoxyhemoglobin is an MRI contrast agent, the MRI signal is enhanced leading to the fMRI blood oxygen level dependent or BOLD signal. Second, glucose consumption does keep pace with the blood flow change, indicating that glucose use outside of oxidative phosphorylation, so-called aerobic glycolysis (Lunt and Vander Heiden 2011), has increased. Glucose used in this way provides energy for membrane pumps and provides substrate for biosynthesis and neuroprotection as schematically illustrated in Fig. 8.19 (for a detailed review, see Raichle and Mintun 2006).

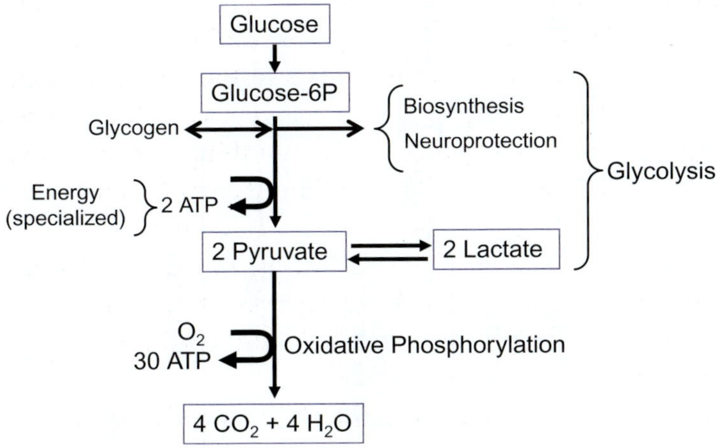

Fig. 8.19 A simplified schematic diagram of the metabolic fate of glucose in brain.

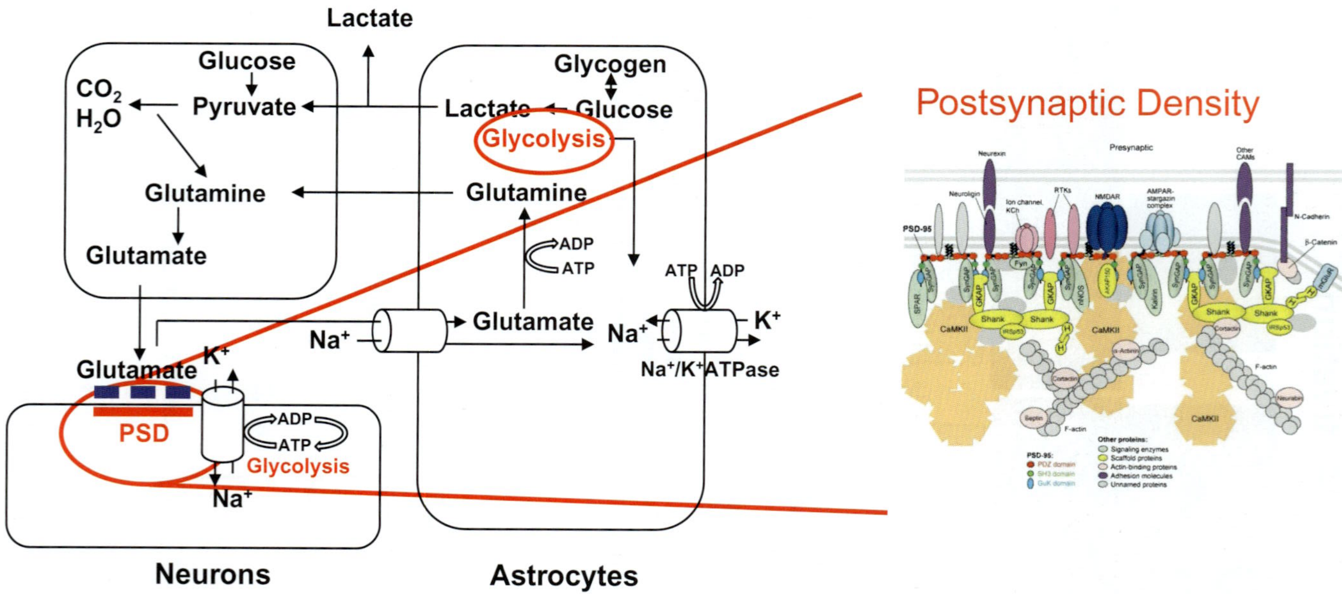

Fig. 8.20 It is important to note that the metabolic demands of the postsynaptic density illustrated on the right (Kennedy 2000; Sheng and Hoogenraad 2007) are largely served locally by glycolysis (Siekevitz 1985) as less than 10% of dendritic spines and filopodia contain mitochondria (Li *et al.* 2004). This coupled with the fact that the elements of the postsynaptic density turn over in minutes, hours, days and weeks (Marder and Goaillard 2006) suggests an important role for the biosynthetic role of glucose at the critical location in the brain. The schematic diagram of the postsynaptic density is from (Sheng and Hoogenraad 2007, Fig. 4).

and action. This challenge is especially daunting if we remember that neurons can live for up to one hundred years but the ion channels and receptors that underlie electrical signaling and synaptic transmission turn over in the membrane in minutes, hours, days or weeks." So, our brain is unbelievably dynamic: what we are learning today, and what we will remember tomorrow, is likely to be stored in a different protein structure than what it was when we took the information in today. And of course this raises some very interesting questions about how such a state of homeostasis of the cell can be effectively maintained. The notion comes to mind that in the course of these activities the brain is not making new cells but, rather, is remodeling the cells that it has.

Now the question is: Does aerobic glycolysis (i.e., that amount of glucose entering the brain not used purely for the generation of energy) vary in interesting ways? This question may be analyzed from many different perspectives. The first possibility is assessing this variation among different brain areas, that is: Is aerobic glycolysis higher in some parts of the brain than in others? Well, the answer is unequivocally so. The prefrontal cortex and the default mode network are at the highest level and surprisingly the medial temporal cortex and the cerebellum are at the opposite end of the spectrum. I consider this to be a clue about how the brain is being used in these metabolic pathways to differentiate the functionality of these systems (as they are clearly in a state of differing functionality). From a complementary perspective it is important to note that the amount of glucose the brain uses outside of energy production *per se* is highly influenced by growth and development (Chugani *et al.* 1987). The brain of a premature infant uses almost no oxygen yet it consumes glucose. Even in newborns, 30 percent of the glucose that reaches the brain is not used for energy generation.

Finally, there is another interesting paper (Madsen *et al.* 1995) looking at whole brain metabolism, and at blood flow, glucose and oxygen consumption and the oxygen/glucose consumption ratio. During task performance blood flow and glucose consumption rose whereas oxygen consumption did not. What was quite interesting occurred following task performance when glucose consumption stayed up for many minutes despite the fact that blood flow had returned to the control state; it was as if the brain was using glucose in some special way to deal with the events just experienced. Could this change be a biomarker of events concerned with learning and memory in the adult human brain? Only further research will provide an answer.

I will conclude by quoting the words of W. James (1890, Vol. 2, p. 103): "Enough has now been said to prove the general law of perception, which is this, that *whilst part of what we perceive comes through our senses from the object before us, another part* (and it may be the larger part) *always comes… out of our own head.*"

DISCUSSION

Wolf Singer: So far, everything suggests that there is a causal chain flow: there is increased demand because of the increase in activity and therefore increased blood flow occurs. However, it seems to me that there is now more and more evidence that it may also be the other way around. That is, increased blood flow determines an increase in the metabolic processes that affect the activity on their own.

Marcus Raichle: These data (Fig. 8.21) from the work of Sirotin and Das (2009) and the concerns of Wolf Singer with regard to these data deserve comment. In this study the investigators recorded directly from the visual cortex in awake monkeys electrical activity (i.e., spikes and field potentials) as well as optical signals that capture the changes in cortical oxygenation underlying the fMRI BOLD signal. In this study the monkeys were given a cue that a visual target was imminent and then a short time later the target appeared and the monkey responded. What they observed when this sequence was followed (denoted "stimulated" in Fig. 8.21) was an abrupt increase in electrical activity (spikes and field potentials) followed by an increase in the optical equivalent of the BOLD signal. In some trials, however, the monkeys received a cue that was *not* followed by the visual target (a standard procedure in such experiments). Obviously, the monkey anticipated a target but on these trials it was absent. Surprisingly, to the investigators at least, a BOLD signal was observed anyway, despite the absence of any change in electrical activity. They concluded that by whatever mechanism the BOLD signal was operating independently of the evoked electrical activity. The flaw in this experiment, which led to this erroneous conclusion, is evident in Fig. 8.21. Note that in recording electrical activity, they ignored the slow

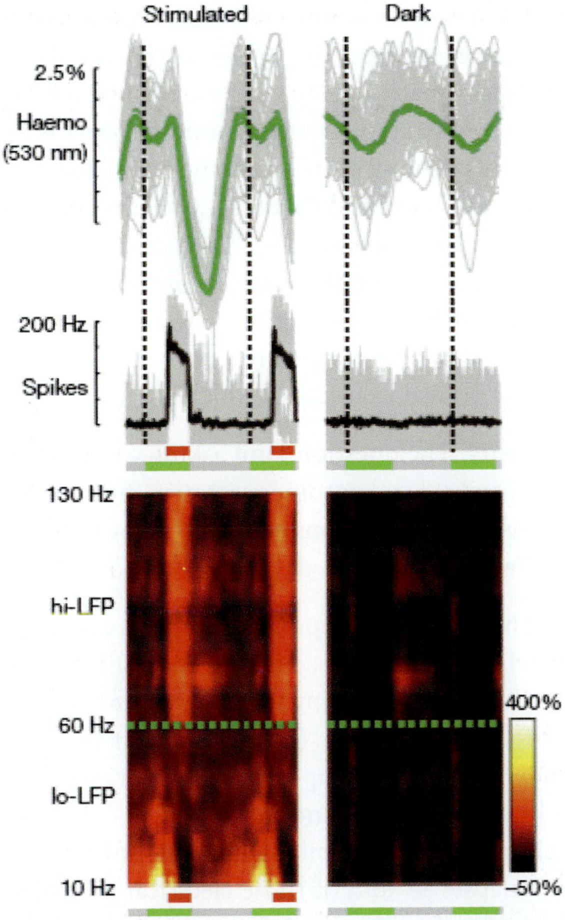

Fig. 8.21 A recent paper (Sirotin and Das 2009) challenges the notion that there is a clear association between the vascular changes associated with the fMRI BOLD signal and underlying electrical changes in brain electrical activity. In their experiment, monkeys were cued to upcoming visual stimulus and in the presence of the stimulus there was an increase in the optical equivalent of the BOLD signal (upper green trace), an increase in spikes (upper black trace) and a broad-band increase in the power of local field potentials. When the cue was not followed by a stimulus there was, nevertheless, an increase in the optical signal at the point in time that the monkey expected the stimulus. Critically, in their minds, there was no change in electrical activity. However, importantly the electrical recordings did not record frequencies below 10 Hz. As shown and discussed in Fig. 8.16, the frequency of local field potentials associated with the BOLD signal fall well below 1 Hz. The only reasonable conclusion to be drawn from this work is that appropriate electrical activity was simply not recorded.

cortical potentials that are the known correlates of the fMRI BOLD signal! (i.e., they recorded no electrical activity below 10 Hz; see also Fig. 8.16 above) This was deeply unfortunate, as it has led to considerable confusion in the scientific community. What is interesting about this work is that it reveals the anticipatory nature of the brain's intrinsic activity, a theme developed in detail over many years by Schroeder and Lakatos (2008) at Columbia University in New York and further developed by me in a recent essay (Raichle 2011). A great example of this, predating the work of Sirotin and Das (2009), comes from the laboratory of David Heeger at New York University (Ress *et al.* 2000). As David Ingvar said so eloquently many years ago, "the brain is in the prediction business," a great example of which is shown in Fig. 8.21.

Jean-Pierre Changeux: The question I am asking you is whether or not you can separate the central control or some vegetative control of the innervations of the blood vessels from the various activities of the brain. I think that this is the very issue that the paper raised.

Marcus Raichle: This paper is not a particularly good example of the relationship of blood vessels to the ongoing activity of the brain, as I have pointed out. Much more complete and compelling work has come from others, beginning with the work of Logothetis *et al.* (2001). In all of these discussions, it is important to understand what components of the brain's electrical spectrum are influencing brain blood flow and metabolism. As we are learning more and more, it is clear that it is not spikes *per se* but rather electrical potentials associated with synapses — and of those electrical potentials it most likely restricted to the lowest frequencies. When approached from this perspective, I am not aware of any data that show discordance in a normal brain of humans or laboratory animals.

Jean-Pierre Changeux: The main reason why I am sceptical about these techniques is that they are not measuring the intrinsic activity of neurons, but the blood flow. This is a relevant point to keep in mind.

Marcus Raichle: I think that the better position to take is that we have techniques that are looking at whole processes, and techniques that are looking at electrical ones. And the electrical processes are not singular; for example, techniques relying on electrical processes look at activities between neurons and astrocytes. Furthermore, the metabolic processes that influence the blood vessels — and hence blood flow — are every bit as much a part of neuronal activity as are spikes and field potentials.

Jean-Pierre Changeux: None of these techniques appropriately distinguish the excitement of the neurons. When you look at the firing of cells, even if you have an absolute measurement of their firing, there is no distinction between excitatory neurons and inhibitory neurons. I think that this is extremely important from the standpoint of processing information.

Marcus Raichle: I agree with that. But it is also the case that the remarkable ongoing activity of inhibitory interneurons is rarely recorded by neurophysiologists, and the influence of these busy cells on brain circulation and metabolism remains to be fully understood, a subject that I explored with Gyuri Buzsaki and Kia Kiala in a *Neuron* review in 2007 (Buszaki *et al.* 2007).

Jean-Pierre Changeux: It would be extremely interesting to have a technique that gives you separate data of the excitation of cells, for instance, by recording the neurotransmitter traffic between cells. But, this is not the case.

Marcus Raichle: The search for the "perfect" technique is, in my opinion, an unnecessary diversion. We do much better by combining the available techniques, models and viewpoints. The problems we face are multidisciplinary and must be approached as such with serious attention to the matter of communication across levels of analysis.

Marc Jeannerod: I have a very short question for information: Do you find lateralization with the default network?

Marcus Raichle: No, it seems very symmetrical; we can split it in two. I did not show you this, but our work in patients with corpus callostomy reveals that they suddenly have two symmetrical default mode networks (Johnston *et al.* 2008).

Marc Jeannerod: I am asking this because, given the function of this default activation, I would expect that at some point you would have more activity in the left brain.

Marcus Raichle: This is an interesting question. We have never formally looked at or done experiments on that. However, I have never noticed anything even remotely asymmetrical.

Jean-Pierre Changeux: What happens to the personality of the patients with corpus callostomy?

Marcus Raichle: Well, you have two individuals. We have published the following: you have two separate defaults networks and they are not coherent from side to side. The behavioral work on patients with section of the corpus callosum began, of course, with the pioneering work of Roger Sperry and his student Michael Gazzaniga many years ago and now represents a large body of work from several research groups. I could not begin to summarize the results of this work here.

Gennaro Auletta: Would it be correct to summarize your point of view by saying that the increase in blood flow prior to the stimulus is evidence of endogenous processes happening in the brain?

Marcus Raichle: Yes. I would posit further that they represent anticipatory endogenous processes of the type recorded by electroencephalographers well before the advent of imaging (e.g., contingent negative variations, bereitschaftspotential). The lovely book *Slow Cortical Potentials* (Rockstroh *et al.* 1989) summarizing this often-overlooked research is worthy of note in the present context.

Jean-Pierre Changeux: Well, we know this thanks to experiments. There exist EEG measurements that show this point.

Gennaro Auletta: I am asking this because I have read some papers of Kandel (in around 2000). He worked with rats, especially focusing on attentional behavior when rats explore mazes in the lab. The experiments show a top-down process of dopamine production in attentional processes. I stress that Kandel himself used the word "top-down." Now, would you agree that this is further evidence of brain endogenous processes?

Jean-Pierre Changeux: Yes, of course.

Gennaro Auletta: I mean endogenous processes in a top-down way.

Marcus Raichle: At least in humans we have such processes.

Wolf Singer: What does "top-down" mean here? I am not sure I understand this expression.

Jean-Pierre Changeux: There is a theory of spontaneous endogenous processes. It especially concerns the capability to anticipate rewards in learning processes that occurs in a top-down way.

Gennaro Auletta: In a general sense, top-down means that one assumes that something is coming from some higher parts of a hierarchical structure.

Jean-Pierre Changeux: Yes, but this depends on whether the monkey is *anticipating* a reward. Well, in this case, for example, you actually have such top-down processes, indeed.

Gennaro Auletta: Yes, I agree.

Marcus Raichle: One thing I did not talk about is that in association with the default mode network, which tends to reduce its activity during externally oriented task execution, humans also have what is referred to as the dorsal attention system (a well-defined bilateral system). These two systems have an "anticorrelated" relationship such that when one goes up the other goes down, a relationship first noted during the performance of externally directed tasks but more recently observed in their spontaneous activity as well (Fox *et al.* 2005).

Jean-Pierre Changeux: I have a question about the evaluation of the metabolism of the brain in the morning and at night. What is the meaning of the difference that you find in these two states?

Marcus Raichle: Giulio Tononi and colleagues (Tononi and Cirelli 2006) suggested that during waking there is a potentiation of existing synapses and possibly an increase in the number of synapses. If this is the case, then because synapses are responsible for the majority of the brain's energy consumption, it follows that energy consumption will rise. We are presently looking at this in more detail with quantitative PET metabolic studies, but I have nothing to report presently.

Jean-Pierre Changeux: So this also means that the consumption of the brain is higher during sleep?

Marcus Raichle: What we know for certain is that blood flow to the brain prior to sleep is about 20 percent higher than on awakening (Braun *et al.* 1997) and that during sleep there are fluctuations in blood flow and metabolism most markedly during slow wave sleep, where both blood flow and metabolism fall dramatically.

Jean-Pierre Changeux: So this means that the view that being conscious costs more energy than not being conscious (e.g., during sleep) is not true.

Marcus Raichle: Maybe, we could say that sleep is not the only non-conscious state in the normal cycle of the day; the difference when the brain sleeps and in conscious awake behavior, or alert, is much more modest than the 20 percent I have mentioned.

Michael Arbib: I would like to return to Marc Jeannerod's question about the symmetry of the default network as opposed to the lateralization of many of our mental and sub-mental processes. This may perhaps suggest that your findings, as exciting as they are, do not bear on your original question in terms of our mental life as a response to external events versus our mental life as a function of the imagination. Indeed, we would imagine, if you will forgive the term, that our imagination would exhibit some of the same patterns of lateralization as our active engagement with the environment. Now, the fact that you do not see the lateralization, and the fact that the default network does not get desynchronized as one ponders different things, would suggest that, although it must play some incredibly important basilar role for our brain function, it is in fact not evidence for intrinsic activity in terms of our thought and imagination.

Marcus Raichle: Let me just answer that. Number one, I did not present all the data that I could have, but I could not let you think that the default network was the only one that exhibits this coherent spontaneous activity. Virtually every system that anybody has looked at in the cortex exhibits this kind of activity. Secondly, it is also the case that the default network can be dissociated in its component parts. For example, if you are engaged in a task involving episodic recall or remembering, it is very clear that particularly the back end of the temporal components are at play and have a very clear role: they show increases rather than decreases. The ventral part of the default mode network in the prefrontal cortex can be clearly dissociated from the activity of other elements of the default mode network when emotions are manipulated.

Antonio Damasio: First, I want to remark that we actually showed differential activity within the posterior medial cortices with different kinds of contents. There is clearly an anterior-posterior relationship, and curiously it exhibits the same kind of pattern connectivity that you find in the monkey relative to the visceral interceptive component versus the muscular skeletal component. This is one remark. Moreover, as Marcus Raichle has said, where certain kinds of emotional/feeling tasks are concerned, increases in activity within the PMC are actually exhibited. So there is actually quite a lot of differential activity. Finally, in relation to Michael Arbib's question, I think that we are barely beginning to look at all these differences inside; probably, we will find some kind of asymmetry there as well. This is just a note, anyway, because this is currently quite a mass of research — and often deflections.

REFERENCES

Anderson CH, Van Essen DC, Olshausen BA. (2005) Directed visual attention and the dynamic control of information flow. In: L Itti, G Rees, J Tsotsos (eds.), *Neurobiology of Attention*, Elsevier, pp. 11–17.

Andrews-Hanna JR, Snyder AZ, Vincent JL, *et al.* (2007) Disruption of large-scale brain systems in advanced aging. *Neuron* **56**: 924–935.

Braun AR, Balkin TJ, Wesenten NJ, *et al.* (1997) Regional cerebral blood flow throughout the sleep-wake cycle. An H2 (15)O PET study. *Brain* **120**: 1173–1197.

Bullmore E, Sporns O. (2012) The economy of brain network organization. *Nat Rev Neurosci* 13(**5**): 336–349.

Buzsaki G, Kaila K, Raichle M. (2007) Inhibition and brain work. *Neuron* **56**: 771–783.

Chugani HT, Phelps ME, Mazziotta JC. (1987) Positron emission tomography study of human brain functional development. *Ann Neurol* **22**: 487–497.

Fair DA, Cohen AL, Dosenbach NUF, *et al.* (2008) The maturing architecture of the brain's default network. *Proc Natl Acad Sci USA* **105**: 4028–4032.

Fox MD, Snyder AZ, Vincent JL, *et al.* (2005) The human brain is intrinsically organized into dynamic, anticorrelated functional networks. *Proc Natl Acad Sci USA* **102**: 9673–9678.

Gazzaniga MS, LeDoux JE. (1978) *The Integrated Mind.* Plenum Press, New York.

Gazzaniga MS. (1998) The split brain revised. *Sci Am* **279**: 50–55.

Gazzaniga MS, Ivry RB, Mangun GR. (2002) *Cognitive Neurosciences: The Biology of Mind.* W. W. Norton & Company, Inc, New York and London.

Gazzaniga MS. (2011) *Who's in Charge? Free Will and the Science of the Brain.* Harper Collins.

Hagmann P, Cammoun L, Gigandet X, *et al.* (2008) Mapping the structural core of human cerebral cortex. *PLOS Biol* **6**: 1479–1493.

Hahn TTG, Sakmann B, Mehta MR. (2006) Phase-locking of hippocampal interneurons' membrane potentials to neocortical up-down states. *Nat Neurosci* **9**: 1359–1361.

Harriger L, van den Heuvel MP, Sporns O. (2012) Rich club organization of macaque cerebral cortex and its role in network communication. *PLoS One* 7, e46497.

He BJ, Snyder AZ, Zempel JM, *et al.* (2008) Electrophysiological correlates of the brain's intrinsic large-scale functional architecture. *Proc Natl Acad Sci USA* **105**: 16039–16044.

Hubel D. (1988) *Eye, Brain and Vision.* Scientific American Library, New York.

Ingvar HD. (1985) Memories of the future: An essay on the temporal organization of conscious awareness. *Hum Neurobiol* **4**: 127–136.

James W. (1890) *The Principles of Psychology.* H. Holt. (Dover, New York, 1950.)

Johnston JM, Vaishnavi SN, Smith MD, *et al.* (2008) Loss of resting interhemispheric functional connectivity after complete section of the corpus callosum. *J Neurosci* **28**: 6453–6458.

Kadecaro M, Craine AM, Sokoloff L. (1985) Differential effects of electrical stimulation of sciatic nerve on metabolic activity in spinal cord and dorsal root ganglion in the rat. *Proc Natl Acad Sci USA* **82**: 6010–6013.

Kennedy MB. (2000) Signal-processing machines at the postsynaptic density. *Science* **290**: 750–754.

Kety S, Schmidt CF. (1948) The nitrous oxide method for the quantitative determination of cerebral blood flow in man: theory, procedure and normal values. *J Clin Invest* **27**: 107–119.

Li Z, Okamoto K, Hayashi Y, Sheng M. (2004) The importance of dendritic mitochondria in the morphogenesis and plasticity of spines and synapses. *Cell* **119**: 873–887.

Logothetis NK, Pauls J, Augath M, *et al.* (2001) Neurophysiological investigation of the basis of the fMRI signal. *Nature* **412**: 150–157.

Lu H, Zou Q, Gu H, *et al.* (2012) Rat brains also have a default mode network. *Proc Natl Acad Sci USA* **109**: 3979–3984.

Lunt SY, Vander Heiden MG. (2011) Aerobic glycolysis: meeting the metabolic requirements of cell proliferation. *Annu Rev Cell Dev Biol* **27**: 441–464.

Madsen PL, Hasselbach SG, Hagemann LP, *et al.* (1995) Persistent resetting of the cerebral oxygen/glucose uptake ratio by brain activation: Evidence obtained with the Kety–Schmidt technique. *J Cereb Blood Flow Metab* **15**: 485–491.

Marder E, Goaillard JM. (2006) Variability, compensation and homeostasis in neuron and network function. *Nat Rev Neurosci* **7**: 563–574.

Norretranders T. (1998) *The User Illusion.* Viking.

Petersen SE, Fox PT, Posner MI, Mintun M, Raichle ME. (1988) Positron emission tomographic studies of the cortical anatomy of single-word processing. *Nature* **331**: 585–589.

Raichle ME. (2010) Two views of brain function. *Trends Cogn Sci* **14**: 180–190.

Raichle M. (2011) The Restless Brain. *Brain Connectivity* **1**: 3–12.

Raichle ME, MacLeod AM, Snyder AZ, *et al.* (2001) A default mode of brain function. *Proc Natl Acad Sci USA* **98**: 676–682.

Raichle ME, Mintun MA. (2006) Brain work and brain imaging. *Annu Rev Neurosci* **29**: 449–476.

Raichle M, Snyder AZ. (2007) A default mode of brain function: a brief history of an evolving idea. *NeuroImage* **37**: 1083–1090.

Ress D, Backus BT, Heeger DJ. (2000) Activity in primary visual cortex predicts performance in a visual detection task. *Nat Neurosci* **3**: 940–945.

Rockstroh B, Elbert T, Canavan AG, *et al.* (1989) *Slow Cortical Potentials.* Urban & Schwarzenberg, Baltimore, Munich and Vienna.

Schroeder CE, Lakatos P. (2008) Low-frequency neuronal oscillations as instruments of sensory selection. *Trends Neurosci* **32**: 9–18.

Sirotin YB, Das A. (2009) Anticipatory haemodynamic signals in sensory cortex not predicted by local neuronal activity. *Nature* **457**: 475–479.

Sheng M, Hoogenraad CC. (2007) The postsynaptic architecture of excitatory synapses: a more quantitative view. *Annu Rev Biochem* **76**: 1.1–1.25.

Siekevitz P. (1985) The postsynaptic density: a possible role in long-lasting effects in the central nervous system. *Proc Natl Acad Sci USA* **82**: 3494–3498.

Sokoloff L, Mangold R, Wechsler RL, *et al.* (1955) The effect of mental arithmetic on cerebral circulation and metabolism. *J Clin Invest* **34**: 1101–1108.

Tononi G, Cirelli C. (2006) Sleep function and synaptic homeostasis. *Sleep Med Rev* **10**: 49–62.

Vincent JL, Patel GH, Fox MD, *et al.* (2007) Intrinsic functional architecture in the anaesthetized monkey brain. *Nature* **447**: 83–86.

Zhang D, Raichle M. (2010) Disease and the brain's dark energy. *Nature Rev Neurol* **6**: 15–28.

Zhang D, Snyder AZ, Fox MD, *et al.* (2008) Intrinsic functional relations between human cerebral cortex and thalamus. *J Neurophysiol* **100**: 1740–1748.

CHAPTER 9

Top-Down Causation in the Brain: Promises for Cognitive Psychology and Challenges for Research

Charles R. Crowell, Joseph R. Moskal, Jeffrey Burgdorf and Jaak Panksepp[1]

9.1. INTRODUCTION

The conference upon which this volume is based was intended to explore the possibility that top-down causation (TDC) might offer a new and interesting way of understanding how neural and molecular brain mechanisms can support the occurrence of certain seemingly endogenous behavioral or psychological phenomena such as goal seeking and voluntary action. The present contribution examines several considerations associated with both the promise as well as the challenge of understanding TDC. In accomplishing this goal, we will first provide a brief review of the key aspects and importance of TDC as the editors of this volume set forth these matters. Second, the paper will suggest a few ways in which a better understanding of TDC might be relevant to a field such as cognitive psychology. Finally, we will outline several potential challenges that exist for achieving a more complete understanding of whether or not and how TDC might actually work in the brain. Due to space limitations, each of these topics will be treated at a summary rather than a detailed level.

[1]This paper was presented during the workshop by C. Crowell alone.

9.2. TDC: NOTABLE ASPECTS AND IMPORTANCE

The characterization of TDC we will use in this paper is a brain-centric version of the original notion outlined in a document prepared for this conference and volume prepared by Jeannerod, Auletta and Colagè, described herein as the "Lyon document" (see Appendix II). TDC has also been discussed in detail in important papers by Ellis (2008) and Auletta *et al.* (2008). Following these sources, TDC in the brain can be conceived of as a process in which events at a higher level of brain organization (e.g., neocortex) have causal power over events at a lower level (e.g., motor output and emotional responding). The notion of levels of brain organization here roughly parallels the idea of "levels of reality" provided by Ellis (2008). As Ellis points out, examples of TDC are very common throughout the natural world, so it is quite possible that this process would operate in the brain as well.

We share the following reflections from a perspective that acknowledges the potential importance of both top-down and bottom-up causation in mind-brain organization (Panksepp 1998). This perspective reflects the fact that: (a) the brain is an evolutionarily "layered" organ where one can still find the imprints of earlier brain-mind functions at the caudal and medial levels, with more recent brain developments generally being added more laterally and rostrally; (b) these levels of organization are interpenetrant with one another, reflected in abundant interacting neural pathways, at all levels of brain organization; and (c) early in development bottom-up influences prevail, with the lower brain areas allowing organization of the higher brain areas, while at maturity, especially with social-emotional education, top-down influences begin to prevail. It is didactically useful to divide brain-mind layering into a tripartite scheme where the evolved mental functions, such as raw emotions and motivations (from anger to hunger, so to speak), can be considered to be *primary-processes* of the brain-mind, the *secondary-processes* reflecting the basic learning mechanisms that are enabled by the primary emotions and motivations (both of which are easily studied using animal models), and *tertiary-processes* reflecting the ability of organisms to think and plan with the aid of higher brain regions, which can only be well studied in humans (Panksepp 1998). This view of brain organization is reminiscent of, though not isomorphic with, the one proposed by MacLean (1990).

Bottom-up control in the brain is highlighted well by the way ancient passions such as primary-process emotions and motivations — from anger and lust to hunger and thirst — allow animals and young humans to learn about the world. In contrast, top-down control is well exemplified by the Aristotelian concept of *phronesis* — for instance, knowing how to master anger as opposed to being the slave of anger. Adults with phronetic emotional regulation can chose with whom to be angry, in what manner, for how long, and with what intensity. This same emotional wisdom is captured in the biblical mandate "Irascimini, et nolite peccare" or "Be angry and sin not" (Old Testament, Psalm 4:5 in the latin vulgate; New Testament, Ephesians 4:26).

A great deal of confusion in discussions of top-down causation, such as the debate over "free-will"(Klemm 2010), is engendered by the failure of many investigators to recognize mind-brain levels of control: lower levels of brain function are automatic, as are many learned procedural habits. At the same time, a well-organized brain surely has enough control over the behavior it supports, especially in the generation of long-term planning, to deserve to be seen as a mind-generating organ that can elaborate self-generated wilful actions that deserve the term "free will." Humans undoubtedly use deliberative thought to plot courses of future actions. Here our discussion is strictly focussed on the kind of TDC that must be considered in discussing mature, well-controlled mental activities.

9.2.1. Notable Aspects of TDC

The notion of TDC as outlined in the aforementioned reference documents has at least three notable aspects. One is that TDC rests on an important distinction between *causal power* and *causal effectiveness* (Auletta *et al.* 2008; Ellis 2008). As one way to explain this distinction, Ellis suggests that the physical force of wind (or fire) in a forest can be modulated by the trees or other geographical features of the land. In this example, the forest is said to have "causal power" over the wind or fire but not "causal effectiveness" per se. That is, the shape or structure of the forest can alter, canalize or change the intensity or destructiveness of wind or fire, but the forest, in and of itself, would not be a sufficient cause on its own to produce any damage or destruction. Viewed in this way, it is only the wind or fire that has causal effectiveness (e.g., strength, intensity or destructiveness), but this effectiveness "collaborates with" (i.e., can be modulated by) the causal power of the forest.

TDC, then, is fundamentally a means whereby one or more events or processes exert an influence (i.e., causal power) through constraining or amplifying the causal effectiveness of other events or processes. When the causal power and causal effectiveness emanate from different levels of complexity, then TDC may occur. For example, Ellis (2008) suggests that a light switch is a common example of a top-down constraint on amplification for electric current. That is, the physical nature of the switch is such that it has an on-state and an off-state, which correspond respectively to the mechanical completion or interruption of an electrically conductive pathway through the switch itself. In essence, the switch acts as a constraint on the flow of electrons in the underlying electric circuit controlling the light bulb. In the on-state, the switch allows the flow of electrons, whereas in the off-state this flow is prevented. In other words, the switch acts to channel the causal effectiveness of the electric current in activating the bulb. Moreover, since light switches and electric currents exist at two different levels of physical complexity, TDC is involved in the constraining influence of the former on the latter. Light switches exist at a molar phenomenological or experiential level of complexity and consist of mechanical parts, wires, etc. that are assembled in certain ways. In contrast, electric currents are made up of electron flows operating at the much lower and far less experiential atomic level. Thus, in this instance, events with causal power at the higher molar level (switch state) influence events with causal effectiveness at a lower atomic level (electron flows). Hence, TDC is operative. From a brain perspective, TDC might be illustrated by higher levels of cortical organization exerting causal power (i.e., modulating or canalizing influence) over the causal effectiveness found in subcortical regions of brain organization.

The light switch example may also be used to illustrate a second key aspect of TDC: namely, the idea of *equivalence classes.* As noted in this example, the state of a light switch constrains the electric current, either allowing or disallowing electron flow. However, at the molecular level of the current flow itself, there are many possible configurations of electrons that could flow in the circuit at any particular moment, all of which would have the same effect with respect to activation of the light bulb. Therefore, it is overall rate of electron flow through the switch, rather than which specific electrons are flowing at that moment, that determines the current. All of the combinations of specific electrons that could flow at any given rate thus would be considered equivalent from the standpoint of the end result (activation of the bulb). These various combinations of specific electrons would be considered an equivalence class with respect to their causal effectiveness in producing light, given the constraint of the switch being in the on-state (Ellis 2008). In a similar way, there are many distributions of individual gas molecules within a container, given a particular temperature constraint, that all would yield the same pressure result. So, with respect to pressure, all of these molecular distributions would be considered an equivalence class. In the brain, too, functional equivalence classes may exist within each level of neural organization (primary, secondary, tertiary) with respect to certain system outputs or goals. For instance, an emotional-affective system has extensive neurochemical substrates, with multiple neurochemical controls, but the experiential end result may be similar at different times even when exactly the same molecular configurations are not present in the underlying neural substrates.

A third notable aspect of TDC is that the downward causal effect of the higher level of complexity on the lower level cannot be accounted for solely in terms of processes operating at the lower level. As Ellis (2008) notes of the light switch example, the "switch is a higher-level entity: it cannot be described in lower level terms (albeit it is constructed from lower level parts)" (p. 6). This third aspect of TDC implies that top-down causal effectiveness cannot be understood purely in reductionistic terms as an epiphenomenon. This criterion of TDC would be met for brain function by the finding that, say, a cognitive or affective process emanating from a higher level of organization could exert causal influence over a lower-level process, but could not be explained purely in terms of lower-level operations. There are many everyday examples, especially when one is just learning to do something new: for instance, when a pianist sits down to practice a new piece. Higher functions have to participate in the sequential organization of how to place the fingers on the keyboard and how to strike the keys [see Klemm (2010)]. Also, with affect, primary-process emotional states can lead to many alterations in cellular, biochemical, and physiological functions within the body of an organism (Panksepp 1998), but those alterations do not constitute the experienced emotional state, which remains a dynamic neural activity of the brain.

9.2.2. Potential Importance of TDC

TDC then is the abstract notion that higher-level outcomes (e.g., lights on or off) can result from members of a class of lower-level events (e.g., electron flows) that are constrained or influenced by circumstances operating at the higher level (e.g., switch states). If this type of process also works in the brain, it may have important implications for understanding many types of behavior exhibited by organisms. For example, TDC may represent a kind of framework for understanding so-called "endogenous processes" in which goal-oriented actions seem to occur spontaneously without obvious external provocation, as if they are being influenced by some internal representation of the desired goal that has yet to be achieved. Since many types of mental activities may qualify as endogenous, TDC may be quite relevant to our understanding of cognitive psychology. Moreover, as the Lyon document (see Appendix II) suggests, mental activity is often associated with the intent to attain certain goals that can be achieved in many different ways (i.e., by one or more members of an equivalence class of different goal attainment means). In human nature, many types of goals appear to be elected, or chosen, by those who seek them. A sudden decision to pick up a writing instrument and make a note would be a case in point. In this instance, someone makes a seemingly voluntary decision to accomplish a desired goal (a written note) using an appropriate means (pen, pencil, etc). Such wilful action fits the description of an endogenous process because (a) there is no apparent external cause, (b) there seems instead to be an internal cognitive process provoking the action, and (c) the focus is on achieving an end, not on executing a specific means. The Lyon document posited that such volitional actions may in fact be by-products of TDC mechanisms in the brain. If so, then a better understanding of TDC will be required to know how higher-level cognitive or mental processes influence, constrain or canalize lower-level neuromolecular events in the nervous system to allow for specific goal-directed voluntary actions.

9.3. TDC: ITS POTENTIAL PROMISE FOR COGNITIVE PSYCHOLOGY

As noted above, a better understanding of TDC mechanisms in the brain might have great relevance for the field of cognitive psychology. It is very common throughout the literature in this field to find notions of "top-down control" being invoked [e.g., Norman and Shallice (1986)]. If you pick up any text or any article on cognitive psychology these days, very likely you will find some reference to top-down control of cognitive processes. In this section, we will touch on four examples of putative top-down control in cognitive psychology. These examples relate to the ideas of executive functions, expectancy or "mental set," selective attention, and volitional behavior.

9.3.1. Executive Control

Executive functioning is a common form of top-down control (Norman and Shallice 1986). The idea here is that a person's cognitive architecture is made up of "mental schemas" that are essentially cognitive representations of our personal understanding of different aspects of the world. Figure 9.1 shows how these schemas are thought to intervene between our sensory-based perceptions and motor-based actions as a way of connecting stimulus inputs with behavioral outputs. Schemas may have to do with many things like world-view, social roles, stereotypes or even aspects of the self. There are numerous different types of schemas that exist, many of which can be invoked from external stimulation in an exogenous way, while others can be invoked endogenously from internal stimulus events.

The executive system at the top of Fig. 9.1 is believed to be a higher-level mental process that monitors a person's cognitive architecture to activate or inhibit schemas as necessary, as shown by the left downward arrow. The right upward arrow indicates that the executive system can be informed by feedback from schemas and/or from the external world. Thus, the executive controller acts as an internal monitor that is believed to free the cognitive system from an excessive dependence on external cues from the outside world, making it more efficient, flexible, and able to respond even when there is

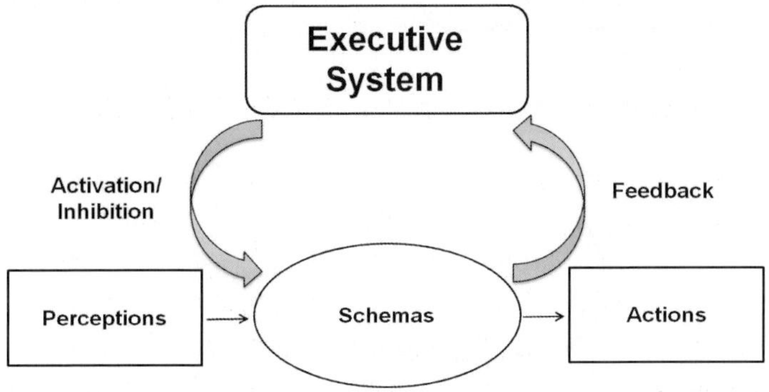

Fig. 9.1 The executive control system as envisioned by Norman and Shallice (1986).

incomplete information from the outside world (Norman and Shallice 1986). Executive control is thus one way to describe top-down influences on human cognition. A better understanding of TDC in the brain may enable a clearer picture of how executive control actually works at a neural level.

9.3.2. Expectancy or Mental Set

Expectancy or mental set is another way in which top-down control is discussed in cognitive psychology. Expectancies or sets are mental states that may be derived from schemas or may arise from the past experiences. Expectancies can exert their effects in many situations, but are especially influential in situations where there is some existing ambiguity. For example, in a classic study reported by Bugelski and Alampay (1961), subjects were first shown pictures of either human faces or animals and were then shown the ambiguous "rat-man" picture depicted in Fig. 9.2 that can be perceived as either a man or a rat. The results indicated that those who were first shown the human faces predominantly identified the ambiguous figure as a man, while those initially shown animal pictures saw the figure as a rat. These findings have been interpreted to suggest that an expectancy or mental set for either humans or animals was induced by the initial pictures, which in turn influenced the perception of the ambiguous figure. In the cognitive psychology

Fig. 9.2 The ambiguous "rat-man" picture adapted from Bugelski and Alampay (1961).

domain, this is also believed to be an example of top-down or executive control over perception.

9.3.3. Selective Attention

A third example of top-down influence described in the cognitive psychology literature is the phenomenon of selective attention, which refers to the fact that humans frequently pay attention to some stimuli while ignoring others (Johnston and Dark 1986). This phenomenon, too, may reflect the operation of some type of executive process that enables selectivity in the focus of attention.

Two common examples of selective attention are the "cocktail party phenomenon" (Cherry 1953) and the so-called Stroop effect (Stroop 1935). The former is illustrated by a person in a crowded room where many conversations are going on who "tunes in" or directs attentional processes to one or more nearby conversations, while blocking out others. The Stroop effect refers to the tendency people have to respond more slowly when asked to name a stimulus word, like "green", when it is presented in a red color than when it is shown in a green color. A common interpretation of this slowing is that it reflects the existence of a competition for cognitive processing resources when the color and meaning of a stimulus word do not match. In these instances, another higher-level process must intervene to resolve the conflict, thereby slowing down the overall response to the stimulus. Thus, both the cocktail party phenomenon and the Stroop effect seem to depend on executive monitors of some sort that can selectively filter our sensory inputs or help to resolve conflicts between the visual and semantic properties of a stimulus. These executive monitors might well reflect the operation of top-down influences in cognitive processing.

9.3.4. Voluntary Behavior

Voluntary behavior also commonly is thought to exemplify the effects of top-down cognitive influences (Haggard 2005; see the Lyon document in Appendix II). The idea here is that voluntary behavior (e.g., picking up a writing instrument to make a note) differs from involuntary behavior (e.g., eye blinking or sneezing) in that the former involves a prior intention or goal leading to a conscious choice, whereas the latter does not. As a result, the mental activity associated with establishing the goal and making the choice is thought to provide for a conscious awareness of intent that is not characteristic of involuntary actions. The seeming intentionality behind voluntary behavior implies a personal control over the activity that is not present for involuntary action (Haggard 2005; Haggard 2008). This control may be another example of top-down executive influence.

Each of the preceding examples seems to involve the operation of higher-level (i.e., executive) cognitive processes that intervene to select inputs, resolve conflicts or elect goal-related behaviors. These executive functions might exert their influences by means of top-down constraints or canalizations similar to how TDC might operate in the brain. Therefore, the promise for cognitive psychology is that understanding TDC could actually shed important light on how executive cognitive processes exert their top-down influences. In other words, it is possible that executive functions operate according to the principles of TDC. However, it is equally important to consider that such executive functions are markedly influenced by past learning and much of that may have developmentally transpired through the influence of bottom-up rewards and punishments, which may psychologically be synonymous with the operations of diverse forms of positive and negative affects. This means that bottom-up controls may have developmentally structured many features of higher mental processes. However, this bottom-up control does not preclude higher functions emerging from the operation of multiple past memories, in diverse permutations, yielding totally new executive configurations in working memory.

9.4. TDC: THE CHALLENGES FOR RESEARCH

A better understanding of whether or not and how TDC mechanisms might work in the brain likely will come at a cost. To achieve this understanding, almost certainly we will need answers to many potentially important questions that remain unanswered at present. In this section, we will highlight only a few of these, with the goal of identifying key research areas that will need further attention as part of a comprehensive effort to achieve the requisite understanding of possible brain mechanisms related to TDC. In the opinion of the present authors, this research will require an interdisciplinary effort among philosophers, cognitive scientists and neuroscientists, because no one of these groups in isolation is likely to be able to provide all the necessary answers. The present volume, and the conference upon which it was based, is an example of just the sort of interdisciplinary dialogue we believe will be required.

However, it is our experience that interdisciplinary research endeavors have their own advantages and obstacles. The principle advantage is that this type of research effort is informed by new perspectives from a diversity of scholars cutting across disciplinary boundaries. An important obstacle is that interdisciplinary researchers often can find themselves operating outside their comfort zones in that the collective work usually requires all researchers to develop knowledge and skills in disciplines different from their own specialties. Nonetheless, in our view, the advantages of interdisciplinary research on a topic such as TDC greatly outweigh the disadvantages.

Now, let us turn our attention to several important research areas that are particularly relevant to our quest for a better understanding of possible TDC mechanisms in the brain. The areas identified here are not intended to form an exhaustive list, but they do represent topics of great relevance to the overall theme of this volume. Space limitations preclude a detailed treatment of these areas from the standpoint of what is already known and what is not. Rather, our goal here is to highlight certain relevant considerations within each area. Readers should note that other contributors to this volume also have provided information pertinent to some of these areas. Moreover, several topics highlighted below relate directly to voluntary behavior because this phenomenon has been characterized as a prototypical type of endogenous process to which TDC might well apply (see the Lyon document in Appendix II).

9.4.1. Consciousness: Unitary or Multidimensional?

As discussed elsewhere in this volume by several other contributors, consciousness is often closely identified with higher-level mental functions in the brain, as well as with the origins of voluntary behavior (Haggard 2005; Haggard 2008). This could suggest that consciousness plays an important role in TDC, which would further imply that bringing more clarity to the nature of consciousness will likely be an important part of achieving a better understanding of TDC.

Some views of consciousness assume it is a unitary phenomenon, but there is reason to think it may be multidimensional (Damasio 1999;

Vandekerckhove and Panksepp 2009). One of the first people to embrace a multidimensional view of consciousness was Endel Tulving, a prominent cognitive psychologist (Tulving 1985). Tulving distinguished three forms of consciousness (anoetic, noetic and autonoetic) and tied each one to a different form of memory (procedural, semantic and episodic, respectively). According to this view, procedural memory is the ability to recall how to do things, how to perform sequences of action or how to accomplish certain things in the world. The acquisition of procedural memories may be guided by sensory-motor associations consolidated by primal affective rewards and punishments that are mediated by a very primitive form of affective experience. Once formed by trial-and-error conditioning, procedural habit structures may become so automatized that they constitute ways of being that are no longer accompanied by the affective arousals/experiences that initially helped consolidate those habits. In contrast, semantic memory is the ability to learn and recall facts, concepts and other information about the world that is no doubt accompanied by its own form of experience. These two types of memory are both different from episodic memory in that the latter is thought to involve the ability to recall personal autobiographical experiences such as remembering who we are, remembering our past, and making projections about the future, all of which constitute yet another unique form of experience.

For Tulving, these types of memory, working in combination, give rise to three different forms of consciousness. The first form, corresponding to the affectively colored formation of procedural memories, is a very elementary form of consciousness, which in Tulving's terms may be called *anoetic* (Tulving 1985). In Tulving's original view, this type of consciousness was thought to support perception and action in response to immediate experiences, without any clear focus on endogenous affective experiences (Tulving 1985), but the idea of anoetic consciousness may be enriched by adding fluctuating "non-knowing" affective experiences as the "reinforcement" that solidifies procedural memories (Vandekerckhove and Panksepp 2009). Anoetic consciousness is "non-knowing" in the sense that it does not rely on experiences of external stimulus states and accompanying affects that are not immediately present, yielding an evolutionarily old kind of consciousness homologous across many species (Panksepp and Northoff 2009). A second form of consciousness was termed *noetic* by Tulving and was thought to be derived from semantic memory allowing for a more "knowing" and flexible form of understanding, perception and action that could be influenced by events not immediately present. Finally, in Tulving's view, the third and highest level of consciousness was termed *autonoetic*, characterized by a self-knowledge of "personal time" extending from the past, though the present, and into the future.[2] Presumably, it is this level of mental organization that allows for "free will."

In Tulving's view, episodic memory was believed to depend heavily upon the presence and integrity of autonoetic consciousness. That is, without the sense of "personal time" provided by autonoetic consciousness, episodic memory for personal experiences was thought not to be possible. As Tulving (1985) described, this interdependency was dramatically illustrated by the case of NN, a person who lost autonoetic consciousness as a result of traumatic brain injury. In the clinic, NN showed intact language skills, had normal procedural and semantic memory, and could learn and remember basic concepts and facts. He had a normal knowledge of the world and could use it flexibly, suggesting an intact noetic form of consciousness, but he had a profound disruption of episodic memory with essentially no sense of "personal time." When asked what had happened to him in the past, NN could not report anything about his previous personal experiences, nor could he say anything about what he was going to do tomorrow or in the future. When asked to describe his sense of future, NN suggested it was a kind of blankness, like going into a room where one expects to find chairs and tables but not finding anything. Tulving characterized NN's plight as a person living in a permanent present with no sense of yesterday or tomorrow. NN's deficits seem consistent with Tulving's idea that there are important distinctions to be made not only among different types of memory, but also among different levels of consciousness. Presumably, NN had relatively intact affective feelings, although that was not formally evaluated or described. It will be important to continue research efforts to delineate these aspects of memory and consciousness [see Damasio (1999); Damasio (2003); Vandekerckhove and Panksepp (2009)] as part of a comprehensive effort to understand the role of TDC in the brain.

[2]Professor Antonio Damasio, a contributor to this conference, has also proposed a multidimensional view of consciousness in several of his influential works [e.g., Damasio (1999)]. His notions of "core" and "extended" consciousness correspond perhaps to Tulvig's ideas of noetic and autonoetic consciousness.

9.4.2. Conscious Intention and Voluntary Action

As noted above, many traditional views regard conscious intentions as a necessary precursor to voluntary behavior (Haggard 2005). These views may be interpreted to suggest that some aspect of a conscious desire to act exerts top-down control over behaviors that are considered voluntary. In this connection, other contributors to this volume have addressed the work of Benjamin Libet calling into question the precedence of conscious awareness as a necessary antecedent to the occurrence of voluntary behavior. In a study by Libet and colleagues (Libet *et al.* 1983), subjects were asked to make a spontaneous motor action with one hand at a time of their own choosing while watching a clock. They were instructed to make a mental note of the time on the clock when they first had the conscious "wish" to act, which presumably would happen just before making the movement. After the action, subjects then reported this so-called "W-time." Also, for each subject, preparatory activity in motor areas of the brain (the so-called "readiness potential" or RP) was recorded using electrodes on the surface of the scalp. Results indicated that the subject's reported W-time did in fact precede the overt hand movement by an average of about 200 msec, which was consistent with the expectation that a conscious intention to act should precede any behavior that was truly voluntary. However, a more interesting finding was that the RP actually started as much as a full second or more ahead of the W-time. Libet interpreted this finding as evidence that the initiation of voluntary action was the direct result of a neural process (i.e., the RP) that must have been unconscious because it started well before the advent of any conscious wish to act. Libet also suggested that the conscious intention to act itself might actually be a consequence of the RP, thus calling into question the primary causal role of intention in the control of "voluntary" behavior.

Although Libet's work has been widely discussed and criticized [see Klemm (2010) for an especially thorough deconstruction], the empirical findings have been replicated and extended (Haggard 2008), but they are clearly open to multiple interpretations. In one important extension, Haggard and Eimer (1999) used a paradigm very similar to Libet's except that subjects could freely move either right or left hands instead of being confined to a single hand, as was the case in the original study. Similar to the Libet study, subjects were to note the W-time prior to the movement and report it after the action was complete. Again, brain activity in the motor area was obtained, including the RP along with another neural signal called the "lateralized response potential" or LRP. The LRP occurs slightly later than RP and is a neural marker of the preparatory motor activity *after* the brain has identified which hand (left or right) will be used for the movement.

This study replicated the Libet *et al.* (1983) finding that neural preparation for movement (RP and LRP) preceded the W-time by well more than a second. In addition, two other interesting findings were revealed by this study. First, as a test of the possibility implied by Libet's work that the RP might be a cause of the conscious wish to act, onset time of the RP was compared with W-time across subjects. If the RP caused the conscious wish to act, then there should be a positive correlation between these two measures such that earlier RPs should be associated with earlier W-times, while later RPs should be associated with later W-times. In fact, Haggard and Eimer (1999) found no correlation, which seems inconsistent with the possibility that the RP causes the conscious desire to act. However, the slightly later LRP was found to be correlated with W-time, suggesting that conscious intent might arise in coordination with the brain's preparation for the specific hand that will be used (indicated by the LRP), rather than in connection with motor preparation that is more non-specific (indexed by the RP). This result does suggest the possibility that, at least in this paradigm, conscious intention may indeed be linked to an unconscious brain process, just not to the particular form of neural motor preparation suggested by Libet.

In reviewing this evidence and other more recent work [e.g., Soon *et al.* (2008)], Haggard (2008) concluded that voluntary behavior very likely was the end result of a series of decision and preparation processes in the brain beginning as early as 8 sec before the final action occurs. Along the way, and most likely as a consequence of one or more neural decisions and preparatory processes, a conscious intention to act may arise. However, it seems likely that this intention is not a cause of the final action, but rather is at best a concomitant of the ongoing neural preparation for the overt behavior (Haggard 2008). But, interested readers should consult Klemm (2010) for a provocative and detailed counter-argument in this regard.

Clearly, there may be reason to suspect that a conscious intention to act, and perhaps consciousness itself, may have a different role to play in voluntary behavior than traditionally has been supposed [see Jeannerod (2006)]. As other contributors to this volume have indicated, there may be evolutionary reasons why conscious awareness of intention occurs *after* the brain has already started motor preparations rather than *before* preparation begins. One reason is that consciousness may best serve the nervous system in the role of "optimizer" rather than "initiator." *Undoubtedly, consciousness is a very "costly" resource in the brain in terms of neural processing and metabolic energy expenditure.* Thus, it makes sense from the standpoint of efficiency to invoke this resource after a final action decision has been made to insure that the act unfolds properly, rather than wasting this resource on monitoring the many preparations leading up the final decision (Haggard 2005).[3] In this view, consciousness has an important role to play, but not until the brain has already decided to act. Further clarity is needed regarding the specific function(s) of such higher forms of consciousness in relation to voluntary behavior as part of an effort to identify the role of TDC mechanisms in the brain. Clearly, however, there is sufficient ambiguity in the existing research on this matter that the role of episodic memories and autonoetic consciousness in the initiation of voluntary action should not yet be marginalized.

9.4.3. A Broader Context for Voluntary Action

There is a broader context for voluntary behavior that has abundant implications for TDC. Patrick Haggard has described many important details of this context in two recent review articles (Haggard 2005; Haggard 2008). Figure 9.3 presents a schematic outline depicting several key aspects of voluntary behavior that are described in more detail in these two important papers. As this figure shows, the origin of voluntary behavior (on the left of the diagram) is thought to involve an immediate goal or desired effect to be achieved in the world, such as the aforementioned intent to pick up a writing instrument to make a note. As Haggard (2008) indicates, this immediate goal may be part of a longer term planning process that may involve what cognitive psychologists call "prospective memory," which represents a mental process for encoding information related to future events (McDaniel and Einstein 2007). So, as in our previous example, the goal of making a written note might well have been part of a longer term plan that had been formulated hours or even days before the note-making behavior was executed. Unfortunately,

[3]See Antonio Damasio's follow up commentary to this paper for a slightly different perspective here.

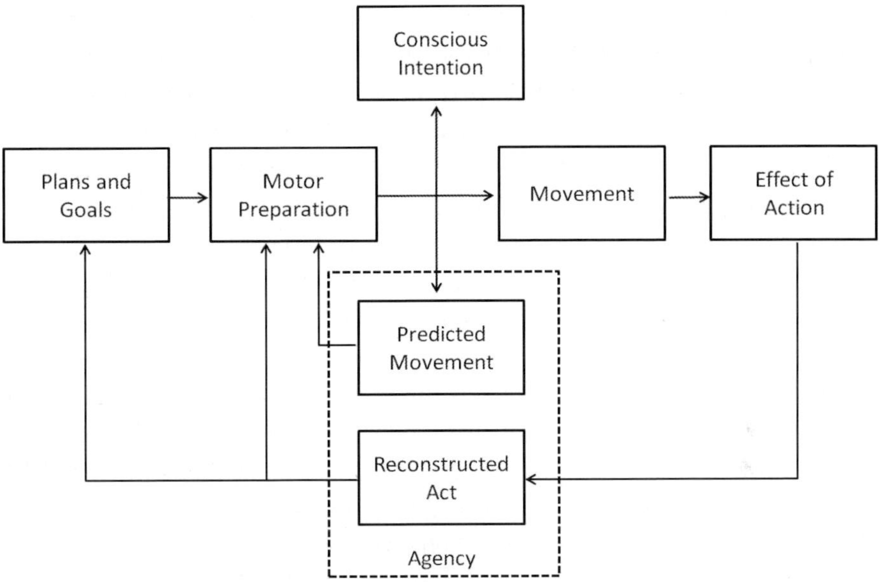

Fig. 9.3 A schematic representation of voluntary behavior (Adapted from Haggard 2005 and 2008).

the brain mechanisms underlying prospective memory are not well known at present, but undoubtedly they will have important implications for TDC.

In a way that has not yet been clearly defined, the immediate goal becomes instantiated as neural preparation for movement. As mentioned in the previous section, this neural preparation (e.g., perhaps prospective experienced planning) may begin in certain areas of the brain (e.g., frontal areas) at least 8 sec before the actual movement occurs (Soon *et al.* 2008). Eventually, as we have noted, this neural preparation results in neural signals like the RP and the LRP that eventuate in the actual overt movement itself. Somewhere in between the onset of the neural preparation for the overt movement and the action itself, a conscious awareness of the intention to act arises, as shown in Fig. 9.3. Prior to that point, the neural preparation for movement seems to be unavailable to conscious awareness (Haggard 2005; Haggard 2008; Soon *et al.* 2008).

Haggard (2005) suggests that the conscious intention to act is closely linked with activation of an internal prediction process in the motor system that somehow extrapolates an appropriate action from the intended goal.

This process, also described by other contributors to this volume, has been termed a "forward model" of movement prediction (Haggard 2005). This predictive process appears to happen quickly, before the actual movement itself, and thus does not depend on sensory feedback from the movement or its effects in the world. As Fig. 9.3 shows, the predicted movement can feedback to the cognitive planning and/or neural preparation process to produce any early adjustments in the movement type or trajectory that might be required. This internal predictive process likely becomes an important component of the conscious intention to act (Haggard 2005; Haggard 2008).

During and after the movement (e.g., picking up the pen to write a note), there is both sensory feedback from the movement itself (e.g., kinesthetic feedback from the arm) as well as from effects of the action (e.g., seeing the note being written or completed). This feedback is thought to contribute to another cognitive process occurring after the fact that has been called a "reconstructive inference" (Haggard 2008). In effect, this inference represents a mental reconstruction of the act derived from a post-hoc awareness of the behavior and the attained goal (see Fig. 9.3). While this reconstruction process can feedback information to the original movement plan, possibly as a way to confirm or modify future goal-directed actions of a similar kind, it also serves importantly to create a sense of personal control over the action and outcome just executed.

The commonly felt sense of personal control over our voluntary action and its effects is thought to result from a "binding" process (shown as the dashed box in Fig. 9.3) in which predicted actions and outcomes become connected to the sensory feedback from the actual behaviors and their effects to form an element of self-knowledge that has been termed personal "agency," or the sense that "*I did that*" (Haggard 2008). To use Tulving's terminology, personal agency may be a form of noetic or autonoetic consciousness that develops when procedural knowledge emerges from anoetic consciousness to become self-referential. Agency thus may represent a more recent adaptation of the evolutionarily older ability to associate behaviors and outcomes that is homologous across many species (Haggard 2005). Undoubtedly, personal agency is related to other similar cognitive phenomena that have been reported such as self-efficacy (Bandura 1997).

If voluntary behavior is indeed a prototypical example of an endogenous, goal-directed process, as suggested by the Lyon document (see Appendix II), then the broader context outlined above needs to be further elucidated as an integral part of efforts to better understand the possible role of TDC in the brain.

9.4.4. TDC and Emergence

The third notable aspect of TDC discussed at the outset of this paper, namely that the top-down influence of a higher level cause cannot be explained solely in terms of processes operating at the lower (influenced) level, is reminiscent of the concept of emergence. As described by Campbell (1974), the principle of emergence is reflected in certain hierarchically organized biological systems when the existence of laws, events or properties operating at a higher level of organization (e.g., law of natural selection operating at the level of species), cannot be explained in terms of, or reduced simply to the operation of, laws or events operating at a lower level (e.g., laws of physics or chemistry operating at the cellular or molecular level within organisms). These higher level laws, events or properties are "emergent" in the sense that they are not deducible (explainable) in any strict sense from the lower level laws. Campbell (1974) also described a potential linkage between emergence and downward causation. "Emergent" higher-level laws (e.g., natural selection of species characteristics over time) can constrain the operation of the lower-level laws (e.g., genetic and molecular) by limiting the genetic variations in existence. For Campbell, this constraint exerts what amounts to a top-down causal influence from higher to lower levels. Bickhard and Campbell (1997) imply that this apparent linkage between emergence and TDC is no accident, but rather results from an inherent complementary relation between the two.

In recent years, the age-old philosophical debate about emergence has received some renewed attention (Clayton and Davis 2006). In the present context, it will be important for philosophers and scientists to have continuing dialogue on the nature of emergence and its relation to TDC. Potentially important aspects of that dialogue have already begun to appear (Andersen *et al.* 2000). The challenge for neuroscientists will be to integrate the latest thinking on downward causation (and emergence) into their research on understanding the layered organization of the brain. To reiterate a previous point in this connection, it is possible that the higher brain functions emerge in development under the aegis of lower brain functions. For instance, as suggested by Panksepp (1998), the reward SEEKING system may have an important role to play in the generation of primal process intentionality (e.g., *anoetic* intentions-in-action), which may be necessary during development for the emergence of higher order executive-cognitive controls (e.g., *autonoetic* intentions-to-act).

9.4.5. A Non-human Model System for TDC

As noted earlier, one goal of this paper was to stimulate thinking about possible ways of investigating the nature and influence of TDC experimentally. What is needed for this purpose are appropriate model systems whose properties (a) are well understood, (b) seem to exhibit phenomena reflective of TDC, and (c) can be studied in flexible ways. For instance, Auletta *et al.* (2008) have proposed a molecular paradigm involving RNA that may fit these prescriptions. On a more molar and behavioral level, human voluntary behavior also may be considered a model system since, as noted above (and in the Lyon document; see Appendix II), this type of action involves seemingly endogenous, goal-directed behaviors that might well be influenced by TDC. While a considerable amount is already known about voluntary behavior, as shown by Haggard (2005; 2008), clearly this phenomenon is complex and must be studied under the constraints imposed by human research protocols. Thus, it would also be useful to have an animal model or analogue of voluntary behavior that would permit TDC to be investigated in a more flexible and less constrained way.

Rough and tumble play (RATP) in animals may well qualify in this regard. Most mammalian species exhibit RATP, which is a form of spontaneous behavior among conspecifics including running, chasing, wrestling and non-aggressive biting usually accompanied by distinctive vocalizations that are indicative of positive affect (Panksepp *et al.* 1984). In rats, for example, 50-kHz "chirping" occurs during play and these ultrasonic vocalizations (USVs) are commonly viewed as an indication that animals are in a positive emotional state (Burgdorf and Panksepp 2006), in contrast to lower frequency 22-kHz USVs indicative of negative emotionality (Blanchard *et al.* 1991). So, RATP and its distinctive accompanying "chirping" behavior appear to be endogenous (instinctual) in nature as would be required for an infrahuman anoetic analogue that could be used to model autonoetic voluntary actions. For example, rats will cease ongoing play whenever emotionally bad things happen, and their vocalizations will abruptly switch from the 50-kHz to the 22-kHz form. Subsequently, however, animals may utilize higher-order brain functions to initiate "voluntary-like" actions intended to reinitiate play. Studying these particular moments during ongoing social activities may allow one to analyze whether TDC processes are involved in the reinitiation of play.

Moreover, there is an apparent goal-directed character of RATP, which would also be important for an animal analogue of human voluntary behavior. Goal-directedness may be shown for RATP by the fact that it is rewarding inasmuch as rats will work for the opportunity to play (Panksepp *et al.* 1984). In addition, rats will exhibit something called "self-handicapping," which refers to the tendency of a dominant partner in a social play dyad to assume an unnatural submissive position, seemingly in an effort to entice the more submissive partner into increased playfulness (Panksepp *et al.* 2002). Self-handicapping, like human voluntary behavior, thus may be guided by an instrumental intent to achieve a desired effect or outcome, which may be a manifestation of the top-down influence of higher brain levels.

The biological basis of RATP and chirping has been studied extensively by Panksepp *et al.* (1984) and others, all of whom have done considerable work to identify the underlying molecular, cellular and neural processes that influence these behaviors (e.g., Burgdorf *et al.* 2010). In view of its behavioral and biological characteristics, RATP seems to offer a model system with animals that may have both bottom-up as well as top-down causal influences that can be studied not only during the natural ongoing flow of behavior, accompanied by both abundant spontaneous positive and negative affects indexed by USVs, but also through a diversity of formal learning paradigms. Such a system might well be useful, alongside the study of human voluntary behavior, as a means to further investigate TDC mechanisms in the brain.

9.5. CONCLUDING REMARKS

In summary, the Lyon document (see Appendix II) has set the stage for the present volume, along with the conference upon which it is based, by suggesting that mechanisms involving TDC may act in the brain to influence or undergird certain endogenous, goal-directed actions such as volitional behavior in humans. The present chapter has argued that the study of TDC mechanisms in the brain is both promising and challenging, and may benefit from clear visions of the evolutionary layering of the *BrainMind* (a monistic term we would suggest for the emerging understanding that mind is a natural function of the brain). For the field of cognitive psychology, the promise in this endeavor is that an understanding of the nature and influence of TDC may help to illuminate the mechanisms underlying such commonly-invoked cognitive phenomena as executive control, expectancy or set, selective attention, and voluntary behavior. Challenges to the study of TDC arise from the need to better understand voluntary behavior and its relation to conscious experience, the intention to act, the sense of personal agency that often accompanies volitional acts, and the developmental affectively-driven BrainMind landscapes — from anoetic to autonoetic — that shape the mature neuro-mental apparatus. Yet another challenge will be to identify model systems that will permit TDC to be studied experimentally in flexible and effective ways. We have suggested that RATP in animals may offer one such system that can easily be transported into standard cross-species laboratory practice offering models where the deep neuropsychological processes that need to be understood, but which cannot be accessed in human brain research, can be rigorously analyzed. When complemented by human studies of autonoetic cognitive play, as in the case of jokes and humor commonly accompanied by laughter, research on the neuropsychological processes involved during play may provide a fuller understanding of the underlying brain mechanisms than the study of a single species (whether mice or men) can allow.

In keeping with the venue for the conference upon which this volume is based (the Vatican), let us conclude with an interesting anecdote about TDC, as possibly reflected in the famous artwork of Michelangelo on the ceiling of the Sistine Chapel (see Fig. 9.4). A common interpretation of the author's intended message in this artwork is that God the Father (figure on right) is infusing life into Moses (figure on left) through the finger touch. However, an American physician named Frank Meshberger has proposed a different interpretation based on the palate upon which the Father and his companions seem to rest, which looks suspiciously like the medial plane of the right cerebral hemisphere. According to Meshberger, it may be that Michelangelo was trying to convey the idea that what God is infusing into Moses is intelligence rather than life. Either way, this would be the ultimate form of "top-down" causation.

Fig. 9.4 A famous painting by Michelangelo on the ceiling of the Vatican's Sistine Chapel.

DISCUSSION

Antonio Damasio: Thank you very much. I just wanted to make a couple of comments, first, on the relation between memory and consciousness. This is very standard and quite well accepted, and I do not think anyone has a problem with that. I do have a problem though with some of the things you said in relation to some specific cases. First of all, since you made a parallel with one of the levels in the proto-self, it is not really quite my proto-self, it is a kind of core, but you did not make a parallel with the other obvious one which is the autobiographical self and autobiographical consciousness which comes with the high noetic level of topic, quite correctly. The other thing is about when you described the patient at hand, and you talked about the fact that the lesions are not well known: this is true. There are actually several well-studied cases, beginning with a case of Boswell that was described by our group around 1985, and the lesions are very well known. So, when you have a disturbance of autobiographical memory, and you do live inside a permanent present, to a certain extent what you have is destruction both of the hippocampus interlateral system, as well as the pole in the lateral aspect of the temporal lobe. And that is a very important issue because if you are deprived of access to all your autobiography, then you are in a very different state as to the way in which you relate to present and anticipate the future; however, you do not have any disturbance whatsoever of consciousness. Only the very top autobiographical core of consciousness and the proto-self are intact, and that leads me to the other comment on Julian Jaynes and his bicameral idea; this is very interesting. Back in the 1970s, when Jaynes made that suggestion, and when he talked about for example the state of consciousness, and the stage of consciousness in which humans would hear voices, he was, by the way, basing all of this in a very beautiful literary connection with Homer's poems. The point is that it is not the kind of situation in which you find the autobiographical deprived patients. The person with no autobiography maintains normal consciousness and is not at all a bicameral person in the early sense of Mr. Jaynes; the patient does not hear voices. The idea of Jaynes relative to consciousness is that there is a lack of integration; not a disintegration but specifically a lack of integration: it is only when you really sew up the two hemispheres in full function that you get the normally organized consciousness that we have today, so that is not the connection that matters. The other thing which I thought was very interesting in your comments about Haggard's work is when you talked about the optimization of consciousness; that is a very dear idea to me, and I think that the higher we go in levels of consciousness, the more optimized we make the mind regulation that I dealt with in my contribution. What I would say is that this is not because consciousness is necessarily more costly, although probably it is, but because evolution just laid it down this way. So, it started first by adding regulation that is totally non-conscious, then by adding regulation that is minded but not conscious, and eventually we do develop consciousness in the course of evolution; so that comes later, and that is probably one of the reasons why it is more costly in terms of economic function. But this is also one of the reasons why it is probably not as abundant as one would wish, and I entirely agree with you that you come to consciousness after you have done lots of other things, and that the general process of decision-making does optimize it; I entirely agree with that.

Chuck Crowell: Thank you very much Professor Damasio for your comments and insights. In our written paper we will clarify the parallel between your view of consciousness and Tulving's. Also, in the paper we will likely omit the reference to Jaynes (originally present in my talk) since, as you so kindly point out, it is not quite an accurate fit here.

Marcus Raichle: The interesting thing to me about the potential relationship between top-down causation and emergence is the question of whether or not there is ever a case of top-down causation without also a case of emergence. So the light switch is an example of a top-down influence from one level (the electrons) to the other, but is it also fair to say then that light is an emergent property? So, is there ever a case where we have a top-down cause without an emergent property that goes along with it? I do not know the answer to that, but it seems to me to be an important question in terms of understanding how top-down causation may actually work. And particularly because there are many people out there, whether it is Chalmers who takes a strong view of emergence and believes consciousness is a strong emergent property, or whether it is Holland who believes that hidden order exists within systems and that when you put individual units together that have certain limited functionality that their aggregate properties become much more dramatically important, so that collections of neurons may emerge with quite different properties. Therefore, I think it is an interesting question about what sort of complementarities really exists between top-down causes and emergent properties.

Wolf Singer: We seem to be very peacefully agreeing that we behave like ontological monists. We have all these beautiful metafunctions seen as emergent properties. Is there anybody here defending an ontologically dualistic position, and if so why do we not have conflicts? Because my perception of the world is that there are about 50 to 60 percent of humanity that would not share our vision of the mental world and the spiritual world, so why are we so peaceful all together?

Antonio Damasio: We are such nice people…

Gennaro Auletta: I think that we should find a sort of middle way between two opposite exigencies; if you will, even between two different modalities of explanation, that is, bottom-up and top-down. I completely agree with you, Wolf, when you spoke of the brain as a distributed system. My point is the following: Would you agree that there are situations in which, it does not matter how, you have complexity emerging, or network emerging, or constraints emerging, and these emerging realities can "back-react" on situations that you had before? In this case, you would have emergence of constraining situations determining in part further developments; this is, for me, really crucial.

So, we have spoken of constraints, but why do we understand constraints only from the outside? I mean, once you have established a network, is that network able somehow to canalize processes also involved in interrelations of single units or not?

Wolf Singer: This was not my problem. My problem was (but maybe there is no one here that defends that position) that in the Asian world, for example, where, in the self-understanding of those people, there is no doubt about another, pre-existing dimension of reality that is not an emergent property of the brain, and that is something interacting with the material world and controls things as they go. So, I am just surprised to see that here, in this environment, it seems to be very peaceful.

Gennaro Auletta: What you are talking about is the mission of the STOQ Project, and for me this is also an important problem that the Catholic Church as such should face, that is, to try to combine in an appropriate way scientific results with the philosophical and theological tradition, which is not dualistic.

Wolf Singer: This is not my problem; we cannot have ontological dualism coexisting with ontological monism at the same time, and not discussing it. This is, I think, something that we need to address at some stage.

Pierre Jacob: I think one way to clarify the discussion would be this way: that relates to both Professor Crowell and Professor Auletta's papers. When we talk about causation, if we want to be philosophically clear about what we are talking about, we are considering relations among events, events instantiating different properties. When we talk about emergent properties, for example about the fact that cognitive or psychological properties (the property of believing so and so, receiving such and such, experiencing such and such), and we say that such properties are emerged from lower level properties, we are not talking about causal relations, I believe. The basic idea philosophers are trying to capture is the kind of relation between emerging properties and their lower level implementation, or the necessary and sufficient conditions at the lower level for the emergence of those higher-level properties. Many philosophers talk about supervenience, the fact that higher cognitive emergent properties supervene on their basic lower-level physical, chemical or neurobiological property. I think one basic issue is whether we want to say that the supervenience relation is a causal relation, that is, whether the emergent property is being caused by the lower level property. I think that the right view to take for a non-dualist is that the emergent properties are not caused by the lower level. Causal relation implies different instance in time: a cause must precede its effect. But when we talk about an emergent property we are not talking about temporal distance between the basic property and the higher-level property, the two are co-instantiating at the same time; so I think it is rather confusing to think of the supervenience relation as an example of a causal relation; it is not a causal relation because the two properties, the higher-level property and the lower-level one, are coinstantiated at the same time. So, I think that one way a non-dualist might think about this is to make a distinction between the causal relation that holds between a cause and its effect (which is temporally related, because the cause precedes the effect) and the way a higher-level property relates to its state's condition (which is not a causal relation at all).

Gennaro Auletta: Let us destroy what is probably a mythology. The mythology is the following: that you can have causal efficient effects between different levels of reality. It does not matter whether top-down or bottom-up: for me this is impossible. Impossible even in the case of bottom-up, for the simple reason that each time that you have a bottom-up effect this can happen only because you have efficient causal processes at the same level, concurring in this action. So, you never have a jump from one level to another, neither top-down nor bottom-up: you need always causation at the same level but you do not have a dynamical action across levels. What you have in both cases (that is, bottom-up and top-down) is: (i) from the bottom, a source of variations that are used at the higher level, that is, conditions of possibilities (for instance, productions of polymers that are used in the cell and so on); (ii) from the top, you have formal constraints that arise and can concur with the final result exactly activated by dynamical processes at the same level. So, for me also the word "causation" is a little equivocal, because we should distinguish between what is dynamical causation, and this can never act across different levels, and what are constraints and conditions of possibilities. The latter are to be meant in a completely different way, but they are not causally irrelevant in my opinion. So, for me the point is simply to understand whether formal constraints exist or not. This is the point. Frankly, this issue of supervenience is a game of words. The issue is only the following: once you have a constituted network, that is, constraints and relations, do these correlations exist or not? Now, if they exist, if they have relevance somehow, then they are able to exert some effects when there are some dynamical causes at play. If you think that they do not exist at all, obviously you think that only the single elements exist. But, you know, we have a huge amount of data showing that formal constraints at the physical level, in quantum mechanics, in complexity, in chaotic behavior, are very relevant; therefore, I do not see any reason for denying this. Already the word "network" implies that we have interrelations, i.e., formal constraints.

Jean-Pierre Changeux: I think that the question that you have not answered is a question that Wolf raised and which is a very important one. A very large population on earth believes that we have a soul apart from our body and our brain, and that there is even an immortality of the soul; this is something that is present in many societies. The question is that science is deconstructing this. However, as a matter of fact, first of all, not all scientists think so: in the French Academy of science, there is the idea of a separate world of mathematics, a Platonic world. I simply do not accept this. So, to convince some scientists that mathematics is a construction of our brain is something that you have to do and we have to do. The second thing is that this belief of dualism is not only present in the Catholic Church. Wolf mentioned the Oriental societies, but I think they are more monist than we are. I had this personal experience with a Japanese scientist in my lab, and he was trying to learn French, and I told him about the separation of Descartes; so, it took seven weeks for me to try to make him understand what the point was, and he said: "We never had this in our traditions. For us it is a matter of physiology of our brain, and we always thought that." So, I think the issue raised is that, as scientists, we think we have no evidence of the dualistic conception, but there is a very important conflict about that. So, you said that you had a plan for us to try to deconstruct this idea: how are you going to do it?

Gennaro Auletta: First of all, let me clarify that I am not a theologian and I am accustomed to talking about things I know, not about things I do not know. Secondly, I would say that the Catholic tradition is on this point rather Aristotelian (monistic) than Platonic.

Jean-Pierre Changeux: Well, I have a different feeling since I have been working with theologians for several years. However, just to shorten the discussion, surely the positions on the embryos that the Catholic Church has now are not inspired by Aristotle.

Gennaro Auletta: I know too little about embryos, but let me tell you something more general. In general, I do not like to avoid possible conflicts, as they are for me always very important in order to understand each other. But I do not think that scientific statements can prove or disprove religious or theological statements and vice versa; these are different orders of questions. So, let me say that you will never see any progress in neuroscience telling us anything about the soul, at least for me; and vice versa, obviously. Moreover, when you speak of the immortality of the soul, you should also consider that for at least a part of our tradition, we speak of the man as a unity of body and soul, and as to such a point we speak of the resurrection of the body, not only of the soul. But this is again a theological question into which I would not enter, because I am not able to appropriately deal with it as a philosopher.

Jean-Pierre Changeux: The question is how are you going to match the Catholic Church, which you said does not have a dualist point of view, with the view of many Catholics, who, I think — most of them — do have a dualistic point of view, for instance in France.

Gennaro Auletta: I cannot deny that, but I think that the source of this view is simply the lack of scientific background; and in my opinion, one of the reasons for doing what we are doing first of all, very elementarily, is to try to improve the level of scientific background within the Church. Anyway, I would mention that in our conference on evolution, the conference we held at the Pontifical Gregorian University at the beginning of March 2009, two important theologians spoke: Cardinal Cottier and a very important French theologian, Jean-Michel Maldamé, from Toulouse. And both of them stressed very much what I am saying now, that is, a monistic view and even a "naturalistic" view. Finally, I think that Cardinal Cottier even spoke about the relevance, to understand more about these issues, of considering the results of neuroscience. So I am relatively confident that further advancements could be made.

Pierre Jacob: Do you think that having a soul can actually have some causal effect although souls are not consistent or compatible with the basic laws of physics?

Gennaro Auletta: I think that one of the main problems here is that it is not by chance that the main opposition comes from the French part of this workshop, because in French the word "mind" does not exist at all. And I would prefer, as scientist and philosopher, to speak of mind because this is something we can deal with at this level. The soul is instead a theological concept.

Pierre Jacob: But do you think there are things which instantiate properties of souls and which do not obey the laws of physics, but can have causal powers?

Gennaro Auletta: Are you asking me an opinion?

Pierre Jacob: Yes, your view.

Gennaro Auletta: This is about my personal opinion, and I would confine ourselves here to a scientific and philosophical discussion, if you like. I am sorry but you are going outside the framework of this workshop.

Jean-Pierre Changeux: But this is a scientific question.

Gennaro Auletta: No, this is exactly *not* a scientific question.

Jean-Pierre Changeux: Yes, the question that Pierre asked is a perfectly scientific question, because if there is an additional effect on our brain which is resisting data and is not taken into consideration, then we are missing the aim of this workshop; so you have to give an answer to this question.

Marc Jeannerod: I think we are now entering into folk theology. Theology is a serious matter; I am not a theologian but I know a few theologians, and I think they use scientific arguments and they try to some extent to match their data with the data of science and particularly neuroscience. There is a gap between the goal of this meeting and the many people who think differently. Maybe more than half of humanity is dualist; but the academic community or the educated community is not.

Wolf Singer: We had a workshop at the Pontifical Academy of Sciences just before your conference on evolution, and the topic was exactly evolution, not only at the biological level. There was also Cardinal Schönborn from Vienna giving an interesting paper; he actually introduced the topic. I think the agreement among everybody at that meeting was that, yes, the process that Darwin described and that modern genetics described has taken place on earth, and the probability that it happened that way is very high. As to when and why this process was initiated or inaugurated, the scientist would simply say: "*ignoramus.*" Then, when we enter the belief systems we may find answers. The Revelation has answers to these things, but these are outside science and we cannot discuss them. Things are a little bit different here as we are talking about top-down causation, and there is a strong position. I know that for many people, what we actually do in the next moment is not only determined by physical processes that preceded that moment, but by something that is called the mind or the soul, in which there are immaterial properties. So, let me rephrase my question: Do we believe that this, whatever it is, mind or soul, is not describable or is not following or is not dominated by the physical laws we know? Is it something ontologically different? And if we believe so, and if this has causal effects, then we need to discuss this seriously and we have a real problem, I think. It is a scientific problem.

Gennaro Auletta: If the worry is about the word "mind," I can avoid using it, because for me it is not a separated ontological reality. So, if you do not like it, I can completely avoid it, and we can speak of consciousness or anything you like. But be sure that when I use the word mind, I am not speaking of the soul or of a sort of external "virus" affecting the brain from outside; it is not the way in which I envisage the word and the issue. The scholastic tradition also does not do that, since St. Thomas speaks of the soul as the form of the body.

Jean-Pierre Changeux: We will need to further discuss that.

Gennaro Auletta: Yes, sure.

Juan José Sanguineti: I think there is a popular tradition among people in the Catholic Church that is dualist because it is very difficult to speak about this higher dimension without a certain ontology. With the appearance of the moral sciences and with Descartes, many Christian and Catholic thinkers embraced the dualistic ontology as a way of explaining things. However, in the Catholic Church there is also the tradition that comes from Aquinas, not from Augustine, and in this tradition I think that there is no dualism. There are some aspects that could be dualist, but the brain, for example, was seen as the organ of intelligence. There was also localization, although mistaken of course. For example, according to Galen, there were drawings with the different parts of the brain associated with practical intelligence, imagination and perception. Such a view was maintained by Aquinas and by his master Albert the Great. Therefore, the brain was seen as the organ of intelligence, of imagination, of perception, of reasoning. This holds true for theologians but was very far from uneducated people. A more dualistic tradition can be found in some writers, following especially Augustine. Now, if you go to the Bible, there is not the typical dualism coming from Plato; in the New Testament the word "spirit" or "soul" sometimes does indeed appear, but in a more phenomenological way and not in an ontological one. Another point stressed by theologians, especially after Vatican II, is incarnation. Incarnation and resurrection are very strong points for a non-dualistic view. I want to add a final point. If you read, for example, *The Acting Person* by Wojtyla, I think that you will not find dualism there. Many issues about the human being are treated therein: behavior, action, emotions, and so on. I think that the theology by Wojtyla — and he was also a phenomenologist philosopher — is a theology and a philosophy of the personal body.

Antonio Damasio: Very short: since we are talking about Descartes, we have to talk about Spinoza too. Especially concerning the fact that he was a central monist, looking at the mind and body/brain as two aspects of exactly the same thing, with a construction that I think is actually quite anticipatory of a neurobiological thought. Quite curiously, however, Spinoza's view allows perfectly well for an inclusion of some kind of force, presence or impulse that would, in fact, come from an external power: this is the interesting thing about Spinoza. His work can have very different readings: one is totally materialist, but another would be compatible with just what you said. You know, as a matter of fact people cannot give answers to what there was before the Big Bang; Spinoza would not be troubled by that, as in one of his famous statements: "There is God in nature, and nature is God." It is only one of the readings that Spinoza was diametrically opposed to Descartes. Incidentally, both of them probably articulated part of their thoughts, having very much in mind possible persecution in relation to the time, historically related to religion and several churches.

Gennaro Auletta: Very briefly, I would only like to add that probably the cultural background of Platonism, of the ancient Platonism, was actually represented by Oriental religions and by Orphism; there are some allusions also in the first Platonic dialogues. Moreover, it is very helpful, I think, the remark that actually in the Church the situation is a little bit problematic, as you, Juan José, have explained before, more on the side of the popular understanding of religion than on the side of theology. This is also the point that Marc Jeannerod was stressing just before as to the differences between folk and professional theology, with which I agree, of course.

REFERENCES

Andersen PB, Emmeche C, Finnemann NO. (2000) *Downward Causation: Minds, Bodies, and Matter.* Arhus University Press, Copenhagen.

Auletta G, Colagè I, Jeannerod M (Appendix II). *Top-Down Causation in Cognitive Science.* Appendix II to this volume.

Auletta G, Ellis GFR, Jaeger L. (2008) Top-down causation: From a philosophical problem to a scientific research programme. *J R Soc Interface* **5**: 1159–1172.

Bandura A. (1997) *Self-Efficacy: The Exercise of Control.* W. H. Freeman and Company, New York, NY.

Bickhard MK, Campbell DT. (1997) Emergence. Retrieved from *www.lehigh.edu/~mhb0/Emergence27Jul97.pdf.*

Blanchard RJ, Blanchard DC, Agullana R, Weiss SM. (1991) Twenty-two kHz alarm cries to presentation of a predator, by laboratory rats living in visible burrow systems. *Physiol Behav* **50**: 967–972.

Bugelski BR, Alampay DA. (1961) The role of frequency in developing perceptual sets. *Can J Psychol* **15**(4): 205–211.

Burgdorf J, Kroes RA, Beinfeld MC, *et al.* (2010) Uncovering the molecular basis of positive affect using rough-and-tumble play in rats: A role for insulin-like growth factor I. *Neuroscience* **168**: 769–777.

Burgdorf J, Panksepp J. (2006) The neurobiology of positive emotions. *Neurosci Biobehav Rev* **30**: 173–187.

Campbell DT. (1974) Downward causation in hierarchically organised biological systems. In: F Ayala and T Dobzhanksy (eds.), *Studies in the Philosophy of Biology: Reduction and Related Problems.* Macmillan, London/Bastingstoke, pp. 179–186.

Cherry EC. (1953) Some experiments on the recognition of speech, with one and with two ears. *J Acoust Soc Am* **25**(5): 975–979.

Clayton P, Davies PCW. (2006) *The Re-Emergence of Emergence.* Oxford University Press, Oxford.

Damasio A. (1999) *The Feeling of What Happens: Body and Emotion in the Making of Consciousness.* Harcourt Inc, New York.

Damasio A. (2003) *Looking for Spinoza.* Harcourt Inc, New York.

Ellis GFR. (2008) On the nature of causation in complex systems. *Trans R Soc S Africa* **63**: 1–16.

Haggard P. (2005) Conscious intention and motor cognition. *Trends Cogn Sci* **9**: 290–295.

Haggard P. (2008) Human volition: Toward a neuroscience of will. *Nat Neurosci Rev* **9**: 934–946.

Haggard P, Eimer M. (1999) On the relation between brain potentials and the awareness of voluntary movements. *Exp Brain Res* **126**: 128–133.

Jeannerod M. (2006) *Motor Cognition: What Actions Tell the Self.* Oxford University Press, New York.

Johnston WA, Dark VJ. (1986) Selective attention. *Ann Rev Psychol* **37**: 43–75.

Klemm WR. (2010) Free will debates: Simple experiments are not so simple. *Adv Cogn Psychol* **6**: 47–65.

Libet B, Gleason CA, Wright EW, Pearl DK. (1983) Time of conscious intention to act in relation to onset of cerebral activities (readiness potential): The unconscious initiation of a freely voluntary act. *Brain* **106**: 623–642.

MacLean PD. (1990) *The Triune Brain in Evolution: Role in Paleocerebral Functions.* Plenum Press, New York.

McDaniel MA, Einstein GO. (2007) *Prospective Memory: An Overview and Synthesis of an Emerging Field.* Sage, Thousand Oaks, CA.

Norman DA, Shallice T. (1986) Attention to action. Willed and automatic control of behavior. In: RJ Davidson, GE Schwartz and D Shapiro (eds.), *Consciousness and Self-Regulation.* Plenum, New York, pp. 1–17.

Panksepp J. (1998) *Affective Neuroscience: The Foundations of Human and Animal Emotions.* Oxford University Press, Oxford.

Panksepp J, Gordon N, Burgdorf J. (2002). Empathy and the action-perception resonances of basic socio-emotional systems of the brain. *Behav Brain Sci* **25**: 43.

Panksepp J, Northoff G. (2009) The trans-species coreself: The emergence of active cultural and neuro-ecological agents through self-related processing within subcortical-cortical midline networks. *Conscious Cogn* **18**: 193–215.

Panksepp J, Siviy S, Normansell L. (1984) The psychobiology of play: Theoretical and methodological perspectives. *Neurosci Biobehav Rev* **8**: 465–492.

Soon CS, Brass M, Heinze H-J, Haynes J-D. (2008) Unconscious determinants of free decisions in the human brain. *Nat Neurosci* **11**: 543–545.

Stroop JR. (1935) Studies of interference in serial verbal reactions. *J Exp Psychol* **18**: 643–662.

Tulving E. (1985) Memory and consciousness. *Can Psychol* **25**: 1–12.

Vandekerckhove M, Panksepp J. (2009) The flow of anoetic to noetic and autonoetic-consciousness: A vision of unknowing (anoetic) and knowing (noetic) consciousness in the remembrance of things past and imagined futures. *Conscious Cogn* **18**: 1018–1028.

CHAPTER 10

Consciousness and the Global Neuronal Workspace Hypothesis: From Bottom-Up to Top-Down Causation and Vice Versa

Gennaro Auletta and Ivan Colagè[1]

10.1. INTRODUCTION

One of the primary aims in the cognitive neurosciences is to represent behavior or a mental process on the basis of a minimal yet realistic description of neural architectures and of brain activity patterns. This aim is ideally achieved starting from the molecular up to the whole-brain level and all the way back downwards, although many possible gaps still need to be filled. Moreover, such a description should be built into mathematical terms as much as possible. In doing this, it turns out to be crucially important to establish causal relationships between a specific behavior or even a mental *subjective* process and *objective* neural measurements that can be tested experimentally.

[1]This chapter was originally inspired by J.-P. Changeux's contribution to the workshop. When he received the first version of the transcriptions (as edited by both of us, i.e., G. Auletta and I. Colagè), he considered the interesting possibility of all three of us writing the paper together. Thereafter, we (G. A. and I. C.) became increasingly aware of the importance of interpolating the original perspective of Changeux about the global workspace with the points of view of A. Damasio and M. Raichle (which are also expressed in this volume), and for this reason widened Changeux's original contribution in this sense. Although the latter was very sympathetic to our work and also carefully revised it, taking into account this widening of perspective, he finally proposed that it should be published with us (G. A. and I. C.) alone as authors.

The human brain has been carefully described since the 16th century by Andreas Vesalius (1514–1564) in his *De fabrica corporis humani* (1543). It is known that the brain is composed of cells, one hundred billion (10^{11}) neurons, and that there are approximately one million billion (10^{15}) synaptic contacts, particularly into the cerebral cortex. The brain, therefore, is a very complex organ, to the point that it might be said that it is "an organism within our organism" (Changeux 2002; Auletta 2011, Chap. 13). Already this makes us aware of the difficulties in trying to build models of the human brain. Notwithstanding the huge number of contacts between cells (the neural synapses), we have to keep in mind that we are dealing with a *discontinuous* network, characterized by a high number of connections, which is nevertheless small when compared with that of all the abstractly possible ones. Moreover, the synapses are chemical in character; the specific class of chemicals called neurotransmitters is used to communicate across the synaptic cleft. Synapses are made up of cell-surface molecules (namely receptors) that are now known at the atomic level, and that have peculiar structural and functional aspects. In such a discontinuous network, synapses and neurotransmitters are involved in chemo-electrical and chemo-chemical transmission. Therefore, even in the case of chemo-electrical transmission, there is a chemical delay in the propagation of the electric signals (mainly the action potentials). Moreover, the postsynaptic signal sooner or later gives rise to action potentials that propagate at a speed lower than that of the sound. There are several phenomena that take place at the synaptic level, and we are now able to describe nearly all of these processes in molecular terms. This holds true for the whole of the flexible brain organization as well, which starts at the synaptic level and is maintained at all levels of the neural architecture. Even the spontaneous activity that we know happens within the brain and that, as we will see, is such a critical component of its functional organization, can be described in molecular terms as well. Such spontaneous activity is essentially due to the spontaneous firing of neurons, and it is worth noting that neurons fire spontaneously throughout the whole system from the very early stages of brain development. This neural activity gives rise to diverse and complex phenomena of neural assembly and differentiation.

The fact that all these processes that constitute the building blocks of the brain organization and function can in principle be described in molecular terms, whatever the level of brain organization concerned, recalls the famous French philosopher G. Bachelard, who considered neuroscience as a kind of chemistry.

Notwithstanding the fundamental relevance of the molecular level, the brain is not simply a "pile" of molecules; it is highly organized. This organization is the result of biological evolution (Fig. 10.1) spanning the millions of years that preceded the appearance of *Homo sapiens*. Besides the evolution of the species, there is another extremely important type of process that one sometimes tends to forget or overlook and that can be regarded as a kind of Darwinian evolution as well: non-genetic or epigenetic development. The latter is somehow nested within the former (the evolution of species) and occurs on a time scale of days to years. As far as the brain is concerned, epigenetic processes involve the phenomenon of developmental *selection* at the scale of synaptic connectivity, the latter being crucial for both the spontaneous and the evoked activity of the neuronal networks; see Changeux and Danchin (1976). There is yet another *nested* level in the context of neuronal activity that is characterized by the time scale of the synaptic exchanges, i.e., hundredths of milliseconds. Such short-time dynamical processes concern neuronal growth and diversification, as well as synaptic stabilization processes (again characterized by a kind of selection). The latter two processes make possible the human-specific phenomenon of cultural evolution expressed in the production of memories and traditions (Changeux 1985; Changeux 2002; Auletta 2011, Chap. 22; Auletta *et al.* 2013). The processes producing cultural memories have a time scale from milliseconds (synaptic activity) to thousands or tens of thousands of years (long-term extracerebral memories: monuments, paintings, books, etc.).

In addition to the processes of selection taking place at these levels of evolution, the different time scales involved and the various kinds of mechanisms at work, we also have to consider that, from the molecular to the cognitive, several nested levels of organization exist within the brain (Changeux and Connes 1989; Changeux and Dehaene 1989). At each of these levels of organization (molecules, neurons, circuits of neurons, or assemblages of circuits) structures are present that have their own functions linked to the *state of activity* of the structure itself, an issue to which we shall return later. We reiterate that this holds true from the simple level of arc reflexes up to the complex cognitive processes at the highest level of organization. Among these

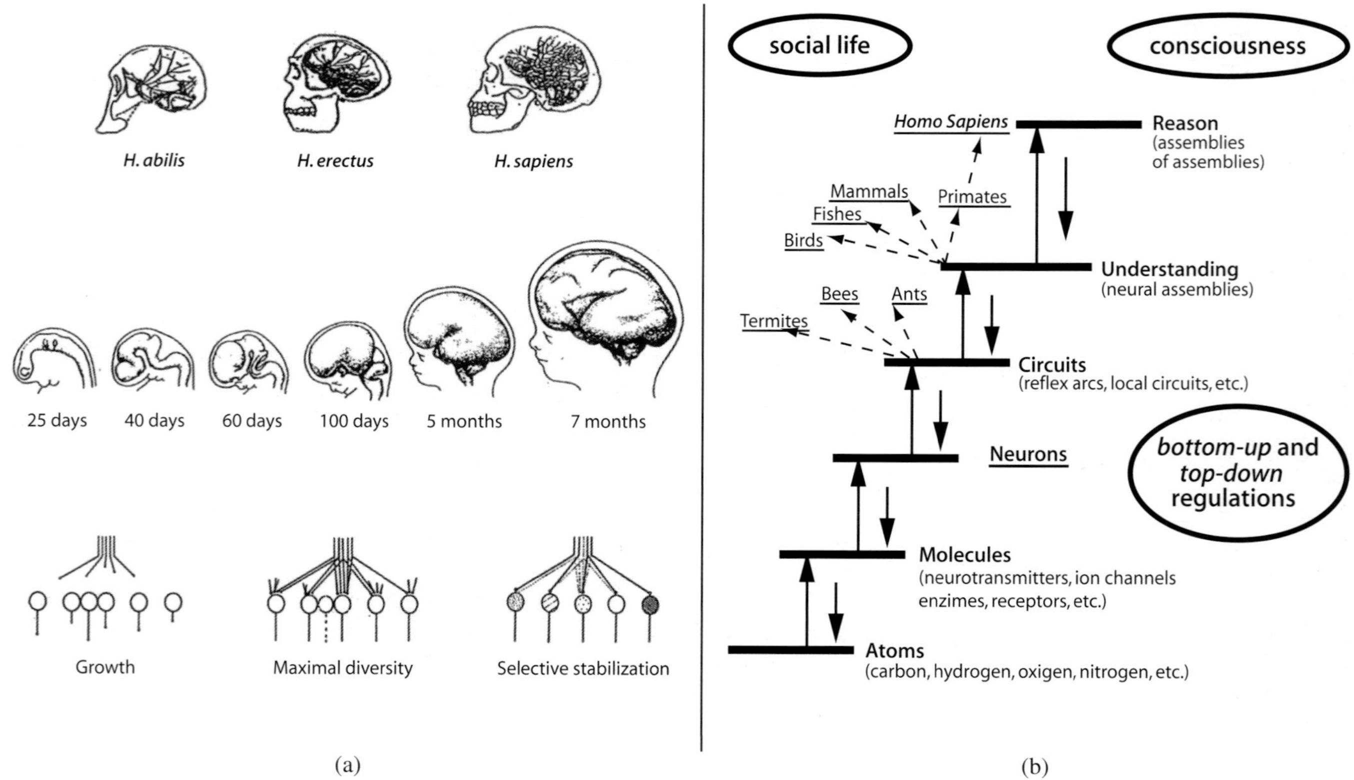

Fig. 10.1 Phylogeny, ontogeny and hierarchical organization of neural/mental processes. (a) *Top*: evolution of species (millions of years); *middle*: Ontogenetic development (days/years); *bottom*: Synaptic processes (milliseconds/seconds). (b) Evolutionary emergence of multiple nested levels of organization from the molecular to the cognitive domain.

levels, both bottom-up and top-down causation processes are engaged. We may even say that the principal aim of neuroscience is to try to understand the complex relationships among these different levels in molecular terms, as well as in terms of networks, in which both parallel and hierarchical processing take place.

10.2. GENOME, BRAIN AND DEVELOPMENT

We should now turn to an important issue that is still far from being understood: the issue of the innate — genetic — universality of the *Homo sapiens* brain. Such a difficult problem can be examined in terms of the nonlinear evolution of the complexity of the genome with respect to the complexity of the brain. From yeast to humans, we have something of a linear increase in the size of the genome, but this increase is not paralleled by an increase in the number of genes. Moreover, if we compare the number of genes in a given species with the number of neurons that it has, the nonlinearity becomes even more striking. The genome size of the mouse is comparable with the human's (the difference being mainly due to repeats) and the number of genes in the two species is nearly the same. As to the number of *neurons*, however, there is a difference of *three orders* of magnitude! The difference becomes even more striking if the structure and, above all, the *function* of the brains of the two species are compared. Therefore, the number of genes cannot explain the increase in the complexity of the brain from the fly (or even from the mouse) to the human (Changeux 2002, Chap. 5). Nevertheless, the relevance of the number of neurons cannot be exaggerated either, since elephants are known to have more neurons than humans.

In facing the above paradox about the relation between complexity of the genome and complexity of the brain, presently known, single-gene brain disorders may be of help. They are caused by discrete changes of the genetic material in our chromosomes, often gene copy number. Let us consider, for example, primary microcephaly that may compromise the size of the brain and result in a situation in which it is significantly smaller than normal. This disorder is due to a decrease in the number of cells, which, in turn, is due to a mutation of a cytoskeletal microtubule-associated protein in centrosomes that affect cell division (namely, neurogenic mitosis) (Bond and Woods 2006). Such kinds of changes may well have had a role in the evolution of our ancestors, as far as it concerns the size of the brain. There are other single-gene deficits that, for instance, cause changes in the general plan of organization of the brain (Arendt and Nubler-Jung 1999), or in the specification of brain areas (Bishop *et al.* 2000). Other single-gene disorders (Wang *et al.* 2002) affect splicing (i.e., the molecular mechanism that allows for a single gene to give rise to several mRNAs or proteins sometimes able to perform different functions or to undergo diverse post-transcriptional regulation). Finally, microRNA-related disorders, such as that which affects the symmetry of the ASE taste neurons, are also becoming increasingly relevant (Johnston and Hobert 2005).

These are a few among several kinds of genetic factors that may immediately affect the complexity and the organization of the brain. Yet the crucial notion is that in general there is *not* a direct relationship between single genes and the brain's functions and performance. On the contrary, these genes *interact* to give rise to the brain phenotype and in particular to neuronal networks. The critical issue, in our opinion, is to try to understand the *combinatorial dynamics* of the products of the genes, which gives rise to complex signaling networks characterized by nonlinear processes stemming from cross-signaling and feedback loops. The complex organization of the brain is the result of such nonlinear dynamics.

One possible model to deal with the nonlinearity paradox is related to the evolutionary increase of certain brain areas, such as the prefrontal cortex, in different species (Fuster 1989; see Fig. 10.2). It is indeed worth mentioning that the shift in the boundaries of expansion of the prefrontal cortex is likely associated with changes in the transcription factors at the developmental level (Mannervik *et al.* 1999). Such processes are relevant to the interaction of transcription factors at the level of the promoters that, during development, orchestrate the gene expression in stem cells and are obviously linked to the developmental production of the general architecture of our brain (Kerszberg and Changeux 1994). However, the combination of transcription factors is already enormously complex in organisms as simple as yeasts (Gavin *et al.* 2002). Likely, this is one of the most pressing problems that we have to face in neuroscience today.

An important point that should be emphasized is the existence of a general brain organization that is *common* within the human species. This is due to networks of gene interactions that give rise to the gross architecture

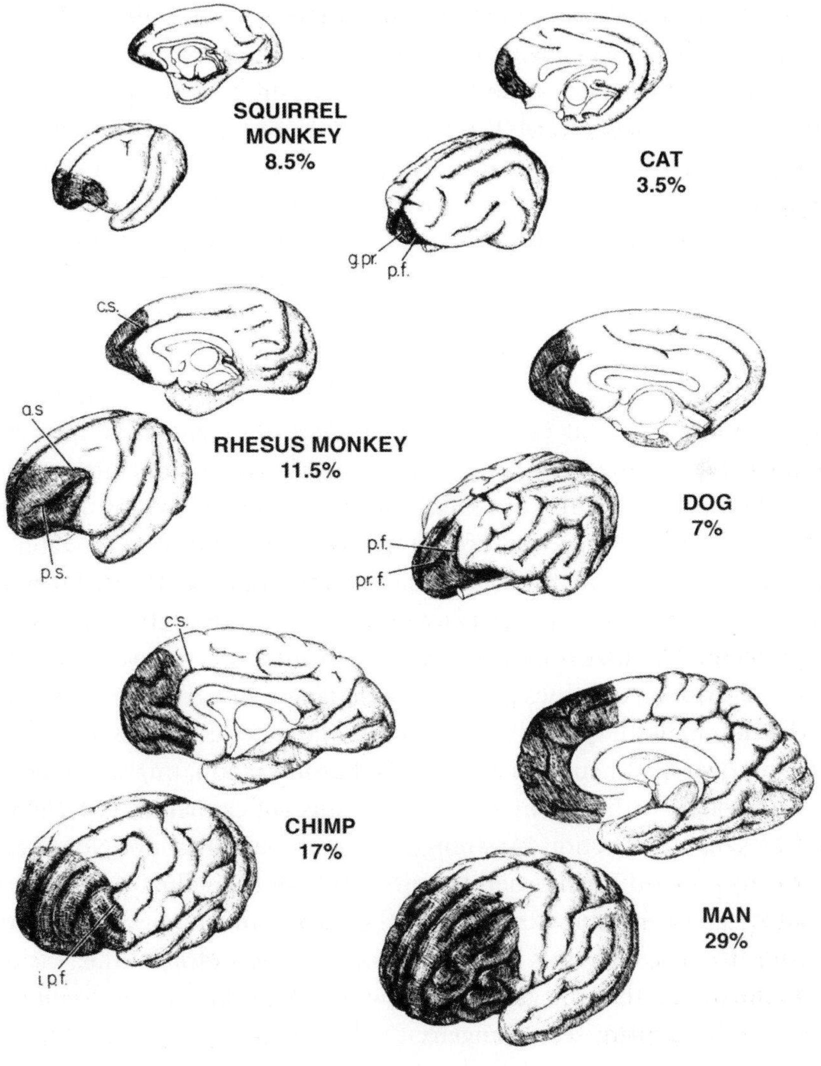

Fig. 10.2 Percentage of the prefrontal cortex on the whole of the brain in different mammal species. (Adapted from Fuster 1989.)

of the human brain and, as a consequence, ensure what may be called the *innate universal knowledge* of the human newborn, e.g., capabilities like breast sucking, recognition of a human face, distinction between living/inanimate objects, primary consciousness, empathy, and many others. Such an innate universal knowledge is part of the biological cognitive endowment of humans but nevertheless needs appropriate environmental stimuli of natural and cultural kinds to be stabilized. Genetic networks only ensure the common gross organization of the brain, thus allowing for genetic-epigenetic variability at the individual level. For example, variation in the number of gene copies is linked to several brain disorders, e.g., schizophrenia (Feuk *et al.* 2006; Redon *et al.* 2006; Cook and Scherer 2008). Nevertheless, since the genome does *not* code for the specific, fine and detailed organization of mature individual brains, these differences between individuals are mainly due to an *epigenetic variability.*

One way to look at the nonlinear relationship between genes and brain organization is to take into consideration the development of the human brain, which takes about 15 years to reach maturity. There is an impressive fivefold increase in the weight of the brain that takes place from birth to adulthood; a comparable increase in the complexity of the brain also takes place. For instance, nearly 50 percent of the total number of synapses is established in the first year of postnatal life. The idea has been suggested that, during this period, critical interactions take place between synapse development and the environment. The theory of the epigenesis of neuronal networks by selective stabilization of synapses proposed by Changeux *et al.* (1973) formalized this developmental process, which can be understood as consisting of the following stages: (i) neuronal and glial cell proliferation, migration and cell death, (ii) outgrowth of connections and transient exuberance of synapses, and (iii) synapse selection by strengthening and elimination. Now, phases (i) and (ii) proceed in an orderly manner in each individual according to a species-specific timetable (Rakic 2000): the main neural pathways are formed according to the constraints imposed by the "*genetic envelope*" (i.e., the cluster of constraints that the genetic endowment sets for the development of the mature organism; see Changeux (2002), Chaps. 3 and 5). At the end of this stage, the maximum variability is reached. Phase (iii) is regulated by activity-dependent mechanisms that, especially after birth, involve individual experiences. Here, the states of activity of the neurons and of the *competition* among neural terminations determine the stabilization of some of the synapses and the elimination of other ones. The process of synaptic stabilization, leading eventually to the adult brain, is strongly influenced by the interactions with the physical, social and cultural environment and the learning process that such interactions bring about or require. This produces the

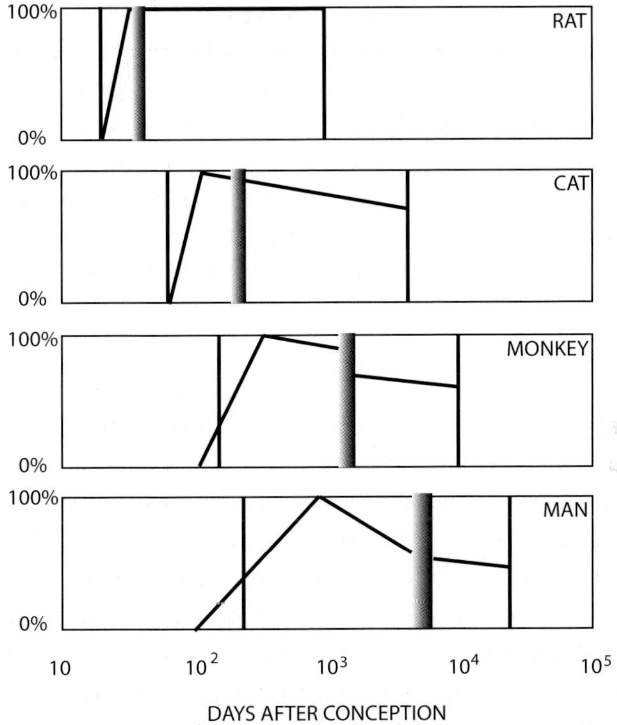

Fig. 10.3 Pictorial representation of the increase in synaptic density in different mammals. The vertical line on the left represents birth, while the one on the right represents death. The large gray bar in between represents puberty. The establishment of topography and synaptic strengthening is more marked in the postnatal period (from the left vertical line to the central gray bar). (Adapted from Changeux 2002, p. 190.)

remarkable difference in the growth of synaptic density in humans relative to other species (Fig. 10.3).

Thus, as to this developmental process of synaptic stabilization by selection, the human being has a peculiar position: this process is indeed significantly prolonged in postnatal life compared with that of any other animal species, primates included [see Bourgeois *et al.* (2000) for a review]. The density of the synaptic connections in the human brain decreases progressively during life and has a high rate at least until the age of fifteen (obviously continuing also thereafter, although at a diminishing rate). It might be worth noting that, since the average lifespan of the early *Homo sapiens* was about

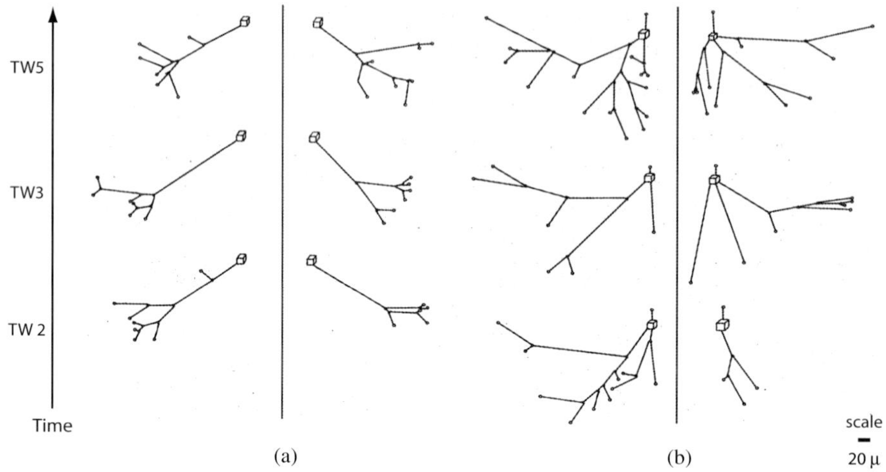

Fig. 10.4 Different experience-dependent branching in two *Poecilia formosa* isogenic individuals. (Adapted from Levinthal *et al.* 1976.)

thirty years, the human being has been selected to spend roughly half of its life developing its brain.

The mathematical consequences of the mentioned model for the epigenesis of neuronal networks (Changeux *et al.* 1973), taken together with data coming from studies on single motor neurons' dendritic branching in distinct *isogenic* individuals (Levinthal *et al.* 1976; see Fig. 10.4), have lent support to the so-called *variability theorem*, which states that different learning inputs may produce different connective organizations and neuronal functioning activity, but give rise to the *same* behavioral activity. Therefore, a particular behavior, or even a functional activity of the brain, represents a behavioral or functional equivalence class (Auletta *et al.* 2008) of different brain structural organizations. Here, we also recall that in certain contexts, the brain can exceptionally rebuild damaged areas and rewire itself without appreciable differences in its functionalities (Hawkins and Blakeslee 2004, p. 61). *Mutatis mutandis*, we can also understand higher mental activities and events (like beliefs, decisions, formulation of plans, and so on) as being possibly implemented in different brain configurations (Auletta 2011, Sec. 24.2; Changeux 1985; Changeux 2002). This is a relevant point in order to better understand the diversity of individual human brains, whose development, as we have seen, starts from a common species-specific topology induced by the aforementioned genetic envelope, and ends up with the individual-specific conformation that strongly depends on personal experience and learning, as well as on the social and cultural environment. The latter diversity is also involved in the generation of the variability of human cultures (Changeux 1985; Changeux 2002; Auletta 2011, Chaps. 11 and 22). As a matter of fact, the human's prolonged postnatal brain development is a necessary condition for culture to develop: it is associated with the genesis and the internalization of culture and, specifically, with the implementation of "cultural pathways" within the brain such as those involved in the acquisition of written language.

The spontaneous activity of the brain is of fundamental importance for the proper course of developmental epigenesis [see also Changeux (1985)]. Spontaneous activity occurs often in the functioning of our brain overour entire life and begins very early. Rita Levi-Montalcini was the first to show that such an activity is present already in the chick embryo (Levi-Montalcini and Visintini 1938). An eloquent example of the relevance of this activity for the process of synapse selection comes from the visual system (Galli and Maffei 1988; Dani *et al.* 2005). If in the retina of a fetus the spontaneous activity of the developing visual networks is blocked by tetrodotoxin, a larger number of synaptic connections is noted with respect to the normal (non-blocked) situation (Stretavan *et al.* 1988). The role of the neuronal activity for the synaptic selection process may also be attested by the analysis of the development of the neuromuscular junction. At birth, around five nerve terminals converge at one neuromuscular junction and only a single one persists in the adult. Moreover, this evolution is regulated by the state of activity of the system (Bourgeois *et al.* 1986). Therefore, the overall pattern of organization of the mature brain does not develop only through the genetic envelope and the evoked activity of the neural networks, but also, and to a certain extent mainly, thanks to the early spontaneous neuronal activity.

10.3. NEURAL ARCHITECTURE IN MAMMALS

Before dealing with higher cognitive functions and processes in humans, it is important to take into account an intermediate level between the local processing of information in neuronal microcircuits and the global level of consciousness. One may distinguish, within these intermediate systems, three main subsystems of the brain: the sensory, the motor-decisional, and the emotional. (i) The sensory system starts bottom-up from the primary sensory cortices, constituted by the somatosensory cortex (Brodman's areas 1–3), the visual cortex (area 17) and the auditory cortex (areas 41 and 42), up to the unimodal sensory association areas, and the multimodal association areas (see Fig. 10.5). (ii) The decisional-motor system takes the top-down path starting from the anterior multimodal motor integration area located in the prefrontal cortex (areas 8–10 and 46), down to the premotor system in

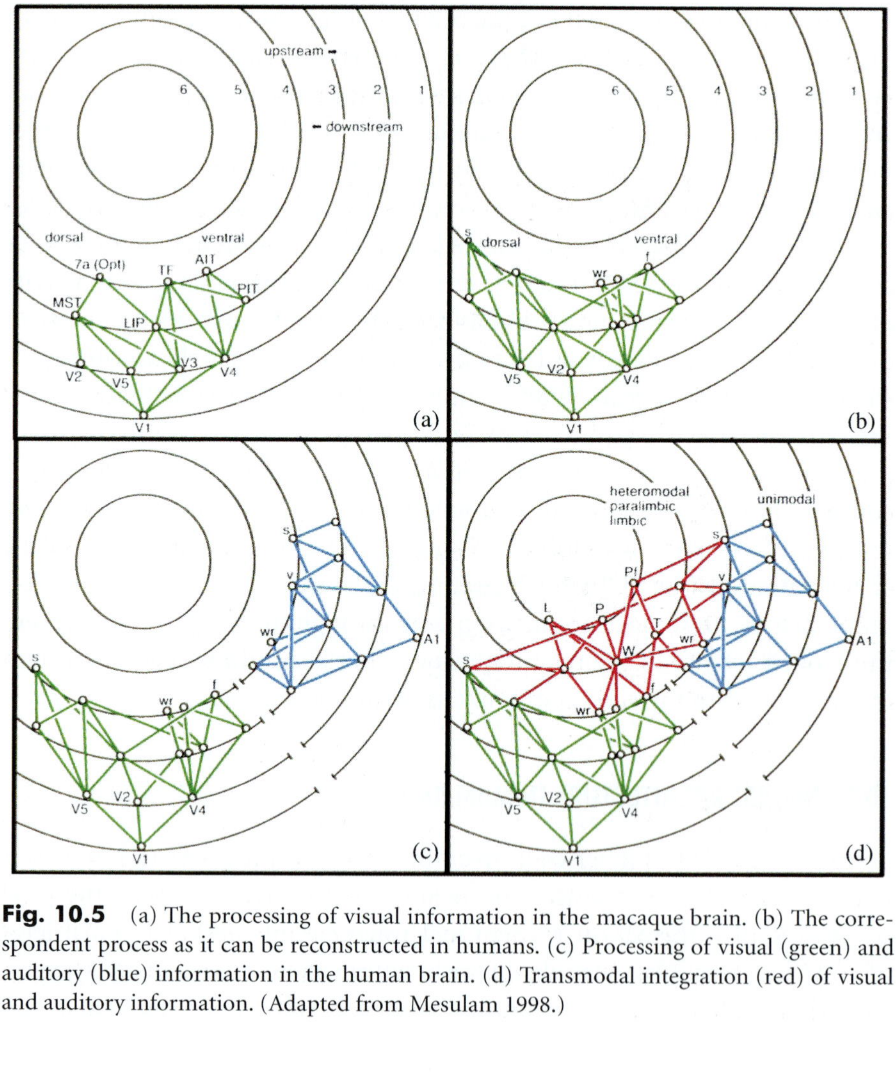

Fig. 10.5 (a) The processing of visual information in the macaque brain. (b) The correspondent process as it can be reconstructed in humans. (c) Processing of visual (green) and auditory (blue) information in the human brain. (d) Transmodal integration (red) of visual and auditory information. (Adapted from Mesulam 1998.)

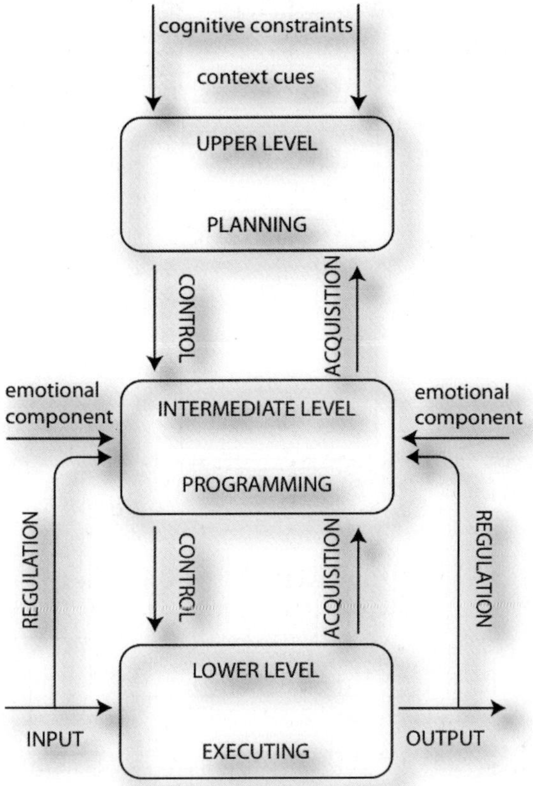

Fig. 10.6 The three levels of the mammal neural organization of the motor system. (Taken from Auletta 2011, p. 446. Adapted from Jeannerod 1999, p. 62).

the frontal lobe (area 6), finally reaching the primary motor cortex (area 4); we may assume that these levels can be roughly associated with those shown in Fig. 10.6, namely planning, programming and executing (see Auletta's contribution to this volume). (iii) The most complicated system is the emotional one, implemented in a circuit that includes three fundamental layers: the brain-stem part represented by the limbic nuclei, the association limbic areas including the cingulate gyrus, the parahippocampal gyrus and the insula. A central role in the emotional system (both for peripheral-autonomic and central-evaluative pathways) is played by the amygdala (Iversen *et al.* 2000). The role of the amygdala in its connections with the other systems is shown in Fig. 10.7.

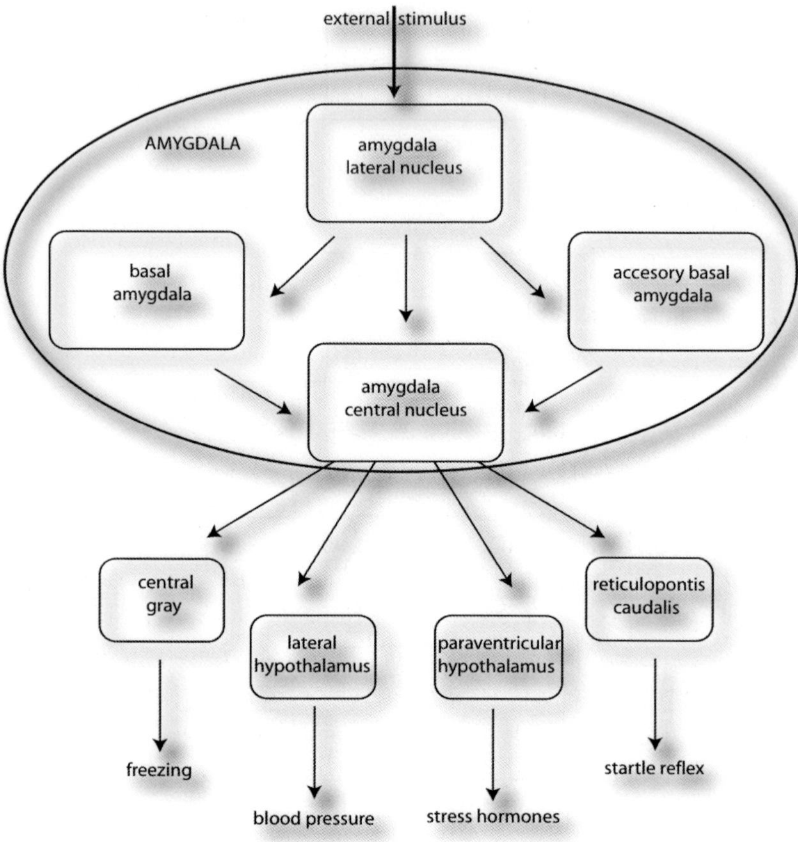

Fig. 10.7 The amygdala's outputs according to LeDoux's model (2003). The outputs are produced by the central nucleus of the amygdala to which both an emotional input (amygdala lateral nucleus) and a motor input (basal ganglia and accessory basal ganglia) concur. The motor system is already informed by inputs coming from the lateral nucleus of the amygdala. (Adapted from Auletta 2011, p. 450.)

At the mammalian (non-human) level, we may assume that the three previous subsystems interact mainly through binary relations (Auletta 2011, Chaps. 13–17). In particular, the sensory and the emotional systems cooperate in producing memories (which are strongly emotionally charged and require sensory inputs at the time of their constitution). The whole temporal lobe (including hippocampus) is likely to be the place where memories are processed before being transferred to more local circuits only after they become robust against disruption. These circuits are relatively well studied at present [see LeDoux (2003) for a summary]. Moreover, we may assume that the sensory and the motor-decisional systems cooperate in learning processes (which require sensory inputs but also selection to provide for generalization). Data (considered below) show that the precuneus (PCUN) takes part in these kinds of connections, since it seems involved in memory and spatial processing, as well as in conscious learning of words. We may further assume that the emotional and the motor-decisional systems concur in constituting behavior. Behavior is often confused with learning, which, for elementary organisms, may also be true, but in higher organisms is markedly endogenously generated and requires both the selection provided by the motor-decisional system and an emotional component.

When we ascend from the mammalian level to the proper human one, crucially characterized by consciousness, clear-cut distinctions such as the ones just introduced cannot be established so sharply (Auletta 2011, Chaps. 18–23). The integration among the subsystems, as well as among the various mental functionalities, seems to be much higher. Memories are integrated in high symbolic systems in which categorical and personal episodic aspects are deeply intertwined, determining an organization based on scripts (clusters of memory schemes showing a temporal dimension). Moreover, human memories have a social-cultural component: intentionality likely plays a fundamental role. Furthermore, consciousness may be the distinctive way in which we humans learn, and it is likely to be a global state that permeates any other aspect of our neural and mental activity. Therefore, especially when consciousness is at play, all the previously mentioned systems and functions might be strictly connected and intertwined in a triadic manner. It is not easy to deal with this level of complexity. We shall first deal with this problem at an evolutionary scale by reference to our next kin, i.e., great apes. Then, we shall formulate a general qualitative hypothesis and try to find out some neural evidence to support and refine it.

10.4. THE EVOLUTIONARY ORIGIN OF CONSCIOUSNESS

Human consciousness is one of the highest and most intriguing functions emerging out of the primate evolution and is involved in the organization of thought and in testing the "*internal*" world. Consciousness, in a preliminary and non-exhaustive definition, may indeed be regarded as a "subjective or mental space" — a "global workspace" (Baars 1988), or an "internal physical milieu," where actions are replaced by simulations, plans, goals and courses of simulated actions [see Dehaene and Changeux (2011); see also Jeannerod (2006)]. Such representations are processed and evaluated in a global manner with reference to the outside world, the self and personal memories, as well as to internalized rules and social conventions that, as we have seen, are internalized in the process of epigenetic development (Changeux 2002, Chap. 6; Dehaene and Changeux 2011; Auletta 2011, Chaps. 22 and 24). It is not possible to understand the constitution of this internal space without considering the way in which other primates deal with their social game. It is here, indeed, that reflexive actions take place (implemented thanks to the mirror neurons), and simulation and understanding of others' behavior becomes crucial for promoting the individual's welfare.

In the course of evolution, the relative surface of the prefrontal cortex increases from lower mammals (e.g., cats) to man (Fig. 10.2), as does conscious processing (Dehaene and Changeux 2011). Significantly, such anatomical change is broadly paralleled by an increase in the sociality of the concerned species. Fission-fusion societies (Goodall 1971; Goodall 1986) emerge late in evolution: in some monkey species, in great apes (notably in chimpanzees) and in man (apart from cetaceans, which have undergone an appreciably different neuroanatomical evolution, especially in that they do not exhibit a marked frontal lobe). Therefore, sociality may be a factor correlated with the appearance of brains capable of supporting conscious processing [see also Auletta (2011), Secs. 18.2, 21.2 and 21.3].

Human reflexive consciousness (or subjective self-awareness) is accompanied by theory-of-mind, sense of personal continuity, theories about oneself, and explicit *reportability* of one's own "internal world," all features indispensable for the highest social forms like humans. Such a higher level of consciousness develops in 3–5-year-old children, and this point is consonant with what we have said above about the importance of postnatal development that, crucially for humans, takes place in a highly complex social environment.

One of the best ways to try to understand the issue of consciousness is to elaborate models for it or for some of its aspects; the next section is dedicated to such an attempt, with special focus on the conscious access, i.e., those processes through which some content reaches the conscious level, thus becoming *reportable* by the subject. The epistemological soundness of this way of dealing with the problem is due to the fact that, when data or formalisms are insufficient, science is forced to proceed with general and qualitative models as an obligatory path of investigation for gathering experimental evidence and so make consistent steps forward.

10.5. THE GLOBAL NEURONAL WORKSPACE HYPOTHESIS AND CONSCIOUS ACCESS

Human consciousness is characterized by several distinctive features (Dehaene and Changeux 2011; Auletta 2011, esp. Chap. 21). First of all it is a discontinuous process that alternates with phases of unconscious processing or states (such as sleep or automatic behaviors). At the brain level, it is a globally distributed and metastable process able to mediate sudden jumps between distant brain information-processes and -processors as well as to establish relations among unrelated contexts. Consciousness is connected with the specifically human way of learning, as it is induced by effortful tasks and is capable of giving rise to lasting and flexible modifications of the executive control on one's own behavior, especially thanks to a top-down molding of spontaneously arising activity.

The global neuronal workspace hypothesis is at the core of a model for consciousness that stems from the early anatomical work of Dejerine (1895) and Ramón y Cajal (1899). In this work, pyramidal neurons with long-range connections were identified and found to be abundant in layers 2 and 3 of the cerebral cortex. Furthermore, these neurons with long-range axons are especially numerous in the prefrontal cortex (von Economo 1929) and give rise to prefrontal-parieto-temporal-cingulate circuits. From this, the hypothesis may be formulated that a reciprocal network of long-axon neurons creates a "global *neuronal* workspace" (Dehaene *et al.* 1998; see Fig. 10.8). On the basis of this model, both the diversity and the globality of consciousness should be emphasized, in the sense that this global workspace is constituted when multiple kinds of information coming from the rest of the brain converge and may give rise to conscious contents. This model can be formalized and, most importantly, can be *tested* experimentally (Dehaene *et al.* 1998). It is made up of two computational spaces. The first one is constituted of several specialized processors that are modular, hierarchically encapsulated, and *automatic,*

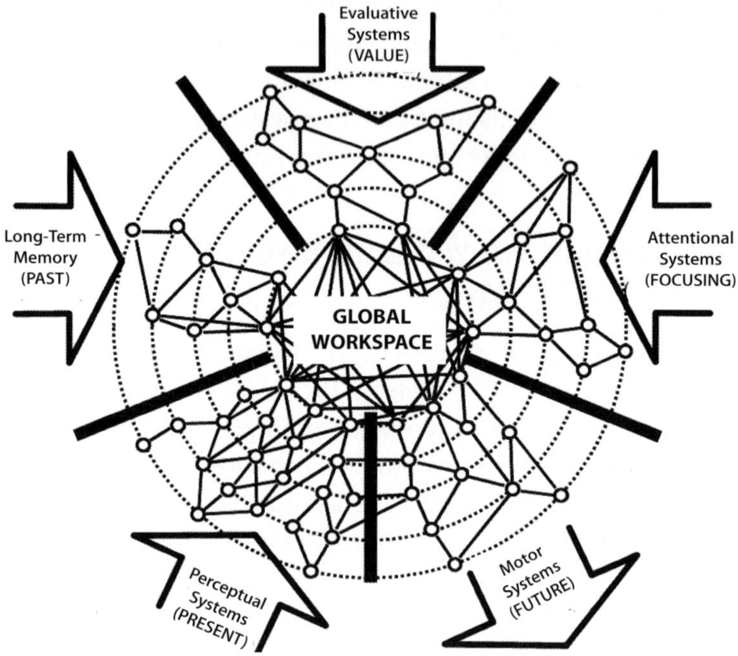

Fig. 10.8 Some of the possible workspace inputs and outputs.

according to the original proposal of Baars (1988). Therefore, these computations are *non-conscious*. The second computational space is the global neuronal workspace. The reverberating and self-sustained top-down effect that follows conscious access (Dehaene and Changeux 2005; Dehaene and Changeux 2011) is cross-contextual and is aimed at searching and ultimately establishing new relations among different contexts. Finally, as we shall see, the late, reflexive consciousness will precede the creation of new contexts or their updating out of this cross-contextual activity (Auletta 2011, Sec. 21.3).

Thus, the global workspace, associated to reciprocally connected, long-range axon neurons, broadcasts signals to multiple areas, thereby yielding to the subjective experience of being conscious, to the storage of future actions and plans, and to *reportability*. Specifically, when a spontaneous mobilization of the aforementioned *automatic* processors and stored memories takes place, such that a sudden activation of a conscious state occurs, this has dramatic *top-down* effects amplifying the operations of the automatic processors constituting the first, non-conscious computational space. A character of the global neuronal workspace that is worth stressing is its capability to work as a "diversity generator" (Changeux 2002, pp. 61 and 64): an endogenous variability of pre-representations is produced that are then *selected* through reward mechanisms [Dehaene and Changeux 2000; Dehaene *et al.* 2003; see also Mesulam (1998)].

For all this to be possible, large-scale brain integration is needed. As a matter of fact, long-range circuits have also been observed in the brains of monkeys (Goldman-Rakic 1988); in humans, however, it appears to be a more elaborated circuitry in which, as mentioned, many functions and processes are interconnected, and the contribution of the prefrontal cortex (Changeux and Dehaene 1998) is particularly important with respect to monkeys. From this point of view, the brain white matter acquires a special importance precisely because of its relevance for the long-range links interconnecting several particular different brain areas. A significant finding by Zilles and coworkers shows the evolution of the rate of white matter over the neocortex from primitive mammals to primates and *Homo sapiens* [Smaers *et al.* 2010; see also Schoenemann *et al.* (2005)]. In particular, the differential increase of white matter in the course of vertebrate evolution displays a significantly sharp rise from prosimians to simians and, even more impressively, from simians to *Homo sapiens*, thus supporting the idea that consciousness relies on such long-range brain interconnections. Furthermore, quantitative studies show that the mammal dendritic field morphology of layer-3 pyramidal neurons — central to the global workspace modeling (Dehaene *et al.* 1998) — displays an increase in complexity from the occipital to the prefrontal cortex (De Felipe and Feriñas 1992; Elston and Rosa 1997; Elston and Rosa 1998). Moreover, there is an increase of such differentiation across evolution, especially from monkeys to humans (Elston 2003).

According to the neuronal workspace hypothesis, conscious access is marked by a sudden *ignition* (Changeux and Dehaene 2008) of a large (prefronto-parietal) cortical network in which the prefrontal cortex is a central hub. This may be correlated with four objective signatures (Dehaene and Changeux 2011): (i) late and distributed event-related potentials in the prefrontal cortex, (ii) large and late increase in high gamma-frequencies, (iii) increase in long-distance cortico-cortical synchrony in beta-frequencies, and (iv) increase in the causal relation among distant cortical areas. As we shall see in the next section, experimental evidence from stimulus-evoked conscious access and conscious access in effortful and attention-demanding cognitive tasks actually emphasizes the role of the prefrontal cortex.

10.6. EXPERIMENTAL EVIDENCE

Let us now turn to evidence in support of the model, starting from stimulus-evoked ignition. The first evidence comes from the cortical processing of conscious versus non-conscious words, assessed thanks to brain imaging records when the subject is presented with visible or masked words (Dehaene *et al.*

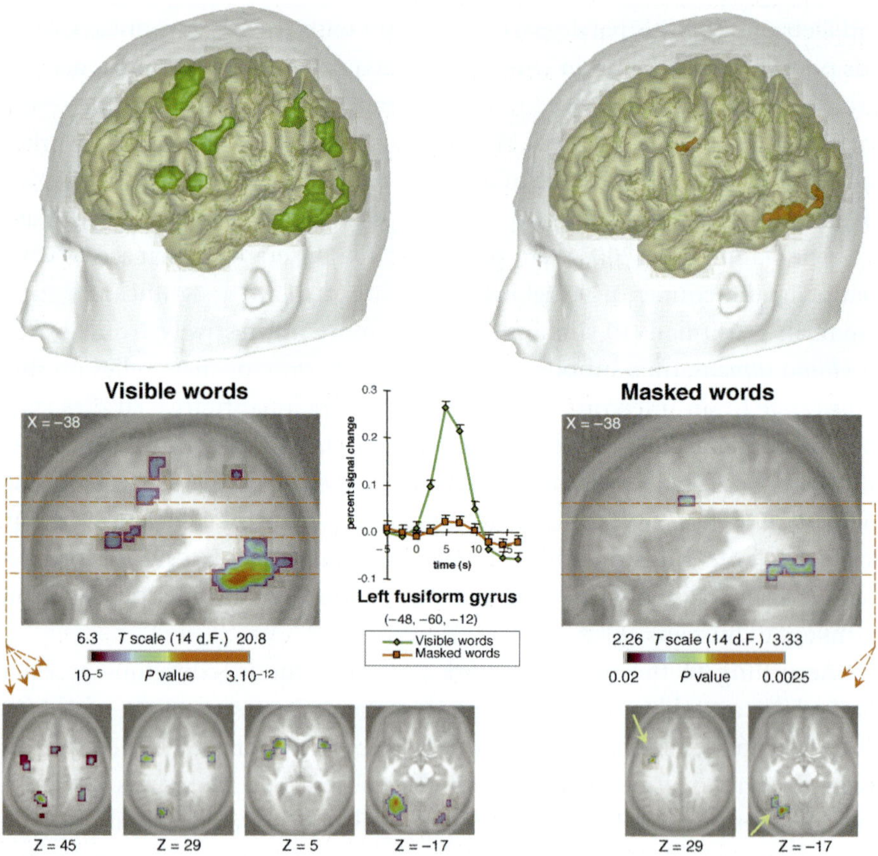

Fig. 10.9 fMRI activation to visible and masked words. *Top*: activations in the left hemisphere; *middle*: activation peak at the left fusiform gyrus; *bottom*: activations in Talaraich space (sagittal and axial views). (Adapted from Dehaene *et al.* 2001.)

2001). What has been observed is a significant increase in the cortical activity (especially high in the left fusiform gyrus) during the conscious processing of visible words. Compared with the case of unconscious processing of the masked words, the previous case shows the activation of several distinct cortical areas, including the prefrontal cortex (Fig. 10.9). This finding is consistent with the hypothesis of the activation of the global neuronal workspace in conscious processing.

A similar experiment has been performed focusing on another task, called the "attentional blink task" (Sergent *et al.* 2005): this effect occurs when an early distracting stimulus, occuring at time T1, prevents access to consciousness of a second stimulus, occuring at time T2, presented shortly (a few tenths of a second) after T1. With regard to the processing of the stimulus at T2, two situations have been distinguished (Fig. 10.10; see also Fig. 10.11): the "not-seen" (subliminal, or unconscious) situation and the "seen" (conscious) one. Event-related potential recordings reveal *early* (300 ms) occipito-temporal activation in both subliminal and conscious trials but a *late* (436 ms) activation in prefrontal areas (namely, dorsolateral prefrontal, as well as inferior frontal and anterior cingulate) *only* in the

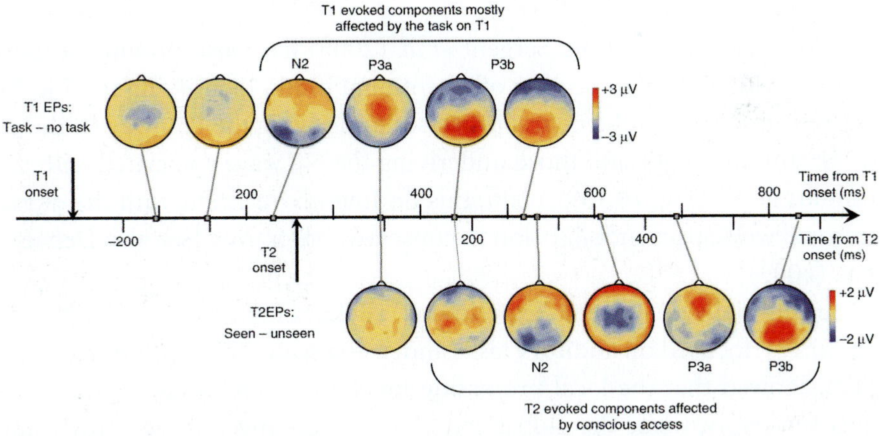

Fig. 10.10 Temporal relation between ERP waves evoked by task occuring at time T1 (task T1 in short), which causes the blink on T2, and event-related potential waves evoked by the stimulus occuring at time T2 (task T2 in short) in the seen-minus-unseen comparison. Voltage maps above the timeline show the difference between trials with and without a task on T1 (short stimulus-onset asynchrony, SOA; T2 absent), over time. Voltage maps below the timeline show the "T2 seen"-minus-"T2 unseen" difference during the attentional blink (dual task, short SOA), over time. (Adapted from Sergent *et al.* 2005.)

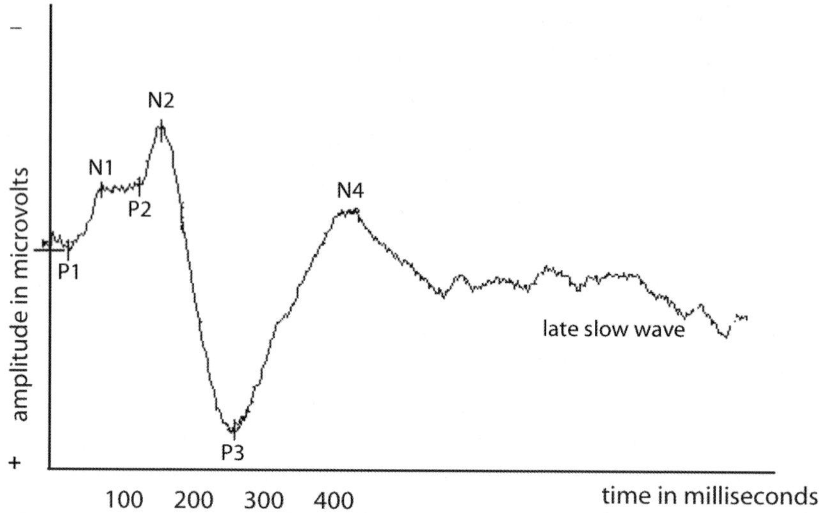

Fig. 10.11 EEG response to sensory stimulus processing (P = positive; N = negative). P1 and N1 waves in sensory areas point out the response to the physical features of the stimulus. P2 and N2 in the temporal cortex may be associated with the brain's detection of the event. P3 and N4 reflect the initial evaluation of a particular sensory event (P3 is further distinguished in P3a, which marks the initial discrimination of an unexpected event, and P3b, associated with the processing of the event's implications). (Adapted from Auletta 2011, Sec 3.5.)

conscious, "seen," situation. Sergent *et al.* (2005) have also provided experimental evidence for the *causes* of attentional blink, identifying it with the competition between the neural processes underlying the P3b wave induced by the stimulus at T1 and those underlying the N2 wave associated with the stimulus at T2 (Fig. 10.10). Again, this finding is consistent with the global neuronal workspace mobilization in conscious processing [see also Dehaene *et al.* (2003)].

Additionally, a recent experiment (Bekinschtein *et al.* 2009; see Fig. 10.12), focused on auditory tasks and, using scalp EEG and intracerebral fMRI, showed that the local processing involving local auditory areas activates very early, while the global conscious processing emerges slowly and late during the task execution. We may say that all these kinds of experiments allow us to treat perceptual consciousness as an experimental variable.

In the previous section, we have seen that conscious access occurs when there is shaking or even disruption of contexts and that the following reverberating top-down effect aims at dealing with the unforeseen situation with a delayed effect resulting in the establishment of new relations among

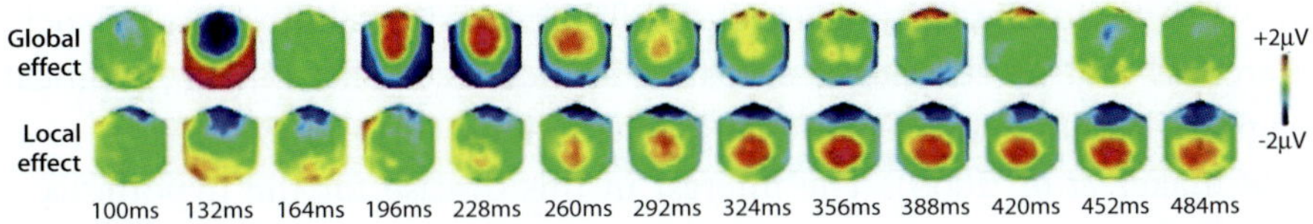

Fig. 10.12 Timings of local and global processing in auditory tasks. (Adapted from Bekinschtein *et al.* 2009.)

different contexts. The neuronal workspace model has been also applied to effortful and attention-demanding cognitive tasks like the Stroop test. In this experimental situation, a subject is asked to identify the color of the ink in which a word is written, but the meaning of the word may differ from the actual color (e.g., the word "green" is written with blue ink), thus creating a conflicting situation. When this happens, the subject is forced to make a conscious effort to solve the task correctly. Such situations have been simulated thanks to the neural workspace model (Dehaene *et al.* 1998; Changeux and Dehaene 2008). The simulations show an activation of the workspace when, after a train of executions in the control situation, i.e., with the non-conflicting set (coincidence of color and meaning), the effortful and conflicting situation (the test) is introduced, and a series of errors (negatively rewarded) occurs as the system keeps applying the routine response. What is interesting in this situation is that the difference between the presentation of the conflicting situation after the non-conflicting one may be interpreted as a very basic instance of shaking or disruption of a previously unproblematic context, i.e., the routine application of the rule that associates the response (color name) to the word meaning. The resulting errors cause the activation of the conscious effort and a phase of search takes place aimed at establishing a new rule for the correct answer, i.e., the rule associating the color name to the color of the word and not to its meaning. This may be interpreted as an updating of the context resulting from the reverberating and self-sustained top-down consequences of the workspace on the lower-level processors.

Additional evidence in support of the neuronal workspace model comes from neuropathology. Frontal lobe dementia is a consequence of single-gene mutations in a growth factor (progranulin) and in a cytoskeletal gene (*tau*), and causes a neuronal loss in layers 2 and 3 of the prefrontal cortex. Frontal lobe dementia results in an alteration of the autonomy of the person and gives rise to apathy, unconcern, disinhibition, distractability, loss of social awareness and of emotional empathy, as well as a shift in, e.g., political, religious, sexual and even dressing orientations (Brun 1987; Neary *et al.* 1988; Baker *et al.* 2006). The interest in this disease for the hypothesis at hand lies in the fact that it involves the prefrontal cortex and precisely the pyramidal cells from layer 2 and 3 assumed to compose the global neuronal network.

Further evidence comes from autism, which also stems from genetic point mutations affecting synapse formation. Autism results in altered social interaction, reduced central coherence, and increased low-level local treatment at the expense of global high-level activity. A hypothesis able to explain the observed phenotype on the basis of the known mutated genotype is that the development of long-axon neurons is differentially altered by the mentioned mutations [Changeux and Edelstein 2005; see also Belmonte and Yurgelun-Todd (2003)].

10.7. PMC AND THE DEFAULT MODE

The evidence just reported clearly shows that the global workspace hypothesis is promising for understanding the functioning of the human brain and the mechanisms underlying high cognitive functions. In particular, we have seen that significant experimental results have come from tasks involving processing of unimodal sensory stimuli (either visual or auditory). Such findings indicate that the prefrontal cortex (PFC) plays a crucial role in conscious access, in agreement with the predictions of the model. Consciousness, however, is a complex human function that cannot be reduced to conscious access alone and that, as we have seen, should be conceived as highly distributed, involving the brain globally, as well as associated to many brain functional subsystems (sensory and motor aspects, along with memory, attentional-decisional and emotional processes, learning and conscious, although delayed, re-elaboration of primary information). The model indeed postulates that the workspace is connected to a huge amount of disparate and specialized subsystems such that the involved neurons are not to be found only in the PFC but should rather be thought of as being widespread throughout the brain so as to give rise to a widely distributed cortico-cortical and cortico-thalamic network. There is mounting neuroanatomical knowledge, especially concerning the connectivity of the brain, that may, on the one hand, provide further support to some aspect of the global neural workspace hypothesis and, on the other, likely suggest additional ways to gain further empirical evidence for its predictions.

There is indeed emerging evidence for the theoretical hypothesis that human consciousness represents a further level of integration, in evolutionary terms starting with the basic neural architecture of mammals and arriving at a new neural organization allowing interconnections among all basic systems and functions, thereby supporting the hypothesis of a global workspace. Let us begin with a recent study by Hagmann *et al.* (2008) (see Fig. 10.13), in which a reconstruction of the whole brain network in the resting and awake state has been provided. Most of the connections go in the anterior-posterior direction passing close to the cutting line between the two hemispheres; the central part, involving in particular the cingulate cortex, is strongly involved. It turns out that several areas provide connections between the motor-decisional system and the emotional system, thus concurring to produce behavior. We stress the roles of three specific regions: the medial orbitofrontal cortex (MOF, Brodmann area 11), the rostral anterior cingulate cortex (RAC, Brodmann area 32), and the caudal anterior cingulate cortex (CAC, Brodmann areas 33 and 24):

(i) MOF is involved in decisions but, because of its function in emotion and reward, is also considered as part of the limbic system. It deals with reward and punishment, as well as with intuitive coherence judgments.
(ii) RAC connects left and right hemispheres and metabolic-emotional functions with decision-making. It is involved in the tasks of error correction and conflict resolution.
(iii) CAC has essentially the same functionalities and tasks as the previous area (i.e., RAC).

Apart from the obvious relevance of these findings for the issue of consciousness, they also represent compelling evidence for the binary connections characterizing the mammal brain. Moreover, this work (Hagmann *et al.* 2008) shows considerable convergence among findings into two relevant research streams in present-day neuroscience: one related to the default mode of the brain function (Raichle *et al.* 2001) and the other dealing with the role of the posteromedial cortex (PMC) (Parvizi and Damasio 2001; Parvizi and

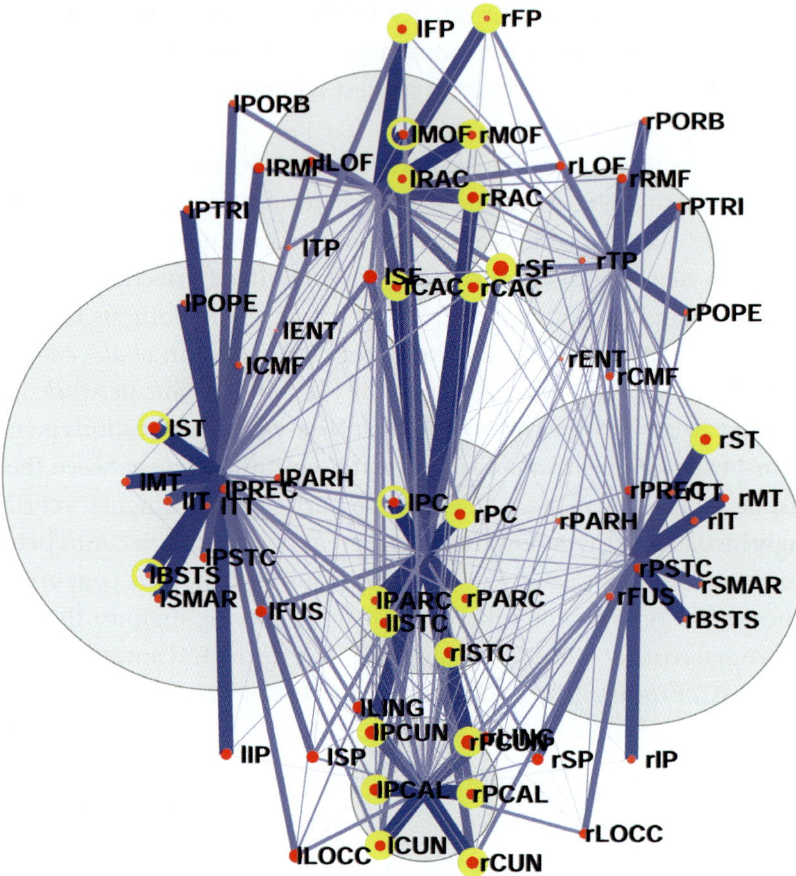

Fig. 10.13 The so-called structural-core neural network as reconstructed in Hagmann *et al.* (2008). Dorsal view of the brain, starting from the frontal part (top) up to the posterior part (bottom). The network shows a relative symmetry between left (l) and right (r) hemispheres. The gray areas are modules. Very schematically, we can say that the two top gray areas correspond to the anterior part of the brain, the three central ones to the center (with the two large left and right circles covering approximately the temporal areas, whilst the small central one, which is in part covered, individuates the central-upper part of the brain); finally the small gray circle at the bottom corresponds to the posterior part of the brain. The red dots individuate specific brain areas. When circled by yellow rings, they represent hubs. The authors distinguish between provincial hubs (showing connections mainly in the same module) and connector hubs (connected also with other modules). The main connector hubs are RAC (rostral anterior cingulate cortex), CAC (caudal anterior cingulate cortex), STC (superior temporal cortex) and PCUN (precuneus). For the meaning of the other acronyms, see the original paper.

Damasio 2003; Parvizi *et al.* 2006). The default mode of brain function is associated with the above-mentioned resting and awake state of the person (e.g., lying with closed eyes) and involves a set of brain areas that are active in conditions of resting, but awake states whose activities decrease in attention-demanding goal-directed tasks and behaviors [see also Shulman *et al.* (1997)]. In fact, many of the key components of the default mode are also found in the structural core individuated by Hagmann *et al.* (2008) (especially the posterior ones): the posterior cingulate (PC), the lateral and medial parietal cortex, and precuneus (PCUN). Interestingly, the medial prefrontal cortex (MPFC), a key component of the default mode, does not count among the main hubs in the structural core defined by Hagmann and colleagues (a point which we will return to later). Posterior cingulate and medial parietal cortices constitute the so-called PMC (namely including Brodman area 31, the posterior cingulate areas 23a–b, the retrosplenial areas 29–30, and the mesial parietal area 7m, i.e., PCUN).

Given these convergent anatomical findings, especially concerning posterior brain areas, let us now turn to some relevant functional considerations about these circuits. The PMC is strongly connected, often bidirectionally, with a great deal of other brain areas (see Fig. 4.5). By virtue of these connections, the PMC may play a crucial role in integrating a huge variety of brain information-processes and functionalities (Parvizi *et al.* 2006). Significant bidirectional connections are present with prefrontal areas 9, 10 and 46 (and to a lesser extent with area 8, the frontal eye field) as well as with the thalamus and the entorhinal cortex (which is crucial for the constitution of multimodal representations and long-term memories that have already been consolidated). In this way, it may be directly or indirectly implicated in processes spanning from memory to decision-making. PMC area 7m (PCUN) also contacts frontal and cingulate structures involved in action planning and execution. PMC retrosplenial areas 29 and 30 receive inputs from the amygdala, thus likely also playing a role in emotions. PMC areas 31 and 7m are interconnected with supplementary somatosensory areas and, although PMC has no direct connection with primary sensory areas, it is, as we have seen, strongly and bidirectionally linked to the thalamus, which is known to be both a relay- and processing-station for virtually all kinds of sensory, motor and emotional information coming from and going to the environment and the body.

All this may suggest that the PMC — and the default mode as well (Gusnard and Raichle 2001) — has a crucial role in self-representation and self-consciousness. As a matter of fact, indeed, this region is the first to increase activity in regaining consciousness from vegetative states (Laureys *et al.* 1999). Interestingly, the PMC is also the most active region in the resting-state default mode (Gusnard and Raichle 2001).

10.8. A LARGER CONTEXT

All the findings reported in the previous section are essentially referred to the resting state, when no goal-directed, attention-demanding task is currently being executed. The question may now arise as to the connection of all this with the empirical evidence reported in Sec. 10.6 as to the activation of the prefrontal cortex in attention-requiring tasks. Moreover, we have also seen that the prefrontal areas are not included in the structural core individuated by Hagmann *et al.* (2008) (Fig. 10.13). However, in our opinion, it is possible to point out some aspects that may turn out to be stimulating for further developing and testing the global neuronal workspace hypothesis. First of all, it must be noted that the report by Hagmann *et al.* (2008) may suffer, at the pure anatomical level, from an underestimation of the long-range connectivity of areas 46, 6, FEF (frontal eye field) and LIP (lateral intraparietal region), which are known both from invasive tracer studies in the monkey and from early human anatomical dissections (Dehaene and Changeux 2011). Secondly, the PMC, together with lateral parietal and medial prefrontal cortices, decreases activity during attention-demanding cognitive tasks *unless* they require reference to the subject's notion of self (Raichle *et al.* 2001; Fox *et al.* 2005; Parvizi *et al.* 2006). Indeed, attention-demanding tasks activate frontal and parietal cortical regions and deactivate posterior cingulated as well as medial and lateral parietal cortices and the medial prefrontal cortex, but the latter deactivation is shown to be appreciably attenuated by self-referential aspects (Fox *et al.* 2005). It seems therefore that the lack of activation of more posterior areas in cognitive tasks not involving explicit reference to the self (which could be called a non-reflexive kind of consciousness) is fully justified.

It is also worth mentioning that the medial parietal cortex (which is deactivated in attention-demanding cognitive tasks) has been proposed to be a nodal structure for self-representation involving emotional aspects related to personal identity and episodic memory, a conclusion also based on transcranial magnetic stimulation (TMS) of this region (Lou *et al.* 2004). As mentioned, medial prefrontal cortex (MPFC) is part of the default-mode brain network [Gusnard and Raichle 2001; Raichle *et al.* 2001; see also Fox *et al.* (2006) for MPFC involvement in attention systems]. Both ventral and dorsal MPFC decrease their activity in various goal-directed behaviours. Dorsal MPFC, however, also shows increased activity in cognitive processes involving the monitoring of one's own mental state or the attribution of mental states to others. Medial prefrontal areas (encompassing Brodmann's areas 32 and 10) are part of a task-negative network that decreases its activity during tasks performance but is involved in self-referential processes (Fox *et al.* 2005). Finally, MPFC (and Brodmann area 32 in particular) is part of a network, also including the hypothalamus and brainstem, seemingly supporting the dynamic interplay between cognitive task performanace and emotions [Simpson *et al.* 2001; see also Bechara *et al.* (1997)].

We may summarize by saying that the PMC is involved when there is spontaneous, self-evoked activation of consciousness, that is, activation not related to an external stimulus or a specific task-context. This may occur in search contexts (Baars 1988, Chap. 6), or when we try to consciously assimilate certain information. It is worth mentioning here that most of the solutions that can be found in the presence of unexpected stimuli or difficult tasks (problems that cannot be solved locally through specialized information processors) are themselves sources of problems, since a solution that is new has not been sufficiently assimilated by the organism and cannot at first be considered as fully reliable (Auletta 2011, Sec. 21.3). Then, after the first stage in which a problem has been solved under the impact of external conditions, a second stage may occur in which this new solution is consciously assimilated. During the latter process, the brain tries to transfer this solution to specialized operators in order to incorporate it in the previous web of neural and mental interdependencies. It is likely that several cycles of this activity must run for a new solution to be fully assimilated. This activity is clearly endogenously generated, is likely to occur in rest state and may be called reflexive consciousness to distinguish it from the stimulus-evoked (non-reflexive) variety. In other words, we may distinguish three different situations: (i) when only reflexive process are involved (mainly the

PMC is activated), (ii) when only non-reflexive processes are activated by external stimuli or tasks (mainly PFC is activated), and (iii) when the two cooperate, especially when some task is difficult enough to demand a second cycle aimed at subsequent error correction or at fine-tuning some newly assimilated solution.

10.9. CONCLUSION

Everything we have discussed (especially in the last sections), on the one hand, provide further support for a general feature of the global neuronal workspace hypothesis, namely, the requirement that the workspace neurons are widespread among the brain and globally share long-distance connections with other local brain processors (Changeux 2002, pp. 87–90). The mechanisms through which the workspace network exchanges information (both top-down and bottom-up) with other brain processors may well be consonant with a hierarchical network of encapsulated convergence-divergence zones (Meyer and Damasio 2009). On the other hand, these findings challenge the model with the possibility of expanding its simulations to even more complex situations encompassing explicit self-referential processes, likely involving, among other things, working and episodic memory, voluntary control and attentional processes (Baars 2002), as well as decision-making and emotional states. Clearly, the variety of brain regions, circuits and processes involved present a serious challenge as to this highly ambitious task, likely resulting in a computational explosion. However, one possible way to deal with this problem may be to resort to a simplification of such a complex network based on the three-subsystem view of the brain mentioned above (Sec. 10.3). Tentatively, we may say that the main hubs involved in the global network, which includes both frontal and posterior contributions, are the PFC, the cingulate cortex, the thalamus, and the PMC.

DISCUSSION

Wolf Singer: Well, my question is about the neural correlates of consciousness: What are we going to look for? It is obvious that once the stimulus has reached conscious processing, the whole sequence of processing looks different from the stimulus that is first processed in producing a response.

So, there is short-term memory involved and, as soon as the stimulus is consciously processed, it turns out to be transcribed into other forms of memory, probably long-term memory in most situations. Now, when we look at the electrographic signatures that you showed, where there were many low potentials; are we really looking at the prerequisite for conscious experience or is something else occurring? I come back to the previous discussion on Libet: access to consciousness must occur within 200 milliseconds. Indeed, eyes move roughly at this frequency, and we have the capability to give conscious reports of this (that is, you know what you have done).

Jean-Pierre Changeux: I have no objection to what I have just heard. I think that the point to be mentioned — that is, where the problem might intervene — is that the singular complexity of the stimulus may introduce bias on the dynamics of the process: in the case of reading a word, of course, the language bias has to be mobilized. If, in contrast, one were looking at a much simpler stimulus, as you just said, it might be possible to have no opposition to that, i.e., no biases to be passed. What is interesting is that it seems that the electrophysiological characters of neurons involved in conscious processes are different from those involved in non-conscious ones. There is a signature of the electrophysiology of neurons involved in conscious processing.

Pierre Jacob: My remark is in the line of what Wolf Singer was pointing out. First of all the experiment considered is about perception of words. Now, working memory is required in order to read words, and attention is needed to record them. The question, then, is whether or not the lack of working memory and attentional resources implies the deprivation of any kind of conscious experience. This also relates to the issue of whether non-human animals could actually have non-reportable experiences, and how we think about these issues.

Jean-Pierre Changeux: There are several ways to answer your question; I will choose a quick one. Looking at the stimulus presentations, one sees that working memory and attention are mobilized in the human workspace; then, you have a very significant effect of them on this kind of processing. And now I turn to what happens in other species. In comparison with other species, it must be noted that the child's brain adapts significantly during development and that we have a much larger contribution of the prefrontal cortex with respect to other mammals. This means that there is a maturation of the human workspace, as well as, of course, during the overall brain evolution. Indeed, different levels of consciousness can be individuated (at least for the purpose of partitioning what is rather a continuum of cases from rats to adult humans). In two-year-old infants there is explicit self-consciousness that develops together with episodic memory. Perhaps chimps can reach this level too. A more basic form of consciousness is present in monkeys. What is important here, however, is that there are techniques through which we can prove whether there is conscious or non-conscious processing.

Antonio Damasio: Well, I would like to start with a comment: Why are you so opposed to the use of the concept of mind? Indeed, when you talk about the global workspace, you are invoking mental events all the time. So, you are just using another name instead of mind. I understand why you may not want to use "mind," but it really does not make a difference. I think it may actually be of help if you want to invoke a notion of self versus non-self. Indeed, in my conceptualizing, there is an add-on to the mind without the self. We do have neurological evidence that there are states in which we have behaviors conforming to the idea that there is mental activity (e.g., maps are being formed, and responses are being made to those maps) and yet there is not a sense of self. Would you care to comment?

Jean-Pierre Changeux: Yes. Instead of using the word "mind," I think I would prefer to speak of a conscious access without reference to the self. It would be less confusing than introducing simplified terms on which we should not insist right now, in my opinion. I have no objections to this, but I think this would turn out in multiplying our definitions, and I would be pleased if we could avoid adding terms in a situation that is already extremely complex.

Gennaro Auletta: My point is only that, obviously, it is impossible to represent mental activity without the brain's excitation patterns and it is also not possible, as I tried to suggest in my talk, to represent mental activity without a sort of combination of physical items (words, gestures, and so on). For me, however, the focal point of human symbolic activity (and possibly of mind or of consciousness) is the fact that we are able to produce a lot of things that do not have a direct representational counterpart. While animals, and primates of course, are able to represent external conditions, the physical environment, events and so on, what strikes me very much in our mental activities is the fact that we are able to produce symbols (for instance mathematical symbols) concerning concepts like the number 2, or "brain" if you will, that do not necessarily have an immediate representational content; sometimes they have no representational content at all, like the concept of infinity, or of a black hole, or mind. I think this is a very relevant point. Obviously, I am not claiming that we are able to fully understand this delicate issue at the moment, but I would suggest to you, as experts in the field, that perhaps we need a further level of sophistication, if you will, to understand this specifically human symbolic activity.

Jean-Pierre Changeux: I agree. I think that the issue of representation is a very imbalanced feature that we have to model more precisely, and this is what we are currently trying to do.

Michael Arbib: In your talk, you presented us three experiments that you argued supported, or at least were consistent with, the hypothesis of the global neuronal workspace. My point is that in each of these experiments the activating areas were different. Now, returning also to Wolf Singer's question at the beginning of this discussion, what is the global workspace as distinct from the areas that are just processing a particular task? Is there, at the present moment, a strong claim as to which areas constitute the global workspace? And finally, what is it about the specific role of these areas in supporting consciousness as distinct from their role in other forms of processing?

Jean-Pierre Changeux: There are indeed areas devoted to different tasks, but what is crucial to say is that, concerning the *same* task, there are features of the task that are conscious and others that are non-conscious, without changing much of the pattern. The task is the same in the sense that we use the same experiment. However, there are only some features of the task that are used to assess consciousness, and they are found to be objectively different from those that are involved in non-conscious processes. In other words, only these specific features are linked with conscious processes.

REFERENCES

Arendt D, Nubler-Jung K. (1999) Comparison of early nerve cord development in insects and vertebrates. *Development* **126**: 2309–2325.

Auletta G. (2011) *Cognitive Biology: Dealing with Information from Bacteria to Minds.* Oxford University Press, Oxford.

Auletta G, Ellis G, Jaeger L. (2008) Top-down causation by information control: From a philosophical problem to a scientific research programme. *J R Soc Interface* **5**: 1159–1172.

Auletta G, Colagè I, D'Ambrosio P. (2013) A critical assessment of transhumanism. *Acta Philosophica*, 2. (in press).

Baars BJ. (1988) *A Cognitive Theory of Consciousness.* Cambridge University Press, Cambridge.

Baars BJ. (2002) The conscious access hypothesis: Origins and recent evidence. *Trends Cogn Sci* **6**: 47–52.

Baker M, Mackenzie I, Pickering-Brown S, *et al.* (2006) Mutations in progranulin cause tau-negative frontotemporal dementia linked to chromosome 17. *Nature* **442**: 916–919.

Bechara A, Damasio H, Trane D, Damasio AR. (1997) Deciding advantageously before knowing the advantageous strategy. *Science* **275**: 1293–1295.

Bekinschtein TA, Dehaene S, Rohaut B, *et al.* (2009) Neural signature of the conscious processing of auditory regularities. *Proc Natl Acad Sci USA* **106**: 1672–1677.

Belmonte MK, Yurgelun-Todd DA. (2003) Functional anatomy of impaired selective attention and compensatory processing in autism. *Cogn Brain Res* **17**: 651–664.

Bishop KM, Goudreau G, O'Leary DD. (2000) Regulation of area identity in the mammalian neocortex by Emx2 and Pax6. *Science* **288**: 344–349.

Bond J, Woods CG. (2006) Cytoskeletal genes regulating brain size. *Curr Opin Cell Biol* **18**: 95–101.

Bourgeois J-P, Toutant M, Gouzé JL, Changeux J-P. (1986) Effect of activity on the selective stabilization of the motor innervation of fast muscle posterior latissimus-dorsi from chick embryo. *Int J Dev Neurosci* **4**: 415–429.

Bourgeois J-P, Goldman-Rakic PS, Rakic P. (2000) Formation, elimination, and stabilization of synapses in the primate cerebral cortex. In MS Gazzaniga (ed.), *The New Cognitive Neurosciences.* MIT Press, Cambridge, MA, pp. 45–53.

Brun A. (1987) Frontal lobe degeneration of non-Alzheimer type I. *Arch Gerontol Geriatr* **6**: 193–208.

Changeux J-P. (1985) *Neuronal Man: The Biology of Mind.* Princeton University Press, Princeton.

Changeux J-P. (2002) *The Physiology of Truth: Neuroscience and Human Knowledge*, Harvard University Press, Cambridge, MA.

Changeux J-P, Connes A. (1989) *Matière à pensée.* Odile Jacob, Paris.

Changeux J-P, Danchin A. (1976) Selecting stabilization of developing synapses as a mechanism for the specification of neural networks. *Nature* **264**: 705–711.

Changeux J-P, Dehaene S. (1989) Neuronal models of cognitive functions. *Cognition* **33**: 63–109.

Changeux J-P, Dehaene S. (1998) Hierarchical neural modeling of cognitive functions: From synaptic transmission to the Tower of London. *CR Acad Sci III-Vie* **321**: 241–247.

Changeux J-P, Dehaene S. (2008) The neuronal workspace model: Conscious processing and learning. In R Menzel (ed.), *Learning Theory and Behavior*. Elsevier, Oxford, pp. 729–758.

Changeux J-P, Edelstein SJ. (2005) Allosteric mechanisms of signal transduction. *Science* **308**: 1424–1428.

Changeux J-P, Courrège P, Danchin A. (1973) A theory of the epigenesis of neuronal networks by selective stabilization of synapses. *Proc Natl Acad Sci USA* **70**: 2974–2978.

Cook EH, Scherer SW. (2008) Copy-number variations associated with neuropsychiatric conditions. *Nature* **455**: 919–923.

Dani VS, Chang Q, Maffei A, *et al.* (2005). Reduced cortical activity due to a shift in the balance between excitation and inhibition in a mouse model of Rett syndrome. *Proc Natl Acad Sci USA* **102**: 12560–12565.

De Felipe J, Feriñas I. (1992) The pyramidal neuron of the cerebral cortex: Morphological and chemical characteristics of the synaptic inputs. *Prog Neurobiol* **39**: 563–607.

Dehaene S, Changeux J-P. (2000) Reward-dependent learning in neuronal networks for planning and decision making. *Prog Brain Res* **126**: 217–229.

Dehaene S, Changeux J-P. (2005) Ongoing spontaneous activity controls access to consciousness: A neuronal model for inattentional blindness. *PLOS Biol* **3**: e141.

Dehaene S, Changeux J-P. (2011) Experimental and theoretical approaches to conscious processing. *Neuron* **70**: 200–227.

Dehaene S, Kerszberg M, Changeux J-P. (1998) A neuronal model of a global workspace in effortful cognitive tasks, *Proc Natl Acad Sci USA* **95**: 14529–14553.

Dehaene S, Naccache L, Cohen L, *et al.* (2001) Cerebral mechanisms of word masking and unconscious repetition priming. *Nat Neurosci* **4**: 752–758.

Dehaene S, Sergent C, Changeux J-P. (2003) A neuronal network model linking subjective reports and objective physiological data during conscious perception. *Proc Natl Acad Sci USA* **100**: 8520–8525.

Dejerine J. (1895) *Anatomie des Centre Nerveux.* Vol. 1. Ru eff et Cie, Paris.

Elston GN. (2003) Cortex, cognition and the cell: New insights into the pyramidal neuron and prefrontal function. *Cereb Cortex* **13**: 1124–1138.

Elston GN, Rosa MG. (1997) The occipitotemporal pathway of the macaque monkey: Comparison of pyramidal cell morphology in layer III of functionally related cortical visual areas. *Cereb Cortex* **7**: 432–452.

Elston GN, Rosa MG.(1998) Morphological variation of layer III pyramidal neurons in the occipitotemporal pathway of the macaque monkey. *Cereb Cortex* **8**: 278–294.

Feuk L, Carson AR, Schrerer SW. (2006) Structural variation in the human genome. *Nat Rev Genet* **7**: 85–97.

Fox MD, Snyder AZ, Vincent JL, *et al.* (2005). The human brain is intrinsically organized into dynamic, anticorrelated functional networks. *Proc Natl Acad Sci USA* **102**: 9673–9678.

Fox MD, Corbetta M, Snyder AZ, *et al.* (2006) Spontaneous neuronal activity distinguishes human dorsal and ventral attention systems. *Proc Natl Acad Sci USA* **103**: 10046–10051.

Fuster JM. (1989) *The Prefrontal Cortex*, 2nd ed. Raven Press, New York.

Galli L, Maffei L. (1988) Spontaneous impulse activity of rat retinal ganglion cells in prenatal life. *Science* **242**: 90–91.

Gavin A-C, Bösche M, Krause R, *et al.* (2002) Functional organization of the yeast proteome by systematic analysis of protein complexes. *Nature* **415**: 141–147.

Goldman-Rakic PS. (1988) Topography of cognition: Parallel distributed networks in primate association cortex. *Ann Rev Neurosci* **11**: 137–156.

Goodall J. (1971) *In the Shadow of Man.* Collins, London.

Goodall J. (1986) *The Chimpanzees of Gombe.* Harvard University Press.

Gusnard DA, Raichle ME. (2001) Searching for a baseline: Functional imaging and the resting human state. *Nat Rev Neurosci* **2**: 685–694.

Hagmann P, Cammoun L, Gigandet X, *et al.* (2008) Mapping the structural core of human cerebral cortex. *PLOS Biol* **6**: 1479–1493.

Hawkins J, Blakeslee S. (2004) *On Intelligence: How a New Understanding of the Brain Will Lead to the Creation of Truly Intelligent Machines.* Times Books, New York.

Iversen S, Kupfermann I, Kandel ER. (2000) Emotional states and feeling. In: ER Kandel, JH Schwartz and TM Jessell (eds.), *Principles of Neural Science*, 4th ed. McGraw-Hill, New York, pp. 982–997.

Jeannerod M. (1999) The cognitive way to action. In: RJ Russell, N Murphy, T Meyering and A Arbib (eds.), *The Neurosciences and the Person: Scientific Perspectives on Divine Action.* Vatican Observatory Publications and CTNS, Vatican City and Berkeley, CA, pp. 57–66.

Jeannerod M. (2006) *Motor Cognition: What Actions Tell the Self.* Oxford University Press, Oxford.

Johnston RJ Jr, Hobert O. (2005) A novel *C. elegans* zinc finger transcription factor, lsy-2, required for the cell type-specific expression of the lsy-6 microRNA. *Development* **132**: 5451–5460.

Kerszberg M, Changeux J-P. (1994) A model for reading morphogenetic gradients: Autocatalysis and competition at the gene level. *Proc Natl Acad Sci USA* **91**: 5823–5827.

Laureys S, Goldman S, Phillips C, *et al.* (1999) Impaired effective cortical connectivity in vegetative state. *Neuro Image* **9**: 377–382.

LeDoux JE. (2003) *Synaptic Self: How Our Brains Become Who We Are.* Penguin, London. (1st ed., Viking, 2002.)

Levi-Montalcini R, Visintini F. (1938) Azione del curaro, della stricnina, dell'esterina, dell'acetilcolina sulla trasmissione dell'influsso nell'embrione del pollo dal 4° all'8° giorno d'incubazione. *Boll Soc Ital Sper* **13**: 976–978.

Levinthal F, Macagno E, Levinthal C. (1976) Anatomy and development of identified cells in isogenic organisms. *Cold Spring Harbor Symp Quant Biol* **40**: 321–332.

Lou HC, Luber B, Crupain M, *et al.* (2004) Parietal cortex and representation of the mental self. *Proc Natl Acad Sci USA* **101**: 6827–6832.

Mannervik M, Nibu Y, Zhang H, Levine M. (1999) Transcriptional coregulators in development. *Science* **284**: 606–609.

Meyer K, Damasio A. (2009) Convergence and divergence in a neural architecture for recognition and memory. *Trends Neurosci* **32**: 376–382.

Mesulam M-M. (1998) From sensation to cognition. *Brain* **121**: 1013–1052.

Neary D, Snowden JS, Northen B, Goulding P. (1988) Dementia of frontal lobe type. *J Neurol Neurosurg Psychiat* **51**: 353–361.

Parvizi J, Damasio AR. (2001) Consciousness and the brainstem. *Cognition* **79**: 135–160.

Parvizi J, Damasio AR. (2003) Neuroanatomical correlates of brainstem coma. *Brain* **126**: 1524–1536.

Parvizi J, Van Hoesen G, Buckwalter J, Damasio A. (2006) Neural connections of the posteromedial cortex in the macaque: Implications for the understanding of the neural basis of consciousness. *Proc Natl Acad Sci USA* **103**: 1563–1568.

Raichle ME, MacLeod AM, Snyder AZ, *et al.* (2001) A default mode of brain function. *Proc Natl Acad Sci USA* **98**: 676–682.

Rakic P. (2000) Setting the stage for cognition: Genesis of the primate cerebral cortex. In MS. Gazzaniga (ed.), *The New Cognitive Neurosciences.* MIT Press, Cambridge, MA, pp. 7–21.

Ramón y Cajal S. (1899) *Comparative Study of the Sensory Areas of the Human Cortex*, Harvard University.

Redon R, Ishikawa S, Fitch KR, *et al.* (2006) Global variation in copy number in the human genome. *Nature* **444**: 444–454.

Schoenemann PT, Sheehan MJ, Glotzer LD. (2005) Prefrontal white matter volume is disproportionately larger in humans than in other primates. *Nat Neurosci* **8**: 242–252.

Shulman GL, Fiez JA, Corbetta M, *et al.* (1997) Common blood flow changes across visual tasks: II. *J Cogn Neurosci* **9**: 648–663.

Sergent C, Baillet S, Dehaene S.(2005) Timing of the brain events underlying access to consciousness during the attentional blink. *Nat Neurosci* **8**: 1391–1400.

Simpson JR Jr, Snyder AZ, Gusnard DA, Raichle ME. (2001) Emotion-induced changes in human medial prefrontal cortex: I. During cognitive task performance. *Proc Natl Acad Sci USA* **98**: 683–687.

Smaers JB, Schleicher A, Zilles K, Vinicius L. (2010) Frontal white matter volume is associated with brain enlargement and higher structural connectivity in anthropoid primates. *PLOS One* **5**: e9123.

Stretavan D, Shatz CJ, Stryker MP. (1988) Modification of retinal ganglion cell axon morphology by prenatal infusion of tetrodotoxin. *Nature* **336**: 468–471.

von Economo CF. (1929) *The Cytoarchitectonics of the Human Cerebral Cortex.* Oxford University Press, Oxford.

Wang Y, Liu X, Bieder T, Südhof TC. (2002) A family of RIM-binding proteins regulated by alternative splicing: Implications for the genesis of synaptic active zones. *Proc Natl Acad Sci USA* **99**: 14464–14469.

CHAPTER 11

Computation, Representation and Physicalism

Pierre Jacob

11.1. INTRODUCTION: TOP-DOWN CAUSATION AND PHYSICALISM

This chapter is devoted to clarifying some controversial claims made by the theoretical linguist Noam Chomsky in the context of issues of top-down causation (selected by Marc Jeannerod and Gennaro Auletta as the focus of the workshop from which this volume stems). As I use the term, "causation" refers to a process that involves a (causal) relation between pairs of events, one of which is a cause and the other of which is (one of) its effect(s). I further assume that causes must temporally precede their effects. Now, both a cause and its effect may have many different properties, which may be exemplified (or instantiated) at many different ontological levels, e.g., organismic level, cellular level, molecular level, atomic level and subatomic level.

As I understand it, the issue of top-down causation is simply the following: As part of a given brain process, could some higher-level (e.g. computational) property instantiated by one event cause another event with some lower-level (e.g. cellular) property?[1] And if so, how? For example, suppose that an addressee's human language faculty is involved in her processing of the phonological, syntactic and semantic properties of the utterance of a sentence she has just heard. Suppose that as a result of this processing,

[1] I focus on higher- and lower-level properties of the brain because the brain is central to the topic of the present meeting. But the issue is general.

she utters a novel sentence in response. The question of top-down causation is: How could instances of some of the computational properties of the language faculty causally modify the molecular structure of some neurotransmitter or perhaps the plasticity of some neural network involved in the addressee's motor system which enables her to pronounce and articulate her verbal response?

The issue of top-down causation so construed only makes sense against two sorts of background assumptions, both of which I find plausible. Among the background assumptions are the following pair of *metascientific* assumptions. First, there is the assumption that all of science cannot be reduced to basic theoretical physics. In other words, basic theoretical physics is not all there is to the scientific enterprise. Instead, the so-called "special sciences" (e.g., chemistry, geology or cognitive neuroscience) are genuine components of the scientific enterprise. By contrast, the physicist Sir Ernest Rutherford once wrote: "There is physics, and there is stamp collecting."[2] The Australian philosopher of science Jack Smart (1963) once wrote: "There is physics. The rest is engineering." I take it that both Smart and Rutherford would question the first metascientic assumption that basic theoretical physics is not the only explanatory science there is. The *second* plausible *metascientific* assumption is that the so-called *special sciences* form a *hierarchy* at the bottom of which is basic theoretical physics. Here, the idea is that all the laws discovered by the special sciences must be consistent with basic physical laws. In this picture, the laws of chemistry are constrained by the laws of physics and the laws of chemistry in turn constrain the laws of biology, which in turn constrain the laws of cognitive neuroscience, and so on.

It seems to me that the issue of top-down causation also requires the background *ontological* assumption of *physicalism*, i.e., that the ontological domain covered by basic theoretical physics (e.g., elementary particles) is both comprehensive and closed. It is comprehensive in the sense that any *event* investigated by some special science or other (e.g., geology or cognitive science) *is* a *physical* event. It is closed in the sense that if any physical event has a cause, or an explanation, then it has a *physical* cause, or a *physical* explanation, in accordance with physical laws. Clearly, ontological *physicalism* so construed is inconsistent with Cartesian substance or *ontological dualism*.

---

[2]Rutherford's famous statement is quoted by Crane (2003), p. 5.

These background assumptions have also given rise to a twin problem much discussed by philosophers: the issue of *supervenient* (or *emergent*) *causation.*[3] Suppose c caused e. Suppose further that c has both basic physical property P and higher-level property M, whereas e has both basic physical property P* and higher-level property M*. Finally, suppose that c instantiates M in virtue of instantiating P, and e instantiates M* in virtue of instantiating P*. (M and M* are said to supervene, respectively, on P and P*.) We could illustrate the situation as in Fig. 11.1, where the vertical lines represent the ontological dependency of higher-level properties M and M*, respectively, on lower-level properties P and P*, and the horizontal arrows depict the causal relations between property instances. P and M are co-instantiated by c at the same time. Similarly, P* and M* are also co-instantiated by e at the same time. But the co-instantiation of P and M by c causes, and thereby precedes in time, the co-instantiation of P* and M* by e (the effect). Now the issue of top-down causation can be depicted in the following graphic way: the fact that c instantiates higher-level property M causes the fact that e instantiates basic physical property P*. This relation can be depicted by the diagonal arrow between the instantiation of M and the instantiation of P* (cf. Fig. 11.1). The very idea of top-down causation raises several ontological questions: is not causation restricted to the more basic ontological level, i.e.,

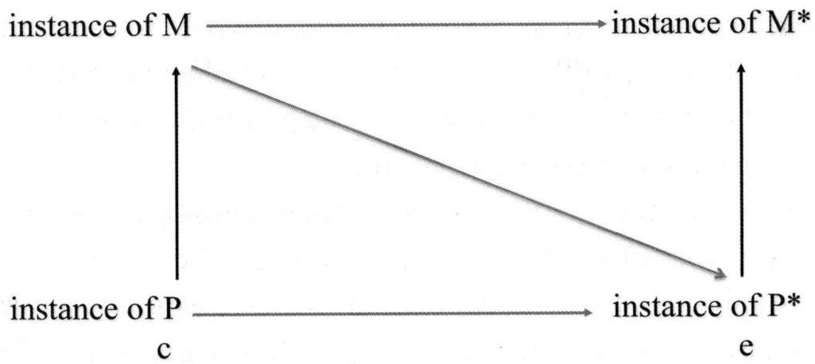

Fig. 11.1 The vertical arrows depict the instantaneous emergence (or supervenience) relation between higher- and lower-level properties. The horizontal arrows depict causal relations. The diagonal arrow depicts a putative top-down causal relation between a higher-level property of a cause and a lower-level property of its effect.

[3]Cf. Kim (1993; 1998) for much needed clarification of the idea of supervenience. In a nutshell, higher-level property M supervenes upon (or emerges from) lower-level property P if and only if anything instantiating P also instantiates M, but not conversely.

between instances of respectively P and P*? Does not causation at the more basic level preempt causation at the higher-level? Can there also be causation between instances of higher-level properties M and M*? Can causal relations hold between instances of properties located at different levels, i.e., between M and P*?[4]

Since I find both the metascientific and the physicalist ontological background assumptions tempting, I have decided to devote this chapter to the views of a very distinguished theoretical linguist and major actor of the so-called cognitive revolution who raises interesting problems for these assumptions, namely, Noam Chomsky. If some of his doubts are correct, then perhaps the issue of top-down causation, as I defined it above, is a kind of a pseudo-problem to be dissolved, not a genuine problem to be solved. The particular way I have chosen to discuss Chomsky's challenge to both the metascientific and the ontological physicalist background presuppositions of the problem of top-down causation is by assessing the gap between some of Chomsky's views and some of the views of a philosopher of mind who has been very close to Chomsky, namely, Jerry Fodor.

Almost fifty years ago, Noam Chomsky laid the foundations for a new scientific approach to the human language faculty, which he called "generative grammar." Briefly, the goal of generative grammar is to provide a finite, explicit and testable characterization of the recursive computational properties of the human language faculty whereby a finite set of lexical items gives rise to an unbounded set of grammatical sentences. In addition, Chomsky's argument that behaviorist explanations of human verbal behavior are inadequate also played a major role in what the psychologist George Miller (2003) recently called the "cognitive counter-revolution" that gave rise to the birth of the cognitive sciences as we know them.

Thus, Chomsky is one of the main architects of a computational approach to the human mind and the human language faculty. Few contemporary analytic philosophers of mind and language would, I think, deny that Chomsky's work has deeply changed and enlarged our scientific understanding of human languages and the human language faculty. Many, however, have challenged one or another aspect of Chomsky's scientific framework for investigating

[4]Kim (1993; 1998) has famously argued against top-down causation and supervenient causation. Cf. Jacob (2002) for a critical assessment.

the human language faculty. In response to his critics, Chomsky has come to make a distinction between two naturalistic theses: methodological naturalism, to which he subscribes, and metaphysical naturalism, to which he does not.

In this chapter, I want to address the question raised by Chomsky's acceptance of methodological naturalism and his joint rejection of metaphysical naturalism. The question is made more pressing by the fact that his former colleague at MIT, Jerry Fodor, has developed one version of metaphysical naturalism, which he calls "the computational representational theory of the mind" (CRTM), and which seems tailor-made for Chomsky's own computational approach to the human language faculty. Chomsky, however, does not endorse Fodor's CRTM. The question is: Why not?[5]

11.2. METHODOLOGICAL NATURALISM

Methodological naturalism is the commitment to apply scientific empirical ("naturalistic") methods to the study of mental and linguistic phenomena. In Chomsky's view, any scientific investigation of the world, in whatever domain, is bound to depart from common sense. There is little doubt that Chomsky's model for linguistics and cognitive science is theoretical physics. As epitomized by what, following Husserl and the theoretical physicist Steve Weinberg, Chomsky calls the "Galilean style," the task for linguistics and cognitive science is to make appropriate idealizations that will lead to the discovery of unexpected principles that will be as remote from both the empirical evidence and common sense as theoretical physical laws are. Thus, the kind of theoretical (or scientific) understanding aimed at by linguistics and cognitive science is both narrow and deep. It is narrow because a simultaneous scientific understanding of all aspects of the world is meaningless. It is deep because it arises from the discovery of abstract principles removed from common-sense beliefs and ordinary observation.

Thus, methodological naturalism stands in sharp contrast with *methodological dualism*, according to which human language and the human mind are *not* amenable to the kind of objective understanding to which other aspects of the world are amenable when subjected to scientific scrutiny.

[5]For a more detailed discussion, cf. Jacob (2010).

Methodological naturalism is just the claim that the scientific investigation of human language and the human mind should not be burdened with conceptual constraints that would be immediately rejected if directed towards the scientific investigation of other domains. In particular, the scientific study of "humans above the neck" should not be subjected to conceptual analyses based upon the epistemic authority of common-sense concepts of, e.g., *language*, *knowledge* or *consciousness* any more than physical and biological theories should be subjected to conceptual analyses based upon the epistemic authority of such common-sense concepts as, respectively, *matter* or *motion* and *life* or *vegetable*.

Theoretical understanding of the world, however, is not the only kind of understanding accessible to humans. The world also offers humans the possibility of artistic (or esthetic) understanding: "the arts may offer appreciation of the heavens to which astrophysics does not aspire" (Chomsky 2000, p. 77). But if the goal is theoretical understanding of the world, then the idealizations of scientific investigation are not dispensable.

Two methodological remarks bear emphasis on the limits of the explanatory scope of generative grammar. On the one hand, Chomsky drew a sharp contrast between what he called grammatical *competence* and linguistic *performance*. This is a distinction between what a speaker tacitly *knows* and what she *does*. One thing is a speaker's tacit knowledge of a system of grammatical rules and principles that enable her to produce and understand any of the grammatical sequences of words that belong to the non-finite set of sentences of her language. Another thing is what a speaker does on the basis of this tacit knowledge: which particular utterance she produces (or interprets) in a particular pragmatic context, at a particular place and time.

This is related to the further distinction between the phonological, syntactic and semantic properties of a sentence-*type* and the further pragmatic properties of an *utterance* of a *token* of this sentence-type. All tokens of a particular sentence-type that belongs to a given human natural language share a number of phonological, syntactic and semantic properties, which they inherit from the phonological, syntactic and semantic properties of their lexical constituents. Each particular utterance consists in producing a particular token of a given sentence, whose type belongs to a particular language. Different tokens of one and the same sentence-type all exemplify the same set of phonological, syntactic and semantic properties. Utterances pertain to language use (or performance): they are composed of words and sentences of natural languages, but they arise from the particular use (or tokening) of words and sentences of natural languages, in particular contexts.

So, unlike sentences (*qua* types), utterances are created by speakers, at particular places and times. Arguably, an utterance exemplifies properties not exemplified by the sentence-type itself: for example, it may be a property of an utterance, not of the uttered sentence itself, that it was addressed to members of the Security Council of the United Nations, in New York City, on February 5, 2003, by the US Secretary of State in a hoarse voice, tape-recorded, broadcast on the French-German channel Arte and perceived by the French president Jacques Chirac, in Paris, on February 6, 2003. Similarly, only the context can help to determine the full content of an utterance of a sentence containing indexical or demonstrative expressions (such "here," "now," "I" or "she"). For example, but uttering the English sentence "I am hungry now," many different speakers will express many different propositions depending on who the speaker is and when he or she uttered the sentence.

One important limitation on the explanatory scope of generative grammar follows from this pair of distinctions (the competence/performance distinction and the sentence-type/utterance distinction): a generative grammar strictly aims at specifying the phonological, syntactic and semantic properties of expression-*types*, not the further contextual properties of *utterances*. Furthermore, it is important to emphasize the distinction between two kinds of possible *explananda* raised by the human language faculty.

On the one hand, it is of course a legitimate task to try and provide a causal explanation of a speaker's linguistic behavior, where the task is to offer a hypothesis about the various causal factors that caused the speaker to utter a particular sentence s̲ of a given human language, in a given context, at a particular place and time. Among the various causal factors involved are the speaker's knowledge of the syntactic and semantic rules of her language and her communicative intention to inform her addressee of some of her own thoughts that she expresses via the use of sentence s̲ of her language. Not unless she intended to inform her addressee of something she had in mind and she knew the grammar of the relevant language would she have uttered sentence s̲.

On the other hand, part of this scientific research program requires explicit computational models of the human language faculty underlying the speaker's knowledge of the syntactic and semantic principles mapping the sound properties and the meaning properties of an open-ended set of expressions of her language. Although this model will be a fundamental component of the causal explanation of the speaker's verbal behavior, the task of the computational model is to describe a cognitive capacity: the human language faculty. Arguably, to describe a cognitive capacity is to offer a form of explanation that is different from a causal explanation: whereas understanding a process (relating events) requires a causal explanation, understanding a cognitive capacity, which is not a process, requires a computational model.[6]

11.3. CRTM: A VERSION OF METAPHYSICAL NATURALISM

Metaphysical naturalism is a family of metaphysical research programs whose defining ontological thesis is physicalism, all of whose versions are committed to the denial of Cartesian substance dualism. Following Quine's (1960) lead, some metaphysical naturalists have accepted the thesis that what I shall call "intentional realism" is simply inconsistent with a physicalist ontology. In other words, they accept the intentional irrealist (or anti-realist) view that words from our ordinary psychological intentional vocabulary, such as "belief," "intention" and "desire," do not pick out real properties, states or events in the world. Other metaphysical naturalists take it that intentional realism is compatible with a physicalist ontology. One particular influential proposal for reconciling physicalism and intentional realism is Fodor's CRTM.

Two features make Fodor's (1987; 1994; 1998) version of metaphysical naturalism uniquely relevant to my present purposes. First, unlike many philosophers of mind and language, Fodor fully endorses Chomsky's methodological naturalistic approach to the study of the human language faculty, which, he thinks, is the most significant contribution to our understanding of human language and the human mind. Secondly, according to Fodor, Chomsky's devastating criticism of Skinner's behaviorism, which he fully accepts, showed that cognitive scientific explanations of human behavior have no option but to appeal to such basic concepts from naive

[6]Cf. Jacob (in press) for further discussion.

common sense psychology as *belief, intention* and *desire.* A conclusion, as we shall see, that Chomsky does not accept.

CRTM involves the five following assumptions:

(i) The first assumption is computationalism (Fodor calls it the Turing picture of mental processes): mental or cognitive processes are computational processes, which could be performed by a Turing machine. On the further assumption that there cannot be computations unless there exist symbols upon which to compute, computationalism amounts to the acceptance of the language of thought hypothesis: thinking is tokening mental symbols with syntactic and semantic properties.

(ii) The second assumption is the representational theory of the mind according to which there could not be any semantic relation between minds (or mental states) and non-mental states of affairs unless there existed mental representations (or symbols), i.e., brain structures with syntactic and semantic properties. Following the language of thought hypothesis, the semantic properties of complex mental symbols depend on the semantic properties of their constituents and the syntactic rules of combination.

(iii) Thirdly, Fodor makes the twin assumptions that scientific psychological explanation is nomic and that psychological laws are intentional because they refer to the contents of an agent's mental states (and thereby to the contents or intentionality of mental symbols in his or her language of thought).

(iv) Fourthly, Fodor assumes that what makes psychological intentional laws causal laws is that they are implemented by underlying computational (non-intentional or non-semantic) mechanisms.

(v) Finally, as made clear by Fodor's (1998) book on *Concepts,* Fodor embraces a *semantic* conception of computation according to which it is constitutive of a computational process that it preserves the semantic relations among mental symbols. As Fodor (1998, pp. 10–11) puts it, "computations are those causal relations among symbols that respect the semantic properties of their relata" and also "a computation is a causal relation that respects the semantic relations among symbols."

On the one hand, CRTM paves the way towards a reconciliation between common-sense or folk psychology and the use of computational models in cognitive science. On the other hand, CRTM offers some hope of reconciling intentional realism with physicalism: the causal and the semantic properties of intentional mental states are coordinated via the computational syntactic properties of mental symbols. Psychological laws can be both intentional and causal.

One might have expected Chomsky to endorse CRTM on the threefold grounds that (i) he is the main architect of the computational approach to one fundamental human cognitive capacity, the human language faculty; (ii) computations apply to representations, and thus (iii) to mental structures with content or intentionality.[7] But this would be a mistake. For two reasons, Chomsky does not endorse CRTM. First of all, Chomsky rejects what makes it appealing to Fodor, i.e., the prospect of reconciling cognitive science with common sense. Secondly, he does not endorse the semantic conception of computation. In Chomsky's view, the semantic conception of computation relies on a *relational* conception of mental representations. Philosophers who subscribe to the relational conception of mental representations are in the grip of a definition of intentionality inherited from Brentano that in turn has generated metaphysical, non-scientific puzzles.[8] According to Chomsky, it is the burden of computational models of human cognitive functions to replace the relational conception by a non-relational, i.e., *syntactic*, conception.

I will now add a short clarification on the scope of the controversy between Chomsky and philosophers, who, like Fodor (1994; 1998) or Dretske (1988), on the basis of Brentano's definition of intentionality, espouse the goal of *naturalizing* intentionality.[9] I just want to point out that one thing is to make intentionality metaphysically compatible with metaphysical naturalism (i.e., physicalism) by unveiling the conceptual links between it and concepts employed in the natural sciences, such as *natural law*, *information* and *biological function*. It would be something else to show that Brentano's concept of intentionality plays a role in computational models and the experimental investigation of human cognitive capacities. Clearly, Chomsky is skeptical about the latter, i.e., the contribution of Brentano's concept of intentionality to a scientific research program in cognitive science.

---

[7] For a forceful development of this line of reasoning, see Rey (2003).

[8] Cf. Jacob (2003).

[9] Cf. Jacob (1997).

11.4. CHOMSKY ON SEMANTIC EXTERNALISM

Intentional realists (e.g., Fodor) assume that the contents of an individual's mental states as conceived by common sense matter to the psychological explanation of his or her actions. They take it that there is a fact of the matter as to what people believe, intend and desire. Unlike Dretske (1988), Fodor (1987; 1994) does not think that the contents of an individual's beliefs and desires are causally efficacious in the production of the individuals' behavior. But he thinks that the syntactic properties of mental symbols are. Now, Chomsky does not think that cognitive science should care at all about intentional realism. In his view, scientific realism arises only if and when some explicit scientific theory has been put forward in a particular domain, as in theoretical physics or in theoretical linguistics. Naive or common-sense psychology, however, is not such an explicit scientific theory. So "to be an Intentional Realist [...] is about as reasonable as being a Desk- or Sound-of-Language- or Cat- or Matter-Realist; not that there are no such things as desks, etc., but that in the domain where questions of realism arise in a serious way, in the context of the search for laws of nature, objects are not conceived from the peculiar perspectives provided by concepts of common sense."

Whether they subscribe to metaphysical naturalism (i.e., to physicalism) or not, many intentional realists accept an *externalist* conception of meaning or content. Fodor is one of them. Semantic or content externalism is the view that the meaning of a linguistic expression or the content of an individual's mental state is a well-defined or determinate property, but it is a relational or extrinsic, not an intrinsic, property of either the expression or the mental state. Semantic externalism applies to the meaning of symbols of public (or natural) languages. Content externalism applies to the contents of speakers' mental psychological states.

Chomsky contends that meaning externalism reflects philosophers' uncritical acceptance of the syntax and truth-conditional (or model-theoretic) semantics of formal languages as models for the syntax and semantics of natural languages. There is no fact of the matter as to which of several extensionally equivalent systems of syntactic rules able to generate the set of well-formed fomulas of some formal (or artificial) language is the uniquely correct one. However, on Chomsky's realistic interpretation, there is a fact of the matter as to which of several extensionally (or weakly) equivalent systems of syntactic rules is the correct description of some fragment of a natural language. It is the task of generative grammar to capture the correct computational recursive procedures in the brain of speakers of a natural language when they produce and interpret sentences of their language.

Now, just as he believes that the syntax of formal languages carries misleading (anti-realist) implications for the syntax of natural languages, Chomsky also believes that the semantics of formal languages carries misleading externalist implications for the semantics of natural languages. In a well-designed formal language, used for some scientific purpose or other (e.g., Frege's *Begriffschrift*), elementary symbols are stipulatively given a denotation and the semantic values of complex expressions depend on the semantic values of their constituents and the syntactic rules of composition. Chomsky, however, disputes the view that such regular word–thing semantic relations hold of expressions of natural languages at all.

For the purpose of a scientific empirical investigation, Chomsky likes to think of linguistic meaning, not on the externalist model of the symbol–thing semantic relation of artificial languages, but rather on the internalist model of the phonological representations of the sound structure of words and expressions. In fact, within his recent so-called "minimalist" approach, he thinks of the syntax of natural languages as the interface between two different human cognitive systems: the human sensorimotor system that controls the articulation and the perception of linguistic sounds and the human intentional conceptual system that performs inferences on the logical forms of uttered sentences. So phonological representations are instructions for the sensorimotor system and logical forms are instructions for the intentional conceptual system. Let M be a mental representation associated with the use of, e.g., the name "London." Linguists take it for granted that there is no scientifically useful *sounding* externalist relation between M and a sound or noise N in the world. By parity or in virtue of the symmetry between meanings and sounds (or semantics and phonology), Chomsky doubts that there is a scientifically useful externalist meaning (or referential) relation between M and some denoted entity E in the world. As Chomsky (2003, p. 271) puts it, "it is not at all clear that the intuitive underlying picture of word–object relations is any more appropriate [in the case of meanings] than in the case of sounds."

In fact, Chomsky accepts the new-Strawsonian (Strawson 1952) view that reference is an *intentional action*, not a semantic relation between words and things. Consider the sentence:

(1) London is so unhappy, ugly and polluted that it should be destroyed and rebuilt 100 miles away.

In an utterance of (1), what the speaker refers to by his use of either "London" or the anaphoric pronoun "it" keeps changing from people (the inhabitants of London, who are said to be unhappy), to buildings (which are said to be ugly), to a place (which is said to be polluted), to a city (whose destruction is being considered), etc. The point of example (1) is to show what complex a scientific understanding of linguistic reference comes to be in view of the fact that (a) reference is an intentional action and (b) speakers' referential intentions keep switching in the course of a single utterance without creating much difficulty for the hearer to keep track of what the speaker is saying.

11.5. CHOMSKY ON PHYSICALISM

I now turn to Chomsky's attitude towards physicalism, which is a defining thesis of metaphysical naturalism. As I said above, one might have expected Chomsky to argue from the explanatory success of his own computational approach to the human language faculty to the truth of physicalism, along the following lines: (i) Generative grammar shows that at least some mental processes are computational processes. (ii) Computational processes are operations that can be performed by a machine working according to the fundamental laws of physics. (iii) Thus, the success of generative grammar vindicates the physicalist thesis according to which some fundamental mental processes are physical processes that can be performed by a physical machine. This line of argument is Fodor's. However, it is not Chomsky's. Why not?

Why does Chomsky not accept the inference from *computationalism* to *physicalism*? All versions of physicalism stand in sharp contrast to Cartesian substance dualism, according to which the ontology of minds is different from the ontology of bodies (including brains). Unlike bodies, minds, which are immaterial, do not follow the basic laws of physics, which Descartes took to be fairly simple mechanical principles, including the laws of collision. Chomsky's recent position on the mind–body problem is both surprising and radical. In a nutshell, in Chomsky's view, the ontological dispute between physicalism and substance dualism did make sense in Descartes' time. But it does not any more. It made sense under the assumptions of the mechanical philosophy, i.e., when basic physical laws were supposed to comply with strict mechanism. But by postulating the law of universal gravitation, Newton introduced action at a distance between physical entities as a major explanatory principle in physics. He thereby undermined the Cartesian mechanical philosophy. In so doing, as Chomsky (2000, p. 84) puts it, "Newton eliminated the problem of [what Gilbert Ryle (1949) called] 'the ghost in the machine' by exorcising the machine; [but] the ghost was unaffected." According to Chomsky, ever since the rise of Newtonian mechanics, the twin concepts of *matter* and *physical object* have been relegated to naive (or folk) physics. They are not part of the physical sciences, let alone of theoretical physics.

If Chomsky's diagnosis is correct, then neither Cartesian substance dualism nor (it would seem) any current version of physicalism makes any real sense. Chomsky's view seems to be that the advent of Newtonian physics has deprived the physical sciences of a physical basis for the ontological reduction of minds to something physical. Since theoretical physics is open and evolving, there can be no *a priori* delimitation of what falls under the predicate "physical." Instead of pursuing the misguided ideal of an ontological reduction of minds to matter, cognitive scientists should aim at "unification" of, e.g., computational accounts of language processing with the core natural sciences, not reduction of the former to the latter. However, as Chomsky (2000; 2002) keeps pointing out, unification of the computational theories of the human language faculty with the core natural sciences might not be achieved until the core natural sciences themselves undergo some major conceptual changes.

Is Chomsky's diagnosis right? Is Chomsky right in claiming that since the demise of the mechanical philosophy, both physicalism and substance dualism have been deprived of intelligible content? The question is too complex for me to offer a definite answer here. However, I wish to make two points in order to deflate some of Chomsky's opposition to ontological physicalism. First of all, suppose we grant Chomsky the point that basic physics does not offer us a stable conception of the physical. Still, it is conceivable that revisions in basic physical theories will not affect our scientific biological understanding of the human brain. Even though our conception of the physical is open, the issue still arises as to how brains made up of neurons (as we know them) can give rise to, or underlie, the computational properties of the human language faculty revealed by generative grammar.

Secondly, Chomsky sometimes gives the impression that he thinks that we should replace the *ontological mind–body pseudo*-problem by the genuine *epistemological* issue of the *unification* between the computational theories of the human language faculty and core natural scientific theories. In a nutshell, Chomsky subscribes to what Jeff Poland (2003) has called *methodological physicalism*. If so, then the questions are: *Why* should we seek theoretical unification at all? Is theoretical unification a purely epistemological (or perhaps *esthetic*) virtue? Presumably, Chomsky agrees that both neuroscientific theories and computational theories investigate the human brain at two (or more) different *levels*. If so, then theoretical *unification*, which is an *epistemic* relation between theories, involves or presupposes a relation between levels of investigation. The notion of *levels* (of investigation), however, is itself an *ontological* notion, not a purely epistemic notion: for example, the chemical level of molecules is distinct from the physical level of atoms because molecules are composed of atoms. Arguably, the distinction between *levels*, i.e., between the level of atoms and the level of molecules, is not expected to collapse under every scientific change in the theories of either atoms or molecules. If not, then the ontological distinction between the level of molecules and the level of atoms does not entirely reduce to the epistemic relation between the best scientific theories, respectively, of atoms and of molecules.

Similarly, the epistemic unification between computational theories of the human language faculty and neuroscientific theories of the human brain may well involve and presuppose an *ontological* distinction between computational *properties* and lower-level *properties* of the human brain. Chomsky (2000; 2002) is willing to refer to the computational properties of the human language faculty as "emergent" (or higher-level) properties of the human brain. Indeed, not unless he accepted the ontological assumption that the computational properties of the human language faculty are emergent properties of the human brain could he, I think, coherently spell out the goal of epistemic unification between computational theories of the human language faculty and neuroscientific theories of the human brain. The ontological distinction between the levels at which, respectively, neurons receive and transmit information and the principles of the human language faculty are computed may well survive many scientific changes in the theories of either neurons or the human language faculty. Thus, my modest point is that (theoretical physics aside) Chomsky's epistemic goal of theoretical unification in cognitive science is better served by monist physicalism than by Cartesian substance dualism. Chomsky's epistemic goal might even be inconsistent with Cartesian substance dualism, i.e., the view that the human language faculty and the human brain do not belong to a single ontology.

11.6. CONCLUSIONS

In this chapter, I have accepted Chomsky's defense of his framework for the scientific study of the human language faculty based on his acceptance of methodological naturalism and his joint rejection of methodological dualism. I have argued that the root of Chomsky's initially surprising rejection of metaphysical naturalism in general, and of CRTM in particular, lies in his entrenched view that the scientific enterprise cannot be reconciled with common sense. The view is that the scientific understanding of such things as the fundamental structure of space and time, the nature of electricity or the nature of the human language faculty requires a sharp break with common-sense concepts. Clearly, Chomsky's model of scientific understanding is that provided by theories from mathematical physics, which have indeed sharply departed from common-sense (or naive) physics.

In these concluding remarks, however, I would like to draw attention to two related points. First, Chomsky might well be right that mathematical physical theories sharply depart from common-sense physics. However, the questions arise whether this is true of scientific theorizing in general and whether Chomsky is right to take mathematical physics (as opposed to, e.g., evolutionary biology) as a model for all kinds of scientific understanding, including scientific understanding of the human language faculty and the human mind. Perhaps Chomsky would not disagree with the Rutherford–Smart view that there is physics and the rest is either engineering or stamp collecting!

Secondly, in Chomsky's Galilean picture, theoretical linguistics and cognitive science must put human common-sense conceptions of human language and the human mind to one side in order to reach a scientific understanding of human language and the human mind — just as theoretical physics removed constraints from naive physics. However, it is also a *higher-level* (*or higher-order*) task of *cognitive science* — which Chomsky labels *ethnoscience* — to reach an empirical scientific understanding of *human common sense* itself.

The ethnoscientific part of cognitive science has no counterpart in other natural sciences, in particular not in theoretical physics. The higher-order task of ethnoscience is to offer a scientific understanding of human common-sense conceptions in various cognitive domains (e.g., naive physics, naive biology, naive psychology, naive arithmetic or naive geometry). The Galilean style emphasizes the gap between common-sense understanding and scientific understanding. Insofar as they both seek to describe and explain the behavior of inanimate physical objects, theoretical and naive physics may be thought of as two competing cognitive perspectives on the same things. Any first-order scientific enterprise (e.g., theoretical biology) may be thought to be in competition with its common-sense counterpart in the same ontological domain (e.g., naive biology). But the ethnoscientific investigation is not a first-order scientific investigation: it is a second-order scientific investigation, since it seeks to provide a scientific understanding of common sense. If so, then the question arises whether the Galilean style is vindicated for ethnoscience, i.e., whether theoretical physics is a proper model of inquiry for the ethnoscientic component of cognitive science, and for cognitive science more generally.

DISCUSSION

Michael Arbib: Let us begin without getting into an endless diatribe. Chomsky is totally irrelevant to the topic in my opinion. From the outset he stated that his subject was competence not performance; he has espoused autonomous syntax looking at the syntactic structure of language divorced from its meaning or its use. The theory is at best a descriptive structure of regularities that can be seen across languages if we look only at the syntactic structure, not at the meaning; as such it does not talk about human performance, and generative grammar may have succeeded in providing a descriptive framework, but it has offered nothing about the acquisition of language, the use of language, the discordance of language, and so on. So, it seems to me that Chomsky disqualifies himself, he pays no attention to data, he pontificates — if you'll forgive the expression in this context — on the brain, but in fact his writings show no attention to the data of acquisition, use or brain function. Fodor has espoused a purely "logical form" approach to semantics that makes no contact with embodied experience, and at least at the meetings I have been to; I have seen him three times, and on each occasion he has ridiculed and been contemptuous of anybody who has suggested that the study of the brain is relevant to cognitive science.

Pierre Jacob: Michael, I take it for granted that you agree with the following five assumptions. First, human natural languages are sets of sentences. Secondly, a sentence of a natural language (e.g., English) is a grammatical sequence of words from the lexicon of that language. Thirdly, the meaning (or semantics) of a sentence or complex expression depends on the meanings of its constituents together with its syntactic structure. Fourthly, there is no syntactic limit on the size of a sentence of a natural language — which means that the set of sentences of a natural language is not finite. But the lexicon of a natural language is finite. So the descriptive task of a generative grammar is to offer an explicit characterization of the finite mechanism whereby a non-finite set of sentences can be generated from a finite set of lexical items. Finally, a speaker who can understand or utter sentences from her language has tacit knowledge of the syntactic rules that enable her to combine the complex meanings of the sentences on the basis of the meanings of their constituents and their syntactic structure. I personally would agree with you that so construed, this description does fall short of providing a full specification of the complete meaning of a speaker's utterance — what philosophers call the truth-conditional content of an utterance, let alone its implicit meaning (or its implicatures). The truth-conditional content of an utterance depends on many contextual features other than what is delivered by a generative grammar. I also agree that this syntactic and semantic description will not be sufficient to explain or predict a speaker's linguistic behavior, i.e., the speaker's next utterance. But I would argue that it is necessary: it seems to me that until you have some initial characterization of the computational properties of a speaker's syntactic and semantic competence, it makes little sense to investigate either language-acquisition (i.e., how this competence arose in the course of ontogenetic development) or the underlying brain structure.

Michael Arbib: With the repression of the autonomy of syntax…

Pierre Jacob: Well, the autonomy of syntax is, I think, an empirical issue. Take Chomsky's example: "Colorless green ideas sleep furiously." There are two basic views one might take of this example. One possible view is that this is a grammatical sequence of English words. However, it is semantically deviant or anomalous: ideas don't have colors; and if so, they can't be green. Nor do ideas sleep — furiously or otherwise. If so, then some sequences of English words are grammatical English sentences, but they have no meaning (or no semantics). That is the interpretation in favor of the autonomy of syntax. Another possible view is that precisely since this sequence of English words has no established meaning, then it is not an English sentence at all. Adjudicating between these two views involves complicated theoretical decisions about whether semantically relevant distinctions (e.g., selection restrictions) should be part of syntactic principles or not. Perhaps we could solve this issue by looking for differences in brain responses to processing this sequence of English words. If we knew exactly how the brain of English speakers responds, respectively, to non-grammatical sequences and to semantically deviant sentences, then we could use knowledge of the brain to solve this question. But clearly neuroscientific progress in understanding the brain structures underlying language processing will itself depend on whether semantic distinctions are relevant to syntactic principles. So in a nutshell, I don't have sympathy for Fodor's polemical claims that knowledge of the brain is irrelevant to our understanding of the human language faculty. But again until you make some progress in what you call a "descriptive" task, you can't meaningfully address theoretical issues about linguistic behavior, language development or brain organization underlying the language faculty.

Jean-Pierre Changeux: I am about to please Michael once more because I agree with him this time, and I must say that what he said in my opinion is fully true and I accept that. Now the question we have in mind is: Why are these very distinguished scientists totally ignorant of neuroscience? And in particular, Fodor is just saying that the brain is irrelevant. So, why is it that in the scientific community we have such a category of mentally blind people? I think this is the question that the philosopher should ask, because I think there is certainly a lot to do, to try to go ahead, and to start from Chomsky: brain sciences have made enormous progress since he proposed his views — we have this brain imaging which is developing, we have a lot of computational neuroscience which is also being progressed; so why are these people, who are famous people, so blind? I use the word "blind," because they are mentally blind. Now, I think that this is unfortunately the case in science, but this is also the case when you look at the behavior of human beings, when you can see the conflicts which exists between, say, Islam and Judaism, orthodoxy and Catholicism, where there are clearly barriers which are impermeable, and human beings consistently do not wish to step further. I think there is something like a system of beliefs around these cognitive scientists that is completely outdated nowadays; and I think this is very unfortunate because we could have very interesting interaction with these people, and that is not happening. Now, I have read the quoted authors in detail, and I am especially concerned by the ontological physicalism of Fodor, which is more or less what most of us think, and the fact that Fodor grounds his ontological physicalism simply on computational theory negating what is the brain. So for me it is a paradox to defend a theory by simply not considering what in fact is the basis of the physicalism of a function. So, the question I am asking to you as a philosopher is: Why are there such kinds of situations, which are not unique to science, and which are at the origin of very bad conflicts in the world?

Pierre Jacob: Allow me to make a basic distinction: in my chapter, I explicitly addressed some of the fundamental disagreements between Fodor's version of ontological physicalism and Chomsky's own view that the traditional dispute between ontological dualism and ontological physicalism makes no sense. Let me stress again that on this ontological issue, Chomsky and Fodor completely disagree. Now, there is a second issue: I fully agree that investigating language use (language processing), language acquisition and also the evolution of the human language faculty is an interdisciplinary enterprise. However, I want to stress that it is (at least) a two-way street: as you emphasize, knowledge of the brain is highly relevant. But so is of course knowledge of syntactic and semantic regularities in natural languages. So, I would not say, as you seem inclined to do, that linguists and psycholinguists are just blind to the accumulation of neuroscientific knowledge about brain organization. There is the converse problem that linguistic and psycholinguistic work on syntactic and semantic structures of natural languages has become highly technical and that it's also hard for neuroscientists to keep up with that research and to give justice to the complexity of syntactic and semantic analyses...

Michael Arbib: But in what way would that leave you to deny...?

Pierre Jacob: No, I don't think the denial is grounded or justified in either direction. And I am saying that until we know much more about the syntactic processes and the syntactic structure of languages, it is hard just to investigate the brain; if we have absolutely no clue as to what the syntactic complexity is, what the syntactic principles are. I mean it makes sense to think that we want the research into the brain structure to be guided by knowledge of the syntactic and semantic structures of natural languages.

Jean-Pierre Changeux: But I think we know enough of this to start to develop a simple system, even if we will not have the whole complexity of the brain. Why not look at some of the basic issues, the simple ones, and look for an experimental system that will be able to show the basic ways in which the brain...

Pierre Jacob: Well, I can understand that one might not be particularly interested in abstract syntactic generalizations about, e.g., c-command in tree representations underlying the syntax of natural languages. But I am afraid that it would be unreasonable to deny the complexity of the syntactic and semantic data...

Jean-Pierre Changeux: And anyway you didn't answer my question.

Pierre Jacob: No, I did not. Maybe the cognitive social scientific investigation of reactions to in-group/out-group membership might be somewhat relevant to the question of the barriers between distinct scientific communities...

Jean-Pierre Changeux: That's exactly my point.

Pierre Jacob: That's an interesting question. But again, I think, it's a two-way street issue.

Gennaro Auletta: I agree with Michael and Jean-Pierre, and I think that probably there is here, by the cognitive school, a sort of abuse of the word "naturalism," or of naturalistic explanations; because when they speak of naturalistic explanations, they jump from the level of grammar and syntax to a sort of computational physical level, bypassing biology completely. Obviously, I agree with naturalism, and I think that we need even a sort of naturalistic monism, so I completely agree with that. But I would strongly suggest, from an epistemological point of view, to consider that we also need a second requirement that is very important when dealing with a true scientific activity; this is the requirement of pluralism, that is, we should always understand that any scientific explanation, in its deep nature, is partial; so we cannot have a sort of hyper-science that explains everything. I mean, science is necessarily regional, so we have necessarily a plurality of different disciplines, and we should even respect different disciplines provided, obviously, that they satisfy some basic requirements of scientific explanations. So, for instance, I see no worry at all with psychology, to give an example. I know there is also a psychology very connected with a neurology which gives fantastic results, but I also think about experiments done on primates by Povinelli and others; they are from an epistemological point of view very much correct; they are science in the true sense of the word. And for this reason, I would suggest a third criterion in addition to naturalism and pluralism, that is, simply a pragmatic criterion. I mean, when we deal with science what matters finally are results; a science should be judged by the results that it brings, not by some hyper- or meta- I-do-not-know-what. I would also strongly suggest — and this is one of the reasons for pluralism — accepting some results even when we do not have sufficient explanations or interpretations for these results. Here, I recall that indeed Newton had no explanation at all for the gravitational force; not only that, he believed that it was somehow acting against his understanding of the world because it was an action at a distance, but nevertheless he formulated the law of gravitation. I mean, this is the way in which we work. I think this is very important to consider, always.

Marc Jeannerod: Let me make a comment as a devil's advocate. I fully disagree with Fodor's position of course, but we should recognize that in spite of extensive investigation of brain functions in 25 years of brain imaging and 100 years of neuropsychology, we still have no idea of how language works, and how the brain implements language. All attempts in neurolinguistics — I remember my mentor Hécaen, who was involved in neurolinguistics in the sixties — completely failed. More recently, there are probably new avenues, in the genetics of language, in language development, which seem promising. But still, could somebody explain why nearly in 2010, we do not know more than Broca and Wernicke about language.

Antonio Damasio: That is an exaggeration.

Marc Jeannerod: That is an exaggeration and it is why I called myself a devil's advocate.

Antonio Damasio: I think we have to close, but I disagree entirely. I think we have far more, we know far more than Broca and Wernicke. Maybe we don't have all the answers, but we have multiple levels of results available, to which different talks contribute. Anyway, I think we are going to close now.

REFERENCES

Chomsky N. (2000) *New Horizons in the Study of Language and Mind.* Cambridge University Press, Cambridge.

Chomsky N. (2002) *On Nature and Language.* Cambridge University Press, Cambridge.

Chomsky N. (2003) Replies. In: LM Antony and N Hornstein (eds.), *Chomsky and His Critics.* Blackwell, Oxford, pp. 255–328.

Crane T. (2003) *The Mechanical Mind: A Philosophical Introduction to Minds, Machines and Mental Representation.* Routledge, London.

Dretske F. (1988) *Explaining Behavior.* MIT Press, Cambridge, MA.

Fodor JA. (1987) *Psychosemantics.* MIT Press, Cambridge, MA.

Fodor JA. (1994) *The Elm and the Expert.* MIT Press, Cambridge, MA.

Fodor JA. (1998) *Concepts, Where Cognitive Science Went Wrong.* Oxford University Press, Oxford.

Jacob P. (1997) *What Minds Can Do.* Cambridge University Press, Cambridge.

Jacob P. (2002) Some problems for reductive physicalism. *Philos Phenomenol Res* **65**(3): 648–654.

Jacob P. (2003) Intentionality. In: *Stanford Encyclopaedia in Philosophy.* http://plato.stanford.edu/ (Revised 2010).

Jacob P. (2010) The scope and limits of Chomsky's naturalism. In: J Bricmont and J Franck (eds.), *Chomsky Notebook.* Columbia University Press, New York, pp. 211–234.

Jacob P. (in press) Meaning, intentionality and communication. In: C Maienborn, K von Heusinger and P Portner (eds.), *Semantics: An International Handbook of Natural Language Meaning.*

Kim JJ. (1993) *Supervenience and Mind.* Cambridge University Press, Cambridge.

Kim JJ. (1998) *Mind in a Physical World.* MIT Press, Cambridge, MA.

Miller G. (2003) The cognitive revolution: A historical perspective. *Trends Cogn Sci* **7**(3): 141–144.

Poland J. (2003) Chomsky's challenge to physicalism. In: LM Antony and N Hornstein (eds.), *Chomsky and his Critics.* Blackwell, Oxford, pp. 29–48.

Quine WVO. (1960) *Word and Object.* MIT Press, Cambridge, MA.

Rey G. (2003) Chomsky, intentionality and a CRTT. In: LM Antony and N Hornstein (eds.), *Chomsky and His Critics*. Blackwell, Oxford, pp. 105–139.

Ryle G. (1949) *The Concept of Mind*. Penguin Books, London.

Smart JJC. (1963) *Philosophy and Scientific Realism*. Routledge & Kegan Paul, London.

Strawson P. (1952) *Introduction to Logical Theory*. Methuen, London.

CHAPTER 12

Overall Discussion

The last part of the workshop was divided into five brief sessions. Sessions II to V are introduced by short summaries of the point of view of the speakers, followed by a discussion. The fourth session included a speech by Michael Arbib summarizing Antonio Damasio's contribution, who left Rome the night before. The fifth session began with a statement by Father Tomasz Trafny on behalf of the Pontifical Council for Culture expressing the utmost interest of the former Archbishop Gianfranco Ravasi, and through him of the Catholic Church, in the workshop. Also Prof. Massimo Stanzione, of the University of Cassino, has contributed to the Overall Discussion.

SESSION I

Michael Arbib: I would like to put a topic on the table. I think we have seen two approaches to the self in the last two days. One is the self as an isolated individual experiencing the world and its interactions with the world. Occasionally, we would see the self as a social being, and I think in some ways it is hard to really imagine a self unless one knew other selves with whom to relate and from whom to distinguish. So we have talked about our agreement or disagreement with top-down causation from, if you will forgive me, the mental level down to the neural level. I wonder if it would be now fruitful to spend a little time talking about downward causation from the cultural level to the habitual personal level. So, if anybody feels that it is an interesting topic to talk about I would be pleased to hear comments.

Antonio Damasio: The question is downward causation from...

Michael Arbib: ... from the social to the personal as distinct from the personal to the neurons.

Wolf Singer: I like the distinction a lot, because it seems to me to be the only way out of problems that we are having. I got very much involved in this discussion of free will in Germany. Well, horribly, somebody who was a naturalist claimed that there is nothing like that and therefore people cannot be guilty; and if you cannot be guilty, you cannot be punished, and then anything goes. Of course, as a neurobiologist, you can ask how come that these attributes are ascribed to you. I think the only way out is to look at brain functions and to consider the brain as a social organ. Throughout ontogeny, knowledge, attitudes and social realities only emerge because brains talk to each other and individuals get knowledge about themselves in the reflection of the perspectives of others. If one recruits this reciprocity then it becomes much less difficult to see that realities can emerge having immaterial connotations like the soul, and all these constructs like free will, intentionality and so forth. They emerge in between the social interactions of people and become realities. Interestingly, one learns about them during childhood amnesia, as Tulving puts it, and therefore they become sort of convictions, as I have said. There is a lot of wisdom that we have learnt from interactions, in this case social interactions, but we ignore the origin of these things because we learnt about them whilst we still had no episodic memory. Thus, we attribute them to us as being just there; we attribute them to our brains. However, we can only understand them if we consider them as emergent properties of social interactions.

Antonio Damasio: The idea that you have a lot of interaction with the others prior to having a fully developed autobiographical self makes perfect sense. And it also makes perfect sense that all that really becomes a part of everybody's implicit background.

Gennaro Auletta: I agree, and I suggest that we should probably distinguish an elementary sense of sociality, in which for instance we have also clusters of cells, kind of multicellular organisms that in a certain sense "talk" during epigenetic processes and so on, and a higher sense of sociality when we speak of interactions between advanced organisms. I would also suggest that something very important for this second aspect, in my opinion, is the emotional system. I think that mammals are very important under this respect.

Mammals in general display a social behavior that is really new because they show, at least at a higher degree, an affective component. For example, newborns can play, as mentioned by Chuck Crowell, and play is very relevant here, as well as the highly sophisticated sexual interaction mammals often display. In the case of primates there is even more, because, as far as I can understand, and I think this was also supported by Michael Arbib's contribution, primates develop a sort of political sense: they have a set of social rules that are not established once and for all. Primates can try to fight each other for the dominant male or the dominant of the hunting group. So, this is a field to explore: how the emotional element plays a relevant role for society.

Antonio Damasio: We did not talk about emotions up to now, but I think it is a very important topic because of its position in evolution. One critical point is that when you go back in evolution, to invertebrates, you can find full evidence of emotional programs. So, I think it is very critical that we make a distinction between emotions and feelings. Emotions as action programs capture a sort of concert of possible actions including movement, endocrine secretion, and all sorts of internal visceral controls. They are, of course, extremely adaptive because they allow organisms to have a full repertoire of connections to situations that can be threatening or that offer survival opportunities. That is obviously and clearly non-conscious and non-minded. Look at *Aplysia*; it is a very stupid animal, but it has a lot of variations of emotions in its repertoire that are very complex, including increases in blood pressure with emission of several endocrine molecules, acceleration of heart rate and behavior of withdrawing into itself. Now, from the structure of its very simple brain, *Aplysia* does not appear to have the possibility of feelings; there is in fact nothing that suggests that it has a platform of the kind that we have in the brain, such as the brain stem and the parabrachial nucleus, and, in the case of mammals, the full insular cortex with this incredible richness of representation of the internal medial visceral components linked with the muscular skeletal ones. Therefore, there is a very interesting progression from very simple adaptive responses of the kind that you can find in multicellular or single cellular organisms without brains, through things that get more complex incorporating the whole package of action programs, and eventually reaching the possibility of representing them in a mindful way, and finally in a conscious way. Of course, the social being, the issue that we are talking about, requires the richness of representation. Let me just add a quick note. We know about a very simple worm with a very limited brain and a limited number of neurons that has clearly different kinds of behavior that lead it to feed together with many other worms if there is scarcity of food products or a threat in the environment as perceived by a smell, and to feed habitually if there is an abundance of food products in the environment. This is very interesting as this aspect hints that sociality can arise clearly without having feelings and mind or consciousness.

Marcus Raichle: I would like to respond to this. Previously, I mentioned something about the ventral medial prefrontal cortex, that is, the part of the ventral medial circuitry that has been very well described. What is so remarkable about that piece of cortex is that its primary direction of change is down. Now, in a fear-generating situation, those who do not experience fear actually attenuate it. So, I very much like the notion that what we call emotion has a feeling component, but it has also a very important *evaluative* component in the context of behavior that is continuously ongoing. It is a part of the way the brain works. It is not as if we turn on the machinery that would be sufficient for the generation of the feeling.

Pierre Jacob: I think that from Jeannerod's contribution, there is a way to think that any agent must be able to make a distinction between itself and the world, to determine the distinction of the observable changes that are completely independent of these actions from those changes that arise from its own actions (I think that this is more or less what he calls the minimal self). Another task is to distinguish the agent from other agents who themselves could produce actions with consequences. And another distinction, it seems to me, is that between a self that ascribes to itself psychological properties and can represent its own psychological states and its own autobiographical memory, and a self able to ascribe also to other agents psychological properties, like beliefs, intentions, and theory of mind. So I think that it is very useful to have these kinds of distinctions.

Massimo Stanzione: I heard your discussion about top-down causation and naturalism, and it is a very haunting and difficult one. I studied this topic in evolutionary epistemology where it was completely clear that it is an effect that can be manifest at the social, the neurological and the biological level.

The problem is that when we talk about the closure of physics, and we support naturalism along with the assumption that there is no scientific explanation that violates the closure of physics, it is difficult to understand many writers of naturalism (like Fodor and Chomsky) without any reference to the importance of the genealogical approach due to the Darwinian scheme. To my mind, the key point is to focus our attention on the best version of naturalistic explanations compatible with the genealogical approach contained in the Darwinian program. That is surely true for language. Chomsky's naturalism is ultimately innativism. But as you surely know, Chomsky is not in agreement with many efforts to consider human language as a natural faculty that evolved as other faculties. He is strongly opposed both to his followers such as Pinker and Jackendorf and to other authors such as T. Deacon, who proposes a more semiotic model of linguistic development. In my mind, these difficulties arise from the fact that if you speak about capacity or ability of language in dispositional terms, you must have a prior notion of what a biological function is in general. This is a very complex and debated question, and we have no definitive solution that enables us to define a biological function, also at a physiological level. Millikan developed the best attempt at entrenching causal explanations with functional and evolutionary explanations, but there are always problems. So I would only say that we can be physicalist in our naturalistic approach to human capacity (from language to consciousness, from self to sociality, and so on), without affirming that there is nothing but physics. According to Campbell's approach, for example, there are many physical determinations that are too complex to be foreseen and that are relevant already in epigenetic phenomena. Do you know the case of *Daphnia*? *Daphnia* is a model organism in embryology studied by Scott Gilbert and others. *Daphnia*, in its early developmental stages, may have two different forms induced by the presence or absence of predators in its aquatic environment. That is an example of social interaction also in such simple primitive forms of life. The case of language is surely a similar one in the following sense: you can consider the mother–newborn relationship as a part of our language-acquisition systems. The Chomskian thesis is that the language-acquisition or other cognitive systems implied in natural languages are not the hardcore of human special languages, that is, the hard core of computability of the special feature of language to be circularly referring to the self. Now, Chomsky seems to assume that this hard core is tied to computability expressed by navigation and orientation capabilities in lower animals. But, against any evolutionary approach to language, he says: a factual physics or a factual chemistry will explicate the jump to our special human language ability, and not a new integrated framework of natural explications which is based on the principle of continuity between the human realm and the animal realm. That is not so physicalist in my opinion, as you assert.

Wolf Singer: I want to get to simpler grounds. Take the example of a community that creates, as an emergent property of a social setting, a set of rules eventually codified either in the regulation or in some legislative books. Henceforth, these rules, which certainly are an emergent property, are not supervenient; they are causally related to the underlying structure. They will be highly influential on the behavior of individuals. Is this downward causation?

Antonio Damasio: I think that it is. I mean, Changeux, for example, was using downward causation at the cultural level.

Wolf Singer: So, actually we have no problem with downward causation. It can be an emergent property of complex systems, and once it has emerged it becomes a reality and so can interact again with its underlying levels.

Gennaro Auletta: This is exactly the point that I was trying to stress in previous discussions. You have emergent complexity able somehow to "back-react" on what it is emerged from.

Wolf Singer: I do not see any mystery here.

Gennaro Auletta: Me neither, indeed!

Wolf Singer: I just think about my brain and consciousness. It just happens that every now and again it becomes creative; something comes together that has not been together before. In that very moment the new entity has emerged from this corporative interaction, whatever it is, and I have that in my long-term memory; and from this very moment on, it will be a cause for everything that continues. It is a bifurcation in my trajectory: it is downward causation.

Gennaro Auletta: I have only a point to make. Once the human brain, which is obviously the substrate of everything we are speaking about, has reached a certain level of complexity, humans are able to manipulate external objects, for instance, or gestures, in a codified way that allows them to express things in a completely new way. So, it is an emergent property that does not devalue at all the brain processes or the physical level, but notwithstanding it allows operations, for instance, manipulation of mathematical symbols, that are novel from an evolutionary point of view.

Antonio Damasio: Well, in essence, and almost the joke is that what comes up must come down.

Jean-Pierre Changeux: Yes. As to the issue of top-down causation, I always use not causation but regulation: bottom-up and top-down regulation. The crucial notion, as Pierre Jacob has been insisting, is the notion of *causality*. The notion of causality seems to me to have a very deep meaning in the sense that it not only causes but it is the only cause, and so, not only is it necessary but also sufficient. I think that in brain science, because it is a very open discipline, it is very difficult to use the notion of causality, and this, by itself, could be a very interesting philosophical issue because it is often difficult to isolate one type of mechanism or process from the context. Indeed, we are dealing with a system that is highly interpolated, and you can obviously take one part and say, "this part is influencing top down." However, it does not mean that it is the core cause or the complete cause of the process we are looking at. I think that in genetics we have the same problem: it has been a long story. Classically, one says that if one changes one amino acid in the active core of a protein, this changes the property of the protein; for example, the enzyme becomes inactive. Since we see that the enzyme is inactive, then we think that the amino acid is the cause of the activity, and there is some kind of causal relationship between this amino acid and the function. There is, I would say, a feeling — a feeling, not a fact! — that the amino acid is the site of the function. *This is completely wrong.* This is true for some cases, but since the discovery of allostery it is also clear that this is not the general situation. I think that the case of the nervous system is essentially the same. As to the question of relating language to genes, for instance, there have been extremely naive statements that have been made by Chomsky himself. He said that the initial stage of learning language is analogous to the development of the visual system: but nobody knows really how the nervous system develops from a definitive point of view, so I will say nothing about, because there is no explanatory power there. Now, there is also a new story about language disease linked to the *FOXP2* gene, and there are people saying that this is the gene for language. I think there is somebody very famous, named Steve Pinker, who has been making a fortune writing books on the existence of a relationship between one gene and one function. This is something that is currently a common view. Therefore, when we talk about causality we have to be extremely careful. Let us take the example of schizophrenia: more and more people have been thinking there is the gene for schizophrenia, but in fact there is not a single gene. And when you look at autism, it is exactly the same picture.

SESSION II

Chairman: Michael Arbib

Speakers: Gennaro Auletta, Chuck Crowell

Gennaro Auletta: The contributions of Pierre Jacob and Jean-Pierre Changeux raised several very important questions, including the relation between explanation and causality. I agree that as long as you are able to show a causal mechanism, you are able to explain something. When I speak of causal mechanism, obviously, I am referring to efficient causation, that is, something having a particular effect on something else. But I would like you to consider with me, now, the fact that during the 20th century there was a growing body of evidence showing that there are relevant correlations in nature also at the physical level that had not previously been sufficiently considered. I am especially speaking about quantum mechanics, as to entanglement and all sorts of correlations, and I am speaking about complex and self-organizing systems. A useful example is given by the Bénard cells, which are able to arrange themselves in such a way that the sense of rotation of each cell is organized in a particular way, turning left or right but in accordance with its neighbors (each cell turns in the opposite direction to the cells immediately around it). I am also talking about networks, as in networks we have a similar relevance of interconnections.

The problem is that when we have correlations, we do not, in general, have causation. In quantum mechanics we have a lot of interconnections and effects without any causal relation; for instance, we have interdependencies at a distance without any transmission of signals, a fact that we can now acknowledge after thousands of experiments. As such, we must ask: How have people dealt with this problem, especially with complexity, networks, and so on? The solution, in general, is simulation, that is, we substitute explanations for simulations whenever we are unable to provide causal explanations. Obviously, simulations are sometimes not very satisfactory, precisely because they are not able to provide explanations in the true sense of the word. So, it would be very interesting for me, and I think even quite wise, to combine these two issues. That is, to understand that perhaps there is now a place for new types of explanations that try to combine the issue of causality in the traditional sense of the word with the issue of the interconnection of formal constraints in the new sense of the word. In which way?

I hope you will agree that it is *impossible* to have a causal relation in the sense of a dynamical relation from an upper level to a lower level. But, what is interesting, in my opinion, is that you may have true traditional causal effects at a lower level and certain constraints from the upper level such that, when they are combined, they produce new, interesting and relevant effects that would not be produced *without* such formal constraints. This is the reason why, in my opinion, it is permissible to speak of top-down causation; when we speak of top-down causation, we are always talking about these two aspects together; namely, formal constraints, as I also used this word in my contribution (and as to this I completely agree with you, Jean-Pierre, that there can be no intervention by a sort of *deus ex machina*: actually, the notion of constraint is very close to your idea of regulation) and traditionally-intended causal processes at the lower level. This seems to make such a way of speaking relevantly meaningful.

Michael Arbib: I have some concerns here. The fact that we apply constraints at one level does not imply that we lack causal explanation at another level, given that these constraints are in place. As far as quantum mechanics is concerned, I think we must not abandon causality. We do have causality at the level of the wave function, but the effect of observation is probabilistic based on the probability amplitudes as they have evolved. One must not conflate causality with determinism. Indeed, even in that most causal domain of Newtonian mechanics, we have the notion of non-holonomic constraints. I should also note that Howard Pattee raised the issue of such constraints in biology in the 1960s — his ideas and related ideas by others may be seen in the series of volumes titled *Towards a Theoretical Biology* edited by C. H. Waddington (Edinburgh University Press).

Gennaro Auletta: I think that you are mixing causality and determinism here. Quantum-mechanical laws are deterministic, but the evolution of any quantum system cannot as such cause any event at all if not interacting with other systems. However, in the general case, we have here resulting events that are not caused by that determinsitic evolution, so that we have a break also of determinism without being such an event caused in the proper sense of the word. This may sound very abstract. We have, however, a quite similar situation (not precisely the same) with complex systems that give rise to events that cannot be forecast even if the laws are deterministic. Why is this point relevant? Because it raises the question of the intrinsic limitation of the traditional causal accounts.

Chuck Crowell: As I read the defining work from Auletta *et al.* on top-down causation, one of the things that struck me was that Gennaro was calling for the evolution of this discussion into a concrete program of research. So what I would like to propose for some discussion is that this group form a research consortium that is dedicated to collaborative work on at least two fronts, one would be a laboratory-oriented front, and the other would be a theoretical-oriented front. It would be an interdisciplinary effort. We have in this room some unbelievable talent, background, and experiences from both the laboratory and the theoretical standpoint. And all of us are now far enough along in our careers that we are afforded the luxury of being able to think about doing something like this, whereas perhaps earlier on it would have been more difficult for us, for a variety of different reasons. So there are two things that I would like to put on the table.

One is my own thinking about what might be a good lab-based model system. Play in animals has promise in terms of a lab approach because it seems to be a kind of endogenous behavior. The cellular, behavioral and neural mechanisms of rough-and-tumble play have been identified fairly well; my colleagues have done a lot of work in that area and we have a lab-device that can be used to study this kind of system. Moreover, play, also has a kind of goal-directed component in terms of self-handicapping. Such an animal preparation would allow us to actually do some fairly sophisticated work to try to identify both the continued bottom-up explanations as well as possible top-down explanations of what may be going on in that particular model preparation.

Secondly, in a more general sense, I would like to have us discussing the research consortium idea. I think that we have an opportunity to put together an international collaboration that may be unparalleled. I think this is a very unique opportunity. We may be able to attract funding in various ways, or other forms of sponsorship from our home universities, other organizations, and possibly even STOQ. Clearly, the logistical aspects of all this need to be worked out, but we have technology that we can throw at this problem in terms of communications, and our ability to do those sorts of things, so I think we can produce some visible products that would be impressive.

Juan José Sanguineti: My first point is epistemological: I wonder about the role of neuroscience in the framework of natural and human sciences. I think there are two extremes: one is Chomsky for example, or Fodor, with regards to the limited relevance of neurosciences to the human sciences. The other extreme would be the total subordination of sciences to neurosciences, which would be reductionist. In my opinion, the neuroscientific explanations are in some way bottom-up explanations in the sense of material explanations, with respect to human sciences like ethics, politics and anthropology. But there is a special relationship between psychology and neuroscience, and in this last point there is a greater importance of neuroscience than the other human sciences.

The second point is to clarify my own position in my talk: I am not a materialist, I believe in spiritualism; I think that this is linked to my personal belief as a Christian. The point of the dualism would be very hard to study, for there are very different types of dualism and monism, and in some sense there is a simplification to see a pure alternative between all dualisms. I think that our spiritual dimension is incarnated, embodied in our animal sensitive body, and the result of this union is the human person. The spiritual dimension cannot be detected separately from other living experiences, but it cannot be reduced to lower levels. I do not think that this metaphysical belief is harmful to neuroscience. I think that it makes sense with many experiences in human life.

The last point concerns top-down causation. My personal conclusion, from what I have heard here, is that we use our body intentionally in normal life, but we know nothing in our common-sense experience about the physical mechanisms that we use; and this is quite fortunate, because it is very disturbing to feel like neurons do all that. So, this dimension is hidden, and this is good. Perhaps it is a good result of evolution. But in relating representations, emotions and the will to the physical, organic level, there is a kind of phenomenological gap, and this does not mean that the physical dimension explains everything; rather it is instrumental for specific scientific investigation that is unable to account for the totality of the human being.

Jean-Pierre Changeux: I think that neuroscientists are not what I would call reductionists. This is not the term I would accept when somebody classifies the work we are doing as neuroscientists. I mean we are exactly opposed to this in that we are actually constrained by our studies to extend our concern, with the aim of progressing with knowledge from the cases that we are studying; and I think that if they were to say that our position was a reductionist one, they are missing certain important aspects. I also think that the most important point is the issue of scientists in general, that is, the progress of knowledge: Is our work going to lead to a better understanding of human beings or not? If a scientist's work is not intended to progress knowledge, he is not a real scientist.

Pierre Jacob: I would like to go back to what professor Auletta said in his discussion of computational frameworks in relation to the issue of top-down causation. In a way, causal explanations in cognitive neuroscience may be the ultimate goal. That is, we want to provide a causal explanation of the causal processes involved in what people do, i.e., for their actions and behaviors. But in many (if not all) cases — unless the behavioral response is a reflex — you cannot directly map an individual's behavioral response to a perceived stimulus without providing a model of what goes on in-between the perceived stimulus and the behavioral response. This goes back to some of my earlier exchanges with Michael Arbib: in the case of linguistic behavior, you cannot directly map a speaker's response to the linguistic or non-linguistic stimulus that may have triggered the response without attributing to the speaker at least knowledge of the syntax and semantics of her language, plus knowledge of contextual features and also such and such a communicative intention towards an addressee. In a nutshell, Chomsky's distinction between grammatical competence (or tacit knowledge) and linguistic performance (or behavior) is just a methodological reminder that a speaker's behavioral response is the effect of a complex network of causal relations. Furthermore, one of the scientific tasks in the causal explanation of the speaker's linguistic behavior is to offer a computational model of her tacit grammatical knowledge.

Gennaro Auletta: Obviously, I would not deny the relevance of computation, even if I would stress that for me it is much more fruitful to approach it in the way that Michael has approached it than in the way that cognitivists like Chomsky have done. That is because, let me say, the network-like approach is much more the true way in which the brain works, and so it is much more interesting to me. Accordingly, important cognitive functions in our dealing with the world can arise due to emerging patterns in a network where no explicit rule is *a priori* and rigidly imposed from the start. By contrast, we are well aware that computationalist models like that of Chomsky suppose a set of rules (a syntax) and that this is even the most qualifying aspect of the model. To a certain extent this may express the specificity of the language system, in which rules, especially syntactic rules, are crucial. Therefore, I will not deny the relevance of the latter approach, but I am simply saying that there are situations in which also the other way could be interesting, and therefore I would warn against too rushed a generalization of this computational approach. So, let us see what the results would be for the language problem: if there are interesting results, ok; if there are not, then there is no problem either.

Marc Jeannerod: I am addressing Gennaro, but also others, about a term that has not yet been uttered in this current discussion, but which is relevant to it, namely that of "emergence." I think that these days there is a renewed interest in emergence. I remember a time when emergence was considered as the pure expression of nonlinear biological systems. In a nonlinear system, something new may happen at some point, just because you jump from one part of the curve to another part. The idea of constraints that you have put forward can also lead to new effects that were not predicted. This is possibly a new way of speaking about emergence. So maybe we should put emergence on our agenda to be discussed further.

Jean-Pierre Changeux: I think you should define what you mean, because the word "emergence" has so many meanings. Well, what is philosophy after all? I think the aim of philosophy is to introduce clarity into the discussion, and not confusion.

Marc Jeannerod: Ok, I will not try to define emergence now, but I think we should discuss this more, simply because it could be an important point. Now, obviously, there are several other definitions, such as expression of nonlinearity, expression of constraints, and so on and so forth.

Gennaro Auletta: It would be interesting to explore the possibility that there are emergent levels of reality that do not violate previous levels, if you will, because, obviously, you never have violations of the objective laws pertaining to those former levels. Nevertheless, upper levels cannot simply be deduced from the previous ones. So, they mean something more. I am aware that this is very general, but I only say this as a possible reference point or as an attempt to start with something in our investigation.

Pierre Jacob: Again, when what we want to explain is something as complex as a speaker's linguistic behavior (i.e., the fact that she uttered a token of such-and-such a sentence of a given human language at a particular point in space and time), it is important to remember that a model of her grammatical competence is only just a small link in the causal chain leading to the utterance, hence one step in the causal explanation of the utterance. There is a long causal chain involving some stimulus, some thought processes and also the mastery of the syntax and semantics of a particular natural language. Now, the computational model of the speaker's tacit syntactic and semantic principles will be part of this complex causal explanation. But we should not confuse the causal explanation and the model that will be part of it. The target of the causal explanation is an utterance or a piece of (linguistic) behavior, i.e., an event. But the target of the computational model is something else: a mental biological capacity.

Wolf Singer: I would like to lower the optimism a little bit here, concerning computation as providing solutions to the problems. As far as I can see, understanding processes in an analytical way with mathematical tools is at present limited to a very specific set of problems, and for many cases of nonlinear systems (but by no means all), you are lost. So what you can do is that you can simulate them, but if you want to have a realistic simulation of our brain processes, then you get into the same trouble as you were in before you started to do that. There are projects like the Blue Brain Project, where one tries to really get a real-time interpretation of your own circuitry in computer systems and find out what they do, but then this attempt at understanding will bring you to exactly the same trouble as you were in before; therefore, I do not think that this will be the key to the desired solutions.

Jean-Pierre Changeux: As to computational models as an intermediate step to causal explanation, I disagree. Because, what is a causal explanation? I think it is the adequacy of the model to things. Therefore, there is no intermediate step. We proceed in building models and confronting them with reality, and this is not an intermediate step to causal explanation. I think we are always proceeding by building models.

Michael Arbib: As computational modelers of the brain, we thus adopt an intermediate solution. To understand the details of small neural circuits, we may still model each neuron in terms of an interlinked set of "compartments," Blue-Brain style, but to gain insights into the neural basis of cognitive processes, we model a number of different brain regions in terms of a reduced set of simplified neurons, thus getting a handle on the performance of the system even though we sacrifice numerical precision with regard to the activity of single neurons.

Ivan Colagè: I would like to spend a moment on the issue of emergence raised previously. The simplest way I have to understand the notion of emergence is that when we try to have a computational model of some situation, what you really have in the model are the elements of the lower level, combined in some way. But the point is that to have a real model of a real situation, you need to restrict the possible range of combinations that you have at the start — as you cannot have all the possible combinations — and when you introduce these limitations, this is what precisely prevents deducing what happens at a higher level.

Gennaro Auletta: I have already said that causal explanations of the traditional type cannot be generalized — otherwise we risk a clash with what current scientific developments have firmly established. We should rather turn the whole problem upside down and say that causal models are rather limiting explanations to which we may attribute mainly a statistical and approximate value when the object of study is relative simple or in very controlled situations when we allow for abstracting certain features in an idealized way. I doubt, however, that this kind of explanation, taken alone and absolutely, will be of help for the kind of problems we are investigating here.

SESSION III

Chairman: Michael Arbib

Speakers: Mark Jeannerod, Wolf Singer

Marc Jeannerod: I have said that there is some sort of methodological duality between different subsystems, and the basic question I am faced with is: How can ideas or beliefs that are conscious expressions of our mental activity directly influence the neural network? We have ideas, we have intentions, we want to do things, and our decisions will influence neural activity. I reject what philosophy and history of science have called the parallelism of conscious activity and neural activity. We know that many cognitive operations occur non-consciously, in that they take place outside our conscious awareness: they are instantiated in the activity of the neural network. But my point is to give consciousness another function, not to cause actions because actions are caused by activity in the neural network that goes on by itself, possibly determined by external or internal events. Rather, I am trying to give consciousness the function of bridging the gap between what happens before and after the action, so as to provide what we could call the cognitive consonance between our intentional states and the final state of our action. So, the influence of conscious states on neural activity would be a delayed influence. Our brain is influenced by the results of our actions, whether positive or negative, and this is reflected in further cognitive activity to a certain degree.

Wolf Singer: I think that at the present stage we cannot give any explanations for the transition from a system's properties to *qualia*. The hard problem continues to be unsolvable, at least at the neural-biological level; and also I think that we have to incorporate social aspects. What we see is that non-conscious processing involves more or less the same structure as conscious processing, so we may accept that there is a correlation, but the conscious processes are distinguished from the non-conscious processes because of a global organized state. So, the self would, in this case, be a dynamical state, or variable. It is not a localizable variable; it is a state variable that distinguishes the conscious self from the non-conscious. Then there is of course, in such a distributed system, a re-entry of the problem of disentangled causal effect.

I think that this is where we are stalling at the moment, because the system completely waltzes along a trajectory whereby the respective next step of the research is to find out the immediately preceding state. Now, it is the global state of the system that also influences the respective local processes in a distributed way. These local processes being influenced by the global state will, in turn, generate the next global state, and so there is a continuous alternation between top-down causation of the global state that influences local processes, and local processes being changed in this way that generate other global states. This is the passage from top-down to bottom-up causation: the local processes L1, having been influenced by a previous global state (of course, through their action), generate the next global state G1 that generates L2, which generates G2. There is, therefore, an intertwined alternation between local and global. So, if we consider this global state as the upper level in the individual brain, the next higher level will be the global state generated through the population of brains, like in this room here; we are all interacting, and my respective next local process will be influenced in a top-down way by what has been going on here in that moment. Also, in this case there is top-down causation: what is happening socially here causes my individual global state to change because of the arguments employed in the discussion, which are then influenced by local processes occurring in my brain which then, in a bottom-up way, generate again the new global state of my brain, which I communicate to you if it reaches consciousness; it will influence your global states, and this is the self-organizing process that goes on and on, whereby top-down and bottom-up are intimately related all the time. So, that is what I feel.

Micheal Arbib: Does this group have a self?

Wolf Singer: No. This group assigns to me a self. This group is distinguished from other groups, so that these can, in turn, join from outside. I would consider this group as a system in which we have a certain identity to share with a bigger structure on the top of what we are as individuals.

Micheal Arbib: And yet it is different from the self, and in understanding that difference we may make some progress in our quest.

Wolf Singer: Yes.

Gennaro Auletta: In order to speak of a self, we always need some biological identity. The self is always related or anchored, if you will, to some biological identity; even when we speak of the narrative self and so on, ultimately we are bound by the fact that there is a single organism. So there is a difference for me, even if I obviously agree with you in saying that this group here is still a system. In my opinion, the sorts of "super organisms," like those formed by ants or bees, cannot be true organisms, because for instance they lack a metabolic system as a group; only individuals are able to perform metabolic functions in general, but this would be a different and long story. I completely agree with Marc Jeannerod that consciousness cannot cause actions, even if it is able to have some constraining or regulating influence, as Jean-Pierre Changeux said previously. And I think it is very insightful what Wolf said about the relations between top-down and bottom-up causation. Indeed, I am of the opinion that the dynamic aspect here is on the side of the excitation patterns, that is, I am absolutely convinced that brain activities are excitation patterns. Even the inhibition coming from the upper or global level needs to go through some excitation pattern and cannot exercise its constraining (or regulative) influence without the contribution of excitation patterns. The difficult point to understand here is that, once some global configurations arise (it does not matter how, by self-organization processes and so on), then they somehow canalize or constrain (they mold, so to speak) further local interactions again and again; so I completely agree, it is a sort of special type of feedback processing, if you like — a special type because it is not a feedback at the same level, somehow, but in a certain way it is the ability to combine in a completely new form, local dynamical processes with formal constraints. I think this is very helpful.

Marc Jeannerod: I would like to ask Wolf: How do you define global state?

Wolf Singer: I wish I knew what a global state is. But, it is obvious that the system is acting as a whole, in a highly coordinated way, or at least is capable of doing so. I think we are still far from having a complete description of such states. But I think we can extrapolate that they do exist, that they can be disrupted, and if they are disrupted, then the system decomposes into many local processes and loses consciousness; and it can still behave in a rather nice way, the behavior can still fluctuate, but it loses a number of functions that you can only have if you are conscious, which is to relate the many parallel processing channels with each other, which allows for abstraction, for recognizing what is the same and what is different, and so on and so forth. So, there is a good reason for having these global states — the ability to use the cortical processes that allow for free exchange between any subprocessors, which is the real charm of the cortex in that once you are beyond layer 4 you are able, in a sort of fluctuating dialogue, to freely exchange all the information that is available to the brain; and if you get into those states where you can relate all those things to each other, then you are conscious. This happens because you see things together that you have not seen together before, and these results can influence the local processes; thus, it is a continuously evolving self-organizing process, part of which is communicable.

Michael Arbib: To say that, when conscious, one can exchange all the information that is available to the brain seems to be too strong a statement. There is much that we are unable to connect — at least when there are time pressures (of which we may be conscious).

Juan José Sanguineti: The gap between consciousness and motor actions seems to me to be a paradox. I think that consciousness is sometimes, or perhaps always, "too conscious" (i.e., too slow) for acting, because if you think too much about what you are doing then you will be disorientated; it is a normal experience. But I would like to shift the domain of our intention and bodily action to the *will*, because I think that the point of all experience is not consciousness, but the will. When I move my body, it is because I want to, not because I am thinking too much, so I am driven by desire. For example: I want to get to the microphone because I have a strong desire to speak now, and I am moving my body, strongly directed by that, and this is a common experience; it is hard to deny that this is a real and effective explanation. Therefore, it is not a problem of consciousness, but it is rather related to the concept of will, and to emotions, of course.

Pierre Jacob: I have a methodological question for both speakers (i.e., Marc Jeannerod and Wolf Singer) about the study of consciousness. Of course, I think we have a common-sense non-scientific notion about what it is to be conscious (or to be free). Wolf recognizes that, so far, there is an explanatory gap between scientific hypotheses about our current knowledge of brain processes and *qualia* — where this latin word (*qualia*) refers to our naive introspectively accessed conscious (visual, or otherwise qualified) experiences. But there should be a gap between a commonsensically described phenomenon and a scientific hypothesis. My question is: could the *explanandum* (that is, what is to be explained) of a scientific explanation be a non-scientific ordinary phenomenon? Should science aim at explaining common-sense introspectively accessible intuitions about our visual conscious experiences? or instead should the phenomena be scientifically re-described before they can be provided with a scientific explanation?

Wolf Singer: I would say no. I think the problem is that we get used to this dichotomy and we will not be worried anymore about it. We have got used to many unresolved problems in the world, but this just shows that we understand more and more about the issue, since more understanding brings also more problems.

Marc Jeannerod: I think that the business of science is first to observe and to describe. There are many concepts here that are still waiting for true descriptions. Then we can look for causes and causal explanations, that is of course also one of the businesses of science, but the first thing is to identify the problems and to describe them.

Jean-Pierre Changeux: I agree with what Marc said, and my feeling is that there is a confusion of language once more here, and I think Pierre has contributed to this confusion by introducing the notion that we do not understand the issue in scientific terms, or that we are using non-scientific words to define consciousness. Is it that you have said?

Pierre Jacob: No, I asked the question: Can science directly explain phenomena as they are naively experienced by common sense and introspection? Or instead should the relevant phenomena first be provided with a scientific re-description before they can be scientifically explained? I do not think this question is confusing.

Jean-Pierre Changeux: Well, it is very confusing because our aim as scientists is to avoid precisely what you have just said. We are scientists not philosophers, and our function as scientists is not to get redefinitions of all of these issues.

And I think there was a little bit of confusion here in Wolf's views about what he called global and local states. What do we mean by states? I think I was careful in distinguishing states of consciousness from the contents of consciousness. Wakefulness, sleep, all these things are what I would call states of consciousness. Now, when we are awake we are in a certain state of consciousness, and then our attention can be more or less intense within this awake state of consciousness. I disagree a little bit with Wolf when he says that there are alternations between states of consciousness, global and local. I would say that *being conscious* in an awake state permits us to have global representations, which may be transitory and flow, like William James mentioned; so these are contents of consciousness that are indeed alternating and transient.

I think Marc raised a very interesting point about having an in-between cognitive consonance, that is, between actions, some kind of cognitive dynamics is taking place, which is slower. Is that what you said? I would like to raise a purely scientific question: can we have, within a given state of consciousness, different kinds of representations in parallel with different time scales? Can we have, for instance, some kind of intentional state lasting, say, for seconds (like the common intentional state that we have during conversation) and having other conscious representations nested *within this intentional state*? I think that this way of speaking perhaps may provide a way to represent these different representations in different time scales.

Marc Jeannerod: I can only give a brief response because this point would require more scrutiny. I think that we always have several actions ready "in our pocket." For example, I am speaking, but at the same time I am conscious of other things going on. So, this could be one possibility for explaining the so-called veto, by which we are able at some point to interrupt an action and to shift to another one, according to Libet's model. We have at the same time two actions ready, for example, and at some point there is a branching between the two, so that we can shift to the other one within a very short time window.

Ivan Colagè: I would like to come back to what professor Singer was saying before about this dynamic interrelation between the local level and the global level, which is somehow cyclical. As to what happens once a global level (for example the level of the cell relative to biochemical reactions) is established, it seems to me that we can somehow reformulate your point as a sort of tendency of this global level to become the local level for a higher global level, for example the level of groups. In my opinion, this could be useful and insightful to understand emergence, because it seems that once you have a level in which you have atoms, the interactions of the elements of this level give rise to a higher level which is global with respect to that level, but can be local with respect to an even higher level; and it seems to me that to behave in this way is a fundamental feature of nature. I would like to know what you think about that.

Wolf Singer: I totally agree. You have to have nodes in your networks, because somehow you have to communicate the result of the local processor to some others that are distributed, and this is based on nodes, and consciousness is one of those nodes, if you consider the social network.

Chuck Crowell: I would like to mention a thread I have heard in comments of professors Sanguineti, Auletta, Jeannerod and Changeux, and it has to do with something cognitive psychologists call prospective memory. The idea is basically this: I decide at time 1 that there is something I need to do at time 2, and time 2 could be in a few minutes, it could be in a few days, it could be in a few weeks. Somehow, I encode into my memory a future action that I need to take at time 2, and then in order for that to happen, presumably I need to recall before time 2 that I needed to do this and then execute the action at that point in time. So, from the standpoint of what we have been discussing, assuming that the action occurs at time 2, my question is: How would you characterize the causal chain that led up to that in relation to decisions that were made at time 1? I think we need to keep that idea in mind, and it relates to something Professor Jeannerod said in his presentation about the long-range planning aspects of motor action.

Wolf Singer: I think that one important aspect is also that the brain, when it is dynamically active, is not always producing a specific result (like taking some decision). I think that one needs to distinguish states that strive to have results in which computations still occur, and those moments in which results occur. We know when we have results, we know when we have finished the computational process, and this state must have a particular signature but we ignore what it is; and our internal evaluation systems know what the signature is because if we have a result we get rewarded, we feel fine; we are striving towards these results all the time. So, the result is always supraordinate; it is more than the computational processes itself, right? And these results influence the search trajectory. I would really like to spend the last years of my scientific career in trying to find the signature of the result in the brain, what it is, because it is such a clear-cut event. It must have a very, very clear signature, because otherwise the internal reward systems would not be able to read it. At present, we do not know what it is, and I think if we had it, it would give us some more answers.

SESSION IV

Chairman: Michael Arbib

Speakers: Jean-Pierre Changeux, Marcus Raichle, Michael Arbib (summarizing Antonio Damasio's positions)

Jean-Pierre Changeux: My first point is the issue of levels of organization. I think, as some of you said, that this is a kind of theoretical aspect, to be related to the modeling work. But I think it is quite essential to try to do it in a fair way, and to try to define these levels of organization (as I think Professor Auletta mentioned), and first of all to be aware of the processes we are using to define these levels of organization and how we view the interrelationships between them.

In the context of the conscious workspace, the questions arose: Are there levels of organization which are nested? And where might we have to use different kinds of modeling to describe these levels of organization and their relationships? I would not endorse the notion of top-down causation, because the downward effect is going to take place at any of these levels. If these levels are nested, then you may indeed have global constraints at the last level, at the level of the cell, or at the level of a given network, or of consciousness. This is why I think it is very important to have a clear idea of what we mean by these different levels of organization.

My second point concerns the brain as a biological organ. I think we should have the same philosophy as we did in the past in biological sciences — perhaps this comes from Claude Bernard. As to notions like structure, function and relationship (and this is something we should clearly distinguish), what is a structure and what is a function? Trying to define function is difficult, I know; I myself spent some time in trying to get a definition of what a function is of an enzyme, of a cell, of a network, of a mass of neurons, and so on. And this is for us, I think again, a way to clarify a situation: to relate these levels of organization, which are structural, to different kinds of functions that are associated with these levels. Because we have functions that are associated with small networks, the micro-sequences, other functions associated with large sequences, and the ones I think that are produced within the global workspace, which is a very basic way, of course, of looking at so-called consciousness, which is a function as well. So, I think that a very important aspect is to clarify what kind of function is associated with what kind of level of organization, knowing that each of these levels of organization is itself integrated within another.

Marcus Raichle: My remarks relate to my trouble with the very thing Jean-Pierre just talked about: What do we mean by this top-down and bottom-up stuff? If I look at the brain and I try to think "How does that relate to how I think about the brain?" I arrive at the question: "What is the brain doing intrinsically, spontaneously and as it regards its environment?" And I have to put the weight onto the former category of intrinsic, spontaneous activity. And so, when I hear words like "top-down" and "bottom-up," I find it very hard to understand how to think about that. On the other hand, I am very concerned about the perspective that one has in thinking about the brain in terms of what we've seen from our conscious awareness, versus what the brain is actually doing. If you are an economist and look at where the energy is going, it is going towards its intrinsic non-conscious operations.

The second point I would like to make is a bit of an issue given what Jean-Pierre said: I think we should be very cautious about suggesting that there is not a sustained and continuous global organization of this system. I think there is evidence for a very large-scale global organization upon which is nested a whole category of processes becoming increasingly local, but reflected globally in various frequency domains which emerge interestingly out of the noise of the brain, out of the so-called frequency spectrum which is reflected in the brain's electrical activity. I think that it's interesting to think about this spontaneous activity. I am not a big reader of abstracts, but I happened to read Jean-Pierre's abstract last night, and it reminded me of a speech I heard that Carla Schatz gave many years ago, which caught my attention about the importance of spontaneous activity in the development of the brain; and I walked away from her lecture thinking, wait a minute, that is what I am looking at; she commented that it might be important in the adult brain. I would suggest that we take from that the fact that this activity may well be involved in the redevelopment of the brain, as it is occurring daily, and in the support of memory and learning.

The final point I would make is based on having recently taken part in the rather large project in the United States sponsored by the MacArthur Foundation having to do with the interface between law and neuroscience. This has now been going on for two years, and we have been trying to figure out how we bring these two very different domains together. I think the answer comes from the program in cognitive neuroscience that was supported by the McDonnell Foundation, and ran for 13 years and basically created an entirely new field. The answer was that you cannot make progress in understanding how the brain works if you do not pay attention to behavior. What is striking to me about the discussions around this table is that we talk about very abstract issues with regards to behavior, but I worry that if we do not get more specific about what we are talking about, behaviorally, we are not going to make progress here; I strongly suspect, indeed, that a model that talks about quantum mechanics and tries to relate it to how the brain works, and does not address specific issues about human behavior in quantitative terms, is not going to get anywhere, and this discussion could be had again in 15 years and we would not have made any progress. Do not get me wrong: I do not walk away from this being pessimistic. I am very, very optimistic.

Michael Arbib (resuming Antonio Damasio's viewpoint): So, firstly I am very glad that Jean-Pierre mentioned Claude Bernard, because I consider homeostasis to be basic to our understanding of the brain, tracing the way in which the brain functions all the way back to the homeostasis of the single cell, and seeing then how it moves forward in large-scale systems. The differences among proto-self, the core-self and the autobiographical-self have been also pointed out. And to some extent that relates to the concern with emotions, where I would distinguish emotions and the role that they play in homeostasis from emotional feelings as part of our consciousness. Turning from that to the recent work in studying the brain, I have been very interested in the recent work of Olaf Sporns, who has applied graph-theoretic tools to analyze the connections between regions of the cerebral cortex, and in particular has looked at the degree of connectivity to look for hubs. The idea is that you take the graph of connections between brain regions, you remove those regions that were connected to only one region, then you remove those regions that are now only connected to two regions, and so on until you get the last group of most connected regions. And the intriguing fact is that in the brain the posterior medial cortex provides the network of regions that seems to be the most strongly connected — it is, as it were, the hub of all hubs. But as a matter of discussion, I think it would be interesting to know whether there are other hubs, and perhaps the ones that Jean-Pierre has talked about as constituting the global workspace of the prefrontal cortex will be a candidate. I think it will be interesting to relate those to the areas that Marcus has seen as providing the default brain's mode.

I think that this idea of large-scale connectivity graphs and what we can learn by looking at connections provide a new window on brain function. It has been asked, if in fact the posterior medial cortex is so well conserved from monkey to human, why is this so important? The suggestion is that the role of connecting many brain regions is conserved, and what it involves in the comparison between monkeys and humans is the nature of the regions that are connected by this hub.

Finally, a historical point about convergence-divergence zones. Some years ago, when first publications came out on this, there was a certain hesitation whether to just call them convergence zones, but it would be a mistake, because that term is not as informative as the term currently used, as convergence-divergence zones denote not only the connections being brought into these hubs, but also those leading outwards from the hubs.

Gennaro Auletta: I will focus on one point raised by Jean-Pierre, that is, the issue of structure and function. In order to speak of true functions, we need a certain minimal level of complexity. And I would then use this word for very large structures and general patterns also at a pure biological level, i.e., with respect to operations like splicing, which involves many steps. And I would stress the fact that, in my opinion, function is not identified by single structures, because you can perform a single function through different structures; for instance, splicing is done in different ways. But, and this is very important to me, there are specific aspects of a structure, let us say the hubs, that are crucial elements for performing that function. So, if these hubs are not present, you cannot have that functionality. Therefore, I would try to establish a sort of middle way, that is, not a complete separation and not a complete identification between structure and function.

Jean-Pierre Changeux: Yes, I think many of these issues such as structure, function, and levels of organization are of course theoretical approaches. Indeed, we have to be aware of this in modeling the complexity of the system. Now, the notion of being minimal in our thinking is very important for us in modeling, that is, to make a minimal model for something. And I think it is very important for us to say whether or not we need a minimal level of organization in order to have what we call consciousness. And if we really try to restrict the definition of consciousness, and try to follow its evolution, I think it would be extremely interesting to try to relate this function to a minimal level of complexity or organization.

Marc Jeannerod: I just want to make a general statement about the relationship between experiment and theories. I myself, a physiologist, am trying to implement observations into experiments, and in my presentation I showed a couple of simple experiments that try to reduce and simplify the problems, so we can try to understand what is happening. Now, what is the relationship between experiments and theory? I see that there is some parallelism, like the one that we have between action and consciousness; sometimes we go from theories to experiments and sometimes from experiments to theories — there is an exchange between the two — and this cannot be done without involving the clinical or the experimental observations of reality.

Pierre Jacob: I have a question for Marcus with regards to his emphasis that it would be bad if we forgot the importance of behavior in the present discussion. Of course, there are many ways in which we can understand this remark, but particularly in the light of the kind of work that you do and presented about the importance of brain imaging, evidence of metabolic processes in the brain, do you think that we should not forget behavior as it is understood by common sense as dealing with actions that people, non-scientists, can see and observe when we walk, when we speak, and so on? Or instead your crucial notion of behavior is related to metabolic processes in the brain, which you find evidence for by brain imaging? Is the relevant notion of behavior the behavior of cells, which is certainly not the common-sense notion of behavior (which is rather exemplified by what people do, not what cells do)? If it is crucial for you not to forget behavior, how do you want to constrain the relevant notion of behavior?

Marcus Raichle: As I understand the charge for the people who take the theological side here, it is to get a better understanding of why we as human beings behave in certain ways — why we make judgements, why we have feelings, why some of us have beliefs of one sort, and others have beliefs of another sort, and how these different beliefs arose, and then to what extent the brain plays a role in this, and how we understand, say, individual differences of many things. So, I could imagine this being addressed at many different behavioral levels, but defining the behavior that we want to understand seems to me to be critical at the outset, and then approaching the brain and trying to understand how that gets implemented in the brain, that is what I was thinking of. Of course, I do think of the behavior of cells, but ultimately I would like to understand the relevant behavior of people and, in this particular context, it may relate to how people believe and how people feel about very critical aspects of their own life. And the legal sense that I have to deal with is how are people responsible or not. Can we determine that? Can we determine whether or not are they telling a lie? How can we understand these things? Today, we have neural economics, neural law, neural marketing, neural literature and you could even say neural theology — you could put it in anything you want to. There is neural ethics, too — there is a whole field of neural ethics. I have got a shelf of these books. I think that at each level where this is introduced, the relevant communities have got to understand one another; this is number one. Then, the question should be: What are the issues on the table that are relevant to that community? We have been struggling for two years to develop the conversation between the U.S. legal scholars and brain scientists, and what is so often the problem is defining the behavior you want to understand.

Jean-Pierre Changeux: I agree with what you said, and there is a future in trying to relate human behavior to present-day cognitive neuroscience; there is a blossoming of all these different approaches such as neuro-economics, neuro-ethics and neuro-epistemology. I think all of this is of dramatic importance. But the issue is also that we have to maintain a critical attitude. Sometimes, there are a lot of abuses in these disciplines, and perhaps the presence of philosophers would be useful to try to contribute with a little more reflection about these things. Now, I would just like to come back to the general issue of modeling in neuroscience. In our definition of models for neuroscientific issues, since our work in 1973, we have the structure, the function *and* the behavior; and in any kind of formal presentation of a minimal biological process, level of organization and of course at a global level, we should have these three different components interrelated with each other. Our aim at the time, which is still our aim at the present time, was to try to build a formal organism; this was the definition we gave. A minimal formal organism is, of course, extremely limited in its abilities, and sometimes even distinguished in its particular function. This was the general epistemological attitude that we had and that we try to maintain. I think we should really have these three main aspects, namely structure, function and behavior, linked together.

Wolf Singer: I would like to make two points. One is on the nestedness of the organization of the brain. There is a fantastic intertwining of processes of different scales, which are expressed in the wave frequencies characterizing our selections, for example, so that the higher the frequencies, the more local the process. It's a very close relation; we see these frequencies going from less than 0.1 Hz all the way up to 30 Hz and more, and all this allows for a very, very complex network of relations; thus, nestedness is a very important aspect.

My second point was mentioned briefly in a previous discussion, and is concerned with the non-dominant, non-speaking hemisphere. It has the PMC, it has all these networks, but it is apparently not enough. Why does the non-dominant hemisphere not become conscious? Could it be that it is a consciousness that is not communicable? I doubt it, because the subjects who would have those problems, can talk about them, and would maintain that they have not been conscious of that process even though their own hemisphere has activated — the PMC, the prefrontal cortex, and everything, except for a few areas in the language part that are not there. So, it is also worth considering, since nobody has talked about the relation between consciousness and linguistic abilities.

Gennaro Auletta: I think it could perhaps be useful to remark that, as far as I can understand, conscious processes, conscious acts come out any time you have some discrepancy; that is, provided that everything goes or works in a satisfactory way, you do not need consciousness at all, because the brain can work in an almost automatic way. But provided that you have some discrepancy, some problem, then consciousness probably comes into play in switching modalities when something unexpected occurs. I mean, you always have non-conscious processes, patterning and non-conscious motor actions, so the job of consciousness cannot be in providing that; but *the switch* from one to another, this, for me, is the role of consciousness. For instance, I am driving a car, and then I see danger, and I switch from one modality to another; then, obviously, my reaction after that is again fully unconscious, that is, I react in a quick way, otherwise it would be too late to avoid the danger; but the switching, this is conscious. So, I think this would be a useful way to consider the problem, that is, in relation to some sort of unexpectedness, of discrepancy and so finally with respect to attention. This is obviously a biological thesis because animals are also attentive; clearly, this is not something specific to humans, but insofar as humans do this in a distinctive way, this could be interesting to further our enquiry.

SESSION V

Chairman: Marc Jeannerod

Speakers: Tomasz Trafny, Pierre Jacob, Michael Arbib

Tomasz Trafny: In modern society, an important dimension of authority is the one based on knowledge, when someone provides the fulfilment of some research and reaches some results, thus achieving epistemic authority. The second dimension of authority is the deontological one, one based on power, which somehow pushes people to obey, to obedience. Now, sometimes people have the impression that the Church has all of the deontological authority, and one of the aims of these projects we are developing is just to try to pass from purely deontological statements to epistemic ones, and in some way to influence the Church; if not the teaching immediately, at least the thinking inside the Church, among the bishops, the cardinals, the leaders, attracting attention to the fact that natural sciences provide us with extremely important results, which we have to consider at least as inspiration for our reflection and our research. Another aim is just to train the new generations of future researchers and scientists, teachers and professors, a process that in the next 50 years could make the difference in how people perceive the Church. I would like to thank you for sharing your knowledge and your capacity with us, and I hope that we can establish some kind of research team that would further discuss various interesting subjects. So, on behalf of the Pontifical Council for Culture (besides that, myself, of course), I thank you very much for your time and I hope to see you again.

Pierre Jacob: I would like to make two basic remarks in relation to the previous discussion and to Michael Arbib's paper. One is a defence of the plurality of explanations, and I shall try to make a few comments, which I hope will clarify, not obscure, the issue — maybe I will fail — and then I shall go back to the issue of the causal explanation of linguistic behavior versus computational models of linguistic competence. Now, Michael will probably disagree sharply with what I have to say, so then we can really focus on our possible disagreement.

The first general comment about the plurality of explanations is that of course there are many different sorts of *explananda*, of things to be explained, and we might want to distinguish what we want to explain; maybe one of the challenges of science is to get rid of pseudo-*explananda* or *explananda* that are not open to deep scientific explanations either now or ever. Arguably, it is also conceivable that different correct explanations could be non-competing explanations of the same *explanandum*. They would not be competing in the sense that the correctness of one explanation does not preempt the correctness of the other. Suppose (contrary to fact) that what is to be explained is the fact that all my friends are Martini drinkers. Well, we could think of two basic explanations, and I want to think about how they might relate to one another. One is that all my friends are Martini drinkers, because I selected from among possible friends only those who drink Martini. Let us call this a "selective explanation." We could also explain the very same fact by looking at the ontogenetic development of the individual life of each of my individual friends, and why in their historical, ontogenetic development, family, whatever, they became Martini drinkers. Let us call this kind of explanation a "developmental explanation." By aggregating the various specific developmental explanations, we would get an explanation as to why all my friends are Martini drinkers. Now, I think that the selective and the developmental explanations are just two distinct explanations of the same fact, and they are not competing explanations in the sense that the correctness of one of them does not automatically preempt the other one from being correct. Both could be correct in providing distinct information, so it makes no sense to ask which of the two is the correct one.

My second point is that, as I already mentioned, when we consider the human language faculty, we ought to distinguish two distinct *explananda*. One *explanandum* is the speaker's linguistic behavior: Why did a speaker make a particular utterance in a particular context, at a particular place and time? Another *explanandum* is: What does a speaker tacitly know about the syntax and semantics of her language in virtue of which she can utter and understand any of the indefinitely many sentences of her language? Whereas the first issue cries out for a causal explanation, what is required to answer the second question is a computational model. Now, the basic computational or even logical issue that Chomsky was concerned with was as follows. We know that every sentence of any natural language is a grammatical sequence of words belonging to the dictionary of the language. We know that the dictionary of a language is a list of words. The problem is: How are we going to provide an explicit account of the fact that the speaker of the language is able to come up with understanding or generating any of an indefinite open-ended set of grammatical sentences? In other words, the task was: How can we provide a characterization of the kind of rule system that must be finitely characterized which will be able to project the finite set of lexical items onto this open-ended set of sentences of the language? And that was the beginning of a new way of doing linguistics. So, it *is* a computational model of the ability of a speaker to do this, to take the finite set, and any child must go through this, that is, go from mapping the finite lexicon of his or her language to generating an open-ended set of grammatical sentences. Now, I think that when (i) we see how this computational path arises, and (ii) we want then to focus on how a causal explanation of brain functions is able to help us understand how speakers actually use their brain tissue to implement those functions, we have a perfectly reasonable scientific program.

Michael Arbib: I think that we can locate Chomsky's original enterprise within the framework of the development of the theory of computation in the 1930s. David Hilbert, a great mathematician, who is the advisor of the advisor of the advisor of my Ph.D. advisor, posed essentially the question as to what is an effective procedure, what is something that can be done by manipulating symbols and a rule in a specified way. And among the different answers, Emil Post came up with rewriting systems, which is basically the idea of taking strings of symbols, and then recognizing them as the left-hand side of the production and replace them with the string on the right-hand side. So the idea is that computation could be described by starting with a stream of symbols (i.e., the initial data), then by keeping on processing, and if you stop, at the end, you can read the result of the computation. What the young Chomsky did — although I am not sure that he explicitly located his work in that way — was to provide us with a hierarchy of such rewriting systems, the finite-state systems, the context-free systems, the context-sensitive systems, all the way up to the general rewriting systems. And working with both linguistic data and the psycho-linguistic stage with George Miller, he convinced us that although the context-free languages were very attractive models of grammatical productions, they were not rich enough to describe a human language, and so the process was on finding those that were.

Meanwhile, Warren McCulloch and Walter Pitts had been interested in relating the ideas of Turing and the ideas of mathematical logic to neural networks, and they came up with a very simple model of the neuron, the threshold logic unit, and basically showed that any Turing machine, which was considered to be a universal model of computation, could have its control program replaced by a network of their neurons. So, this idea of a neural network as in some sense a logical construct, but rich enough to control any computation given appropriate storage facilities to hold initial and intermediate results, was the framework of computation that we had in the 1940s. And I understand that John von Neumann, who designed the first stored program electronic computer, was in fact influenced by all these trends, and had been one of the advisors of Turing when he was at Princeton.

So, when we turn to the brain in the 1960s, at the time of Chomsky's early work on grammars and automata, the initial temptation for those with a computational background was to see the brain as a McCulloch–Pitts network and to consider the world as in some sense the tape which would hold the information in which to interact. But then people noted a very crucial difference between computation as they conceived it and the brain as we understand it, namely, that whereas in a logical system you have a closed set of rules and see whether these would lead to all possible logical results or not, we, as humans behaving in the world, make mistakes and learn from those mistakes. So, the notion of a brain with a fixed structure is not of interest to us; we look for an adaptive structure and, with the perceptron and the Hebbian synapse and so on, we moved from the McCulloch–Pitts network to networks in which the knowledge accumulated within the network was contained in these plastic connections which change over time.

What I showed within my general framework of these different levels of society, person, brain, neuron and further details, was that the story has been complicated in two ways. One is that going down from this model of the adaptive network, we find ourselves confronted with an ever increasingly rich database of molecular details for synaptic plasticity, dendrite structure and so on, even in mathematical terms, and we are challenged to try to understand how these complexities below the level of the single neuron will change our understanding of computation in terms of something that the overall network of the brain does. The other thing that I have suggested, is that no matter how sophisticated our models of the individual neuron, simply combining them as neurons to form a brain region, and then combining brain regions, in other words using *structure* as our only guide to a theory of brain function, is inadequate.

In terms of neurons, we look at adaptation, excitation and inhibition, and I suggest that we introduce a functional mediating level to complement the structural mediating level, namely, this theory of schemas; and I suggested that in this case, unlike the traditional computer in which the data is separated from the processor, schemas were at the same time stores of knowledge and processors of knowledge, as indeed are neural networks, and that we could look at these as units of distributed computation which would both compete and cooperate, so that in a particular situation the schemas that represent our ongoing default state would interact with the motor schemas determining our actions and the perceptual schemas determining our construction of the world around us, and that these processes would frame the way in which we would move forward as embodied persons in the world.

Chuck Crowell: I have some comments on Professor Jacob's remarks. First, with regard to the Martini drinker issue, there is a third explanation in that you are a great Martini drinker and all your friends became influenced by that, and that is why. But the more basic point here is what we call an empirical question in science, and the only way we are going to distinguish whether one explanation is correct as opposed to another, is to do systematic observations, and that is what science is all about. A more important point is that I see a parallel between language and motor activity. In basic motor activity, as Professor Jeannerod has demonstrated in his work, if I want to get a drink, I have a goal in mind, and this implies a sequence of actions that I execute in order to pour the water out of this container and then drink it. As a matter of fact, there are equivalence classes with respect to how I go about doing that, and that brings us to a top-down causality terminology. In language, we have something similar; I use words because I have a goal: I want to express an idea of some sort, and then there are a whole lot of equivalence classes about the actual way in which I choose to express it, particular words and so on, and they can be seen together in a variety of different ways, but they are all in the service of what I really want to say. So, I do not think we should lose sight of that parallel, because it seems to me to be fundamentally important. We might be able to draw in models that already exist for understanding how motor behavior actually happens in order to understand language.

Gennaro Auletta: First, I would try to avoid confusion here, that is, for me we can speak of competing theories when they deal with problems in the same field at the same level of complexity; but when we have different explanations dealing with different levels of complexity, we no longer have two competing theories. I mean, all of us agree that the molecular structure comes from a more basic physics, and that biochemistry comes from another lower level, but I think none of us would try to explain biochemistry in terms of quantum mechanics or whatever. Somehow, this would be completely unscientific. Also when we try to establish a connection between neurons and language — and this is a very important approach — obviously there are linguistic items that must be explained in linguistic terms; we cannot explain linguistics in atomic terms or in physical terms. About Michael's issue, I think that this was very insightful. I think that this approach can indeed provide us with very important insights into the way in which representations can arise. I just have one point that comes much before any representation, that is, at the basic level of the first unicellular organism, a single bacterium. You already have, at this level, a sort of activity; namely, this entity is somehow able to actively refer to something else, for instance, to energetic sources. So, the point is how we are to combine this idea of networks with this idea of being somehow referred to something else, or of the use of some external signals pointing to something else (for instance, some chemical gradient) as meaningful for me as a unicellular organism since it is related to the presence of food. I think that this has to be approached clearly at the most elementary level. For instance, Jean-Pierre has spoken about the minimal level of complexity, and I think that it should be accounted for at the minimal level of possible complexity.

Jean-Pierre Changeux: Michael gave us a historical account of how this kind of modeling developed through the years; but he did not mention what I will mention now, namely, that there has been progress in the understanding of the objects that we have studied in the brain. And I must say I disagree with Pierre on two points. Firstly, what he called the plurality of explanations and second what he considered as a computational model. The computational model of Chomsky is, in my opinion, very useful for progressing in the description of language, so I agree that there is a positive aspect. But I must say that the progress is very limited, and what I would like to introduce as a general issue is the need to try to find what the progress of knowledge is. And I would say that it is not a question of correct explanation, but rather we have to use models in order to deal with explanations subject to a selection. I think this is not simply producing models; it is rather to see what the useful models are, and where we go with these models. And I think that now we need to have more and more multidisciplinary models, or models that put together more and more things. Today, we have to deal with a lot of new facts, a lot of new information, a lot of new insights, and, particularly in neuroscience, I think our aim, now at this stage, is to see where new knowledge is going to take place, and try to put our efforts together. Perhaps here is the need for a multidisciplinary effort, for a single person would perhaps not be able to undertake that endeavor, and we need to have a team contributing to the modeling of these issues. And my wish is now to put a team together where we have brain imaging, linguists and computational people all together, so that they can start with simple systems first, to progress in the understanding, and then have a definition of progress.

Pierre Jacob: First of all, I want to thank Michael for a useful and wonderful account of the development of computational ideas. One thing that we started discussing with Michael is really the kind of (formal or informal) constraints that he wants to put on the notion of the schema, and what counts as a schema in his way of building computational models of various brain functions. I would like to state three points in addition to that. To Professor Crowell, I think that the comparison between grasping the glass and the equivalence class of sentences (whereby I as a speaker could express what I have in mind, where my basic goal is just my communicative intention) is an interesting way of thinking about it. The fact that I have to know the grammar of the language, whereby I can put together those words in a grammatical structure and a grammatical sequence in order to make myself understandable by other English speakers, is in fact related to the linguistic program to which I was referring, and was trying to account for and provide explicit models of, and so on.

With regards to the question raised about competing explanations (and the points about competing explanations that were brought up by Jean-Pierre, Professor Auletta and Professor Crowell), I just want to reflect on something that Wolf provided in his presentation. For example, he was interested in how we can account for the fact that, say, some very distant species in terms of brain structure, homologies and evolution are able to understand pointing gestures. He mentioned the fact that humans have this particular ability to point, and very early on human children can understand pointing gestures as a kind of ostensive communicative behavior, and chimps, which are very close to humans in terms of brains and genetic structure, fail to understand this completely, whereas dogs do not. Dogs, whose brain structures are far more distant from the human brain structures than those of chimps, and who are in fact in the evolutionary tree much further away from the development of primates, manage to understand it. And he said that we do not know this exactly, but we have this idea that there was a co-evolution that can provide some insights into how dogs manage to understand pointing gestures and chimps do not. What I would like to suggest is that when we try to explain things by complexification of brain structure, on the one hand, for example, and when we want to explain something by converging evolutionary solutions to evolutionary problems given some specific evolutionary constraints, we have very definite kinds of explanations and the merits, as I think Professor Crowell was saying, is a question of empirical evidence in favor of one explanation. But we may have to accept that a historical kind of explanation is fairly different from a more molecular one, so we may have to deal with both.

Juan José Sanguineti: Just a remark that may be completely different from what has been said; I am a philosopher, and we are sometimes dangerous people. But for all these problems of Chomsky, representations, frameworks, I really acknowledge that we humans have a capacity to go beyond pure representation. So, the capacity to change, shift paradigms or comparative explanations, go from one grammar to another grammar, all of this is, for me, a kind of glorification being endowed with a very strange power compared to representations. So, there is a higher level, but the problem is that this level is not itself a representation; it is something that we cannot speak too much about — it is, in some ways, the whole language. I know that this is very difficult not only to explain but even to conceive of. But I think that there is something in our human thought that in my view explains all what is done in science and in human thought, and even in human behavior.

Jean-Pierre Changeux: I would like to respond to Pierre as to the question he mentioned about historical explanations in the case of the chimps he was concerned with. I think one should not have a naive explanation for these things (and I believe that evolution is a linear process), so that it is simply the context of the brain that is able to explain everything here. I think the examples that you gave are challenges for scientists, and one day we may accept the idea that some species can have performances that are much better than humans' ones. Let us look at hummingbirds and the way they move; let us look at the way birds that are predators have vision and are able to catch their prey, something that humans are unable to do. There are many examples in the animal world showing that performances of non-human species, in very specific domains, are much more elaborate. So, it is a challenge for us, as neuroscientists, to try to find out what the features in the brains of these species are which let them have these kind of advances, and how there is an effect of domestication as a selection, so that there are some performances which are selected by domestication. Now, I am very bothered by the attitude that says that there is something in human beings that is beyond representation, beyond language, and which explains everything. Is this what you say, Professor Sanguineti? You start to say that it is not accessible to scientific research. Is that what you mean?

Juan José Sanguineti: It is not treatable by scientific research. It does not mean not accessible but not treatable.

Jean-Pierre Changeux: But why do you say that? This is exactly the aim of science, to try to understand what we do not know. The aim of science has always been to try to define the limits of our knowledge, and I agree with you that we have a lot to understand there, but to say that it is beyond scientific approach is, for me, extremely dangerous, because this is a…

Juan José Sanguineti: How are you able to compare different scientific approaches?

Jean-Pierre Changeux: I am not going to compare scientific approaches. I am trying to say, as a scientist, that our aim is to try to progress in the understanding of what we do not know, what resists our understanding; and for us there is no barrier, there are no constraints, no limits. The only limits are represented by what we cannot, at the present stage, grasp.

Juan José Sanguineti: I think that science must build a kind of special framework in the extremely important task of discovering things that we do not know. But when you change to another scientific view, there I see the power of going beyond representations because scientific facts are working within a representation…

Jean-Pierre Changeux: But I am saying that the aim of science is to try to understand and progress in all the aspects as much as possible…

Marc Jennerod (Chairman): I think it is important to have this discussion on the record, as it is worth reintroducing it on further occasions. We do not have time to go on with it at the moment. And now Michael will make some final remarks to close this session.

Michael Arbib: Ok, let me work backwards. I think that shifting paradigms involves a very strange power that may lie beyond languages. This is certainly provocative, but I think that if we look at the history of science we can trace some particular changes; so for example, the biography of Einstein shows the ways in which ideas of 19th-century physics were transformed by many people until finally, yes, Einstein developed his relativity theory by an act of genius we cannot explain directly, but at least the framework can be understood as a combination of ideas. I think about something that Jean-Pierre is somewhat hostile to, and something that I think Pierre is more sympathetic to, that is, we do need a phenomenology, a historical treatment and so on, in some cases, before we can make the precise definitions to tune the scientific effort; and there may be at any stage a gap between that phenomenology and science, but I would not always think that there is an irreducible gap. This is the first point. To Pierre, I think that I would agree that Chomsky's work has been valuable in providing a description of syntax, but my concern is, as we move from getting some idea of what syntactic structure considered by itself does to looking at how it operates in the brain, I think we have now reached the stage where Chomsky is an impediment rather than a help. Now, Chuck talked about relating motor equivalents to communicative equivalents. I would just like to advertize that I have a readably large number of papers on developing — from the inspiration I received many years ago from Marc Jeannerod and later from Giacomo Rizzolatti — an attempt to define language in a way that builds upon, but does not exhaust the underlying sensory mechanisms. What is the direct relation of the grasp to the hand shape in sign language, and what does that tell us about the more abstract relationships between sound and meaning that we have in spoken language?

Finally, let me just talk about the issue of chimp pointing. Chimps in the wild have never been seen to point, which may be an observational gap but so far has proved robust, whereas chimps in captivity do point. So what is interesting about this is that it seems to be a social consequent. The chimp behind bars will see a banana outside and will reach towards it; he is blocked from getting to it, so the open extension of his hand is now perceived by the human as a request for the banana, which is then given. Eventually, a ritualization of failed reach becomes a way of drawing the keepers' attention towards a desired object. This never turns into declarative pointing, which seems to be a human attribute, but it is instrumental for the chimp. And this mix of what is in the genetic structure versus what is only revealed when that genetic structure is confronted with a suitably rich social structure is a very important point. One of the ways that we have been approaching the view of language in relation to motor experience, in a 2008 paper in *Current Anthropology*, is to analyze non-human primate vocalization, gesture and human language. One interesting point there is that chimps do make some use of attention, so if you make a hand gesture, it may be purely visual or it may have an auditory component. And when a chimp realizes that another chimp is not in its visual field, the odds are that it will use an auditory gesture or a tactile gesture rather than one that purely has a visual component. So, there are various ingredients that suggest that the mixture of human biology and human culture may help to begin to realize Chuck's dream.

Marc Jeannerod (1935–2011) A Personal Note by Giacomo Rizzolatti

Marc Jeannerod, one of the most influential cognitive neuroscientists of our days, died on July 1, 2011. Nobody has been as close to me in science for the last 30 years as Marc was. We met for the first time in Pisa about 50 years ago. He went there from Lyon with François Michel to present some new data on paradoxical sleep to Prof. Giuseppe Moruzzi, the great Italian physiologist and discoverer of the role of reticular formation in sleep regulation. Marc, a medical doctor, had already abandoned clinical medicine for research and was carrying out experiments under the supervision of Michel Jouvet on the neurophysiology of sleep. The similarity of our careers, both medical doctors and both attracted by experimental research, created an initial bond between us.

Subsequently, we met almost every year at the meeting of the Neuropsychological Symposium Society (more a club than a formal learned Society). The Society collected together the most brilliant clinical and experimental scientists interested in neuropsychology. We were both very young when admitted to the Society and had the fortune and privilege of meeting people like Brenda Milner, Hans Lucas Teuber, Norman Geschwind, Ennio De Renzi, Mortimer Mishkin, just to mention only some of them. One of the members was Henry Hécaen, an outstanding neuropsychologist and Editor-in-Chief of *Neuropsychologia*, at that time the most prestigious neuropsychological journal. Hécaen was very impressed by Marc. Later, he proposed Marc as his successor as Editor of *Neuropsychologia.* The proposal was accepted and Marc directed *Neuropsychologia* for many years with great success.

Marc, after his initial research on sleep, became interested in movement physiology. His fundamental studies in this field included the study of reaching and grasping, and the analysis of eye movements and of eye-hand coordination. In these studies, Marc and his colleagues, notably Prablanc and Pelisson, demonstrated the key role of anticipatory correction and proposed a fundamental role of corollary discharge in motor control. His book *The Neural and Behavioral Organization of Goal-Directed Movement* (1988) represents a milestone in the field of movement control.

In the 1980s, I became interested in the motor system and decided to visit Marc (already Professor of Physiology at Claude Bernard University of Lyon), for the purpose of learning more about motor control and improving the quantitative aspects of my laboratory. I went to Lyon together with Maurizio Gentilucci. After this visit, Maurizio and I set up a kinematics laboratory in Parma. This visit had, however, other consequences. Both Marc and I were rather disappointed at the predominance of British empiricism ideas in the study of neurophysiology and the attempts to understand how the mind works based on those ideas. Marc was philosophically more advanced than myself and his love and knowledge of the work of Main de Biran had a profound influence on me as well as on other neuroscientists. Today, many neuroscientists accept the view that the basis for our understanding of the world is the motor activity originating from our internal energy rather than sensory information and its subsequent processing, but this was not so in the 1980s, when Marc was heralding this view.

In those years a new international program for research support was launched: "The human frontier science program (HFSP)." This program required original interdisciplinary projects and a consortium formed by an international team. Marc decided to apply to join the consortium and asked me to participate as well. Together we contacted Michael Arbib and Hideo Sakata, who willingly accepted the proposal. Our application was successful (as were other joint applications to HFSP and the European Community subsequently) and from that time on a strict collaboration was started that lasted for more than 10 years

The idea behind our initial application to HFSP was that there must be a link between objects and motor programs necessary to interact with them.

Behavioral studies by Marc were pointing to this direction. In particular, the idea was that when we observe an object we encode simultaneously, in largely independent parallel "channels," the different intrinsic and extrinsic properties of the observed object. These channels transform the objects extrinsic and intrinsic properties in proximal and distal movements, respectively. The studies by Hideo of the AIP area of the parietal lobe as well as studies by this author (G. Rizzolatti) of the frontal areas, F4 and F5 fully confirmed this view. Simultaneously, Michael made formal models of our findings, proving their conceptual validity and indicating the path for further studies. A summary of these experiments showing a specific link between objects and motor program was published in a highly cited review article in *TINS* (Jeannerod *et al.* 1995).

It is difficult for me to describe the atmosphere of those years. Besides the scientific achievements, there was friendship, a desire to share new ideas, a willingness to communicate with others. This unique camaraderie was shared by many post-docs and students like Boussaoud, Luppino, Fadiga, Murata, Gallese, Decety and Fogassi, who participated in our periodical meetings in France, Italy, Japan and California. Marc later wrote me that the "human frontiers" years have been the most rewarding years of his scientific career, an opinion that all who were involved in the research of those years share.

In the 1990s, Marc studied, among other things, mental imagery. He made the important distinction between the mental imagery in which one imagines objects or visual scenes like in a movie, and motor imageries in which one imagine himself acting. As he nicely put it: "Mental motor imagery is a *window* into the representational stages of action" (Jeannerod and Decety, *Current Opinion in Neurobiology*, 1995). Here is again his idea that the self is related to the motor activity, the "internal energy," and not to the integration of proprioceptive and exteroceptive sensory modalities.

Marc spent a sabbatical year in Parma in the early 1990s. This was the period in which we discovered and studied mirror neurons. Marc was excited about this discovery which indicates that the motor system plays an important role in understanding others. With his critical capacity, he helped me a lot in examining and eventually establishing the possible functional roles of mirror neurons. Many years later he wrote what I think is the best formulated explanation of the function of the mirror neurons: "The observation of actions performed by other agents generates in the brain of the observer representations similar to those of the agents. This circular process, from the self to action and from action to other selves, has as a consequence that action representations can be shared by two or more people. These new findings have radically changed the traditional view of the motor system as an executive system that merely follows instructions elaborated somewhere else. Instead, the motor system now stands as a probe that explores the external world, for the purpose of interacting with other people and gathering new knowledge."

Marc was an excellent writer. He always impressed me with his ability to write papers in a clear, "Cartesian" way, which revealed the outstanding clarity of his thinking. He liked writing, as is shown by the many books and book chapters he had written. His books also show the broad span of his interests, ranging from psychophysics to neuroscience, from neuropsychology to philosophy and psychoanalysis.

In the 1990s, Marc struggled to create an Institute of Cognitive Sciences, in France as he felt that European cognitive neuroscience was considerably behind that of the United States. The Institute, funded by CNRS, was officially established in 1997 and inaugurated in 1998 in a beautiful building in Bron. The idea underlying this enterprise was well explained by Marc: "Seule une association étroite entre sciences du cerveau, psychologie, linguistique, informatique, anthropologie et philosophie [...] peut apporter des réponses nouvelles, c'est-à-dire issues de recherches empiriques, aux questions traditionnelles concernant la nature de l'esprit humain" (M.J. "La creation de l'Institute de Science Cognitive"). Pierre Jacob, a philosopher, was appointed as co-director of the Institute. The Institute, the establishment of which gave joy and enthusiasm to Marc, was also a source of great disappointment in the last years of his life. Stupid, mean egoisms destroyed the unity and the originality of the Institute. Its "balkanization" was felt as a wound by Marc.

The last scientific achievement of Marc was devoted to the problem of agency and to the role of sensory-motor system for language understanding. My last meeting with him was in Lyon in a cold winter of the year in which he died. We spent an entire morning at the Institute of Cognitive Sciences discussing various topics and especially language and its relation to the motor system. Before driving me back to my hotel, Marc asked Tatjana Nazir to join us in discussion. At the end he proposed to organize a meeting on language and its relation to the sensory-motor system. This meeting never took place and I never saw Marc again.

If I have to define what Marc was both in science and in his private life in my own language, I will say that he was "un Signore." The English term gentleman conveys, I am afraid, only partially what I mean. He was conscious of the great contributions he made to neuroscience, but he never praised himself. His opinion of others was quite definite, and often negative, but he never expressed his views openly in order not to offend or diminish others. He was also modest about the public recognition he received. He was a member of the French Academy of Sciences, of the "Pontificia Accademia delle Scienze" and Knigth of the Légion d'Honneur. I learned about these important recognitions a long time after he received them. He was also a deeply religious man, but he never discussed these private matters even with his best friends.

My sketch of the personality of Marc Jeannerod covers only a part of his achievements. A comprehensive summary of his life and achievements has recently been written by Yves Rossetti — a former student of Marc and his successor as Professor of Physiology at Claude Bernard University of Lyon (*Neuropsychologia* **50**: 351–356, 2012).

The History of this Volume

Gennaro Auletta and Ivan Colagè

We have not known Marc Jeannerod for a long time. One of us (G. Auletta) contacted Marc in 2005, asking him to do some lectures on cognitive neurosciences at the Pontifical Gregorian University (he accepted and lectured there until spring 2010). G. Auletta had indeed already read his famous book *The Neural and Behavioural Organization of Goal-Directed Movements* (OUP, 1988) and a contribution ("A Neurophysiological Model for the Directional Coding of Reaching Movements") in a volume edited by Paillard (*Brain and Space*, OUP, 1991), remaining impressed by the freshness of the research of the French school, but in particular by Marc's extraordinary brightness and competence. Moreover, G. Auletta knew his contributions ("The Cognitive Way to Action" and "Are There Limits to the Naturalization of Mental States?") published in a volume collecting the proceedings of a Vatican Observatory conference (Russell *et al.*, *Neuroscience and the Person: Scientific Perspectives on Divine Action*, Vatican Observatory, 1999) and was surprised by his dealing with important philosophical questions, but always with respectful understatement.

When we met him personally, among the first things that he told us was that he had tried to establish a dialogue with both philosophers and theologians, but he had not yet fully succeeded. On the other hand, he was amazed by our goal to build a bridge between science and philosophy. In our personal interaction with him, we remarked that he was not only interested in philosophy, but he himself had serious and deep philosophical ideas and concerns. In particular, he was very familiar with the philosophy of Maine de Biran (a French philosopher who is unfortunately not much studied today), not being extraneous to other classical philosophers like Hume and Kant. In particular, we soon became aware that Marc's "philosophy" of motor cognition (he was the founder of this field of study) was an interesting alternative to both mechanistic associationism and irreducible vitalism.

In 2008, G. Auletta, together with G. Ellis and L. Jaeger, published a paper (for the *Journal of the Royal Society Interface*) attempting, among other things, at translating the concept of top-down causation from a purely philosophical concept into a concrete scientific research program by focusing on notions like functional equivalence classes and information control. We discussed a draft of the paper with Marc, and the proposal struck Marc to a considerable extent.

In February 2008, we met Marc in Lyon. During this intensive meeting, the idea came up of organizing a workshop on the issue of *Downward Causation in Neurosciences* (later renamed *Endogenous Processes and Causation in Brain and Behavioral Sciences*), and we immediately became engaged in this enterprise. To clarify the topic of the workshop, we began writing a three-page document in which we tried to combine his ideas on motor cognition and the developments on top-down causation contained in the aforementioned paper by G. Auletta and colleagues. Meanwhile, we established a list of invited speakers to the workshop. The final version of that document (which Chuck Crowell emphatically labeled "The Lyon Document," and which we have reproduced as a further appendix to this volume) was sent to the speakers to allow them to understand the focus of the event.

The workshop was eventually held on April 2–4, 2009, in the Vatican. During the workshop, it was again in a short private meeting with Marc that the idea of publishing the proceedings came to our minds: we could not allow the high quality of the papers delivered and the incredibly lively and in-depth discussions that were taking place to be lost without a trace. We then presented the idea to the participants.

From there, we started to gather the contributions. Fortunately, the workshop's sessions were recorded, and after six months or so of very hard work, we were able to provide the transcriptions of the talks and, most importantly, of the discussions, to the participants. They used the transcriptions to write their chapters (usually updating their views with current findings) and each of them revised both the discussion that followed his own talk and the final Overall Discussion. Michael Arbib took special care of the Overall Discussion, also prompting further revisions by some of the participants. By November 2010, we had collected nearly all the parts of the volume and begun a last cycle of editing and revision. Marc worked very hard on the project, demonstrating the importance that he attributed to the volume, but always with his unique style characterized by understatement and perseverance.

March 3, 2011, is an important date for this volume. That day, a few months before passing away, Marc sent us the first complete version of the entire volume in a single file, in which almost only the introduction was still missing. Marc also expressed his interest in writing the introduction with G. Auletta, who soon started writing a first draft, which he sent to Marc on May 10. Marc approved the text, giving interesting feedback and some suggestions. Although he was fully aware of his approaching death, he endured this enterprise with an incredible stoicism. Unfortunately (see the note by G. Auletta at the beginning of the Introduction), Marc never got to see the last version of it.

The work that we have done on this volume, which without Marc would have been simply impossible, is the highest tribute that we can offer to his memory. We have a particular regret, however. When we met Marc in 2005, he immediately made clear his desire to discuss the issues with philosophers and theologians. Subsequently, he had asked us several times to involve theologians in our discussions and meetings. In the period that elapsed while preparing this volume, we organized, together with Marc of course, a second workshop (*Human Brain, Human Mind, Human Being*, at the Gregorian University, October 2010), where the presence of theologians would have been entirely appropriate according to Marc's own view. Our feeling was that he considered this possible interaction as possessing an appreciable cognitive value. Marc was really unique in the deep respect that he always showed for any field of knowledge. Our regret is that we were ultimately unable to fulfill Marc's wish and involve really competent theologians with an interest in these matters. In praise of Marc, we hope that this volume may contribute to strengthening the interactions between neuroscientists and philosophers and ultimately encourage theologians to engage in this challenge as well.

Top-Down Causation in Cognitive Science?

Lyon, February 25, 2008

A. TOP-DOWN CAUSATION: SOME GENERAL CONSIDERATIONS

The issue of this project is to ascertain whether it is possible to speak of Top-Down Causation (TDC) in cognitive processes. Since there are a lot of ill-posed questions about TDC, we need clear definitions to start a research program that must be first theoretical and eventually, in a second step, also experimental. We also need to avoid infinite regress in explaining such processes, which is one of the major dangers in this field.

It seems that there are different fields in which TDC applies: physics, biology and cognitive sciences. In human cognition, it is reasonable to assume as a starting hypothesis (to be submitted to subsequent verification) that there are *chosen* goals. We define *goals* as states to be attained by a certain agency in view of its needs. For this reason, goals must be distinguished from targets. To take a pen (e.g., for writing) is a goal; the pen is a target. In other words, a goal has no meaning and cannot be understood without the "perspective" of a certain agency, and represents therefore a specific connection between the agent and the target.

Also in the biological domain there are goals. It is very well known that animals have affordances, e.g., a bird can consider a certain tree as good for nesting. However, we cannot speak here of conscious or chosen goals. On the contrary, goals here are, in general, inbuilt and inherited. Moreover, we wish to stress that we do not have goals in evolution, for goals are always related to a single organism or at most to a cooperation of individual organisms (cooperation of social animals).

Instead, we think that, within the physical domain, there is no TDC in the proper sense of the word but only some analogue. As a matter of fact, we can speak here simply of physical constraints and of attractors. For instance, the disposition of the trees in a forest may block or canalize some physical agency, like wind or fire. In this sense, this configuration acts as a formal constraint on such physical agency; however, no goal can be assigned to it. To reach conceptual clarification about TDC, two notions appear fundamental[1]: (i) information control by feedback circuits and (ii) equivalence classes.

Information control consists in the ability of a controlling instance (which may also be non-localized) to drive certain operations performed by a lower executing instance thanks to the ability of a third instance (a comparator) to verify the mismatching between the output of the operation and the input goal (fixing an expected outcome). In living beings, for example, feedback control underlies homeostatic processes (maintaining blood pressure, body temperature, etc.). This is the way that goals (abstract relations) control dynamic causes, thus becoming causally effective.

In order to understand the concept of *equivalence class*, let us make a comparison with a mechanical device, like a thermostat. In this case, there is also a feedback circuit. However, we wish to stress that such a circuit is designed and built by humans and has no goal. This can be seen by the fact that the "information control" (maintaining a certain temperature) coincides here with a *single* path of physical interactions (the switch mechanism is turned on and off according to the temperature level attained). Instead, when we consider true goals, there are, in general, several — physically or chemically different — operations that can lead to the same desired outcome (within a certain tolerance window). This is already true at a biological level, where, for instance, "splicing" (the cell process leading to the maturation of mRNA) can be performed in very different ways (i.e., through different operations). These different operations constitute, from the perspective of a certain goal, an equivalence class, since they are able, despite physical-chemical differences, to produce the same outcome. This ability is the basis

[1]See Auletta G, Ellis GFR and Jaeger L. (2008) Top-down causation by information control: From a philosophical problem to a scientific research programme. *J Roy Soc Interface* 5: 1159–1172.

of an important behavior of any living being: error-correction and learning. Indeed, the potential gap between an expected outcome and the occurred one (relevant because of the presence of a goal) already implies the ability to correct and optimize the operation via feedback mechanisms. Therefore, in general, equivalence classes are established and maintained by information control in a way that cannot be considered automatic, and this is what constitutes the difference between the case of a thermostat and a true goal-directed information-control mechanism.

Equivalence classes under information control are the result of formal constraints. They represent a confinement of possible efficient causal processes, i.e., pathways of physical and chemical interactions (operations), without violating the laws of physics and chemistry. In this way, formal constraints cannot be understood as a direct causal intervention into the lower levels of reality, but in a way that needs to be clarified, they "make use of" causal processes from below. We can again consider the analogy with the physical world. The spatial disposition of trees is not by itself an active agency able to produce certain effects. However, when a true physical agency, like wind or fire, comes into play, then these formal constraints influence this agency (e.g., by blocking or canalizing it) and decisively concur to the final outcome. When we have true information control, the controller is able to *select* (that is, to either admit or reject) a certain operation (or a certain element) as pertaining or not pertaining to an equivalence class, and therefore, through the specific modality of action of the performing instance, to give rise or not to give rise to a certain causal process, and to eventually correct it during execution.

It is important to stress that the physical states or processes aiming to achieve a certain goal are not in general the most probable from a physical point of view, because they either consists of, or imply, some form of order that is usually far from thermodynamic equilibrium. We are particularly interested in TDC where there is an *endogenous* process, that is, a process occurring in an organism in ways that are not dependent on environmental inputs, since, in this case, TDC cannot be explained away by having recourse to such external inputs. An output without an input seems to represent a violation of the energy conservation principle; but we want a theoretical framework for TDC that does not violate the fundamental physical-chemical laws, as mentioned previously. When endogenous processes occur, i.e., processes where there is no starting external input, an important question must be posed: What is the cause for such processes to get started? A good example here is that of a gene starting to be expressed. This question becomes even more important where cognitive processes involving will are concerned.

B. TOP-DOWN CAUSATION IN COGNITIVE SCIENCES

The general considerations made so far fit very well with the mind–body relation and the problem of *will*. For example, when one wants a coffee, brain constraints must be produced to move one's body and finally have a coffee. The point is that we do not know how this happens and where the controlling instance is, even if this does not need to be localized. In fact, what we are focusing on is not a behavior induced by environmental changes influencing some receptor "device," but a state of affairs where some output effects (brain constraints, and then movements and behaviors) occur without any input, that is, without being a consequence of a process triggered (caused) by some input (coming from either the environment or some internal physiological state).

We have seen that equivalence classes rule out pure automatisms in an organism's information-control mechanisms. In the case of intelligent beings, pure automatism is ruled out from the start, since the goals are not inbuilt but chosen and imposed by the being itself. According to J. Searle, one can define two main kinds of causal relations: (i) world-to-mind causation and (ii) mind-to-world causation. The latter is especially relevant in the context of motion and the brain's motion control, and it is the domain where endogenous cognitive processes take place. With regard to world-to-mind causation (i.e., the causal action of the environment on some receptor), the underlying mechanisms are now quite clear; the mind-to-world domain is much more problematic. It is, in fact, concerned with the way we transform a cognitive state into actions. Many low-level mechanisms are now well known, but the point is how such mind-to-world relations get started. In other words, the question is whether the mind (or whatever brain instance) can be considered as a cause. As we have said, this must happen in a way that does not violate physical-chemical principles. From this viewpoint, it is important to realize that mind cannot act directly on the material world and that it is very likely that the mind's influence is a constraining one. That is what makes us think that the TDC approach might prove helpful here.

A final point should be stressed. TDC with regard to endogenous cognitive processes requires that the *representation* of the goal must come before the action aiming at achieving the goal. This is proved by studies showing a hierarchy of the body's motion, such that the subsegments are nested in larger motor segments (like Russian dolls).[2] This also implies that a behaviorist approach cannot work here. Suppose one wants to take a pen; in general, several body movements are needed to achieve this aim. Once these movements are not considered (or removed or blocked), one is still left with an internal state setting the goal to be achieved. Today, it is possible to experimentally prove the existence of internal mental states even if there are no external consequences, by scanning the blood pressure in certain brain areas, for instance, or by measuring the energy expenditure (which can be very high) during action repression (think about a bluff during a poker game). However, this does not solve our problems at all, since our point is to try to establish how such mental states emerge — especially considering that we are focusing on *endogenous* processes.

C. FINAL SUGGESTIONS

The research project we are proposing here has a truly explorative nature, and our framework is conjectural at this stage, although we have shown many elements that make it plausible and promising. Hard theoretical work is needed in order to propose a sound model for endogenous cognitive processes using TDC to be possibly tested in further steps.

Finally, we think our research might have relevant consequences along three main directions:

- to help fill the gap still present between the biological-neurological explanation of cognitive processes and the psychological approach to them;
- to propose an alternative general model of scientific explanation in which endogenous TDC plays the central role;

[2]Consider the French school and in particular Jeannerod's works since *The Neural and Behavioural Organization of Goal-Directed Movements*, Oxford University Press, Oxford 1988, 1990.

- to clarify a possible root of many difficulties in AI. Indeed, we have seen that error-correction is dependent on goal-imposition. This, in turn, demands full autonomy of the system, something that could run against the exigency to build a universal Turing machine able to faithfully pursue any required task.

Gennaro Auletta, Ivan Colagè, Marc Jeannerod

AUTHOR INDEX